Brief Lives and
Memorable Mathematics

George F. Simmons
Professor of Mathematics
Colorado College

With portraits by
Maceo Mitchell

McGraw-Hill, Inc.
New York St. Louis San Francisco Auckland Bogotá Caracas
Lisbon London Madrid Mexico Milan Montreal New Delhi
Paris San Juan Singapore Sydney Tokyo Toronto

This book was set in Times Roman.
The editors were Maggie Lanzillo and John M. Morriss;
the production supervisor was Richard A. Ausburn.
The cover was designed by Caliber/Phoenix Color Corp.
Project supervision was done by Keyword Publishing Services, Ltd.
R. R. Donnelley & Sons Company was printer and binder.

CALCULUS GEMS
Brief Lives and Memorable Mathematics

1 2 3 4 5 6 7 8 9 0 DOC DOC 9 0 9 8 7 6 5 4 3 2

ISBN 0-07-057566-5

Library of Congress Cataloging-in-Publication Data

Simmons, George Finlay, (date).
Calculus gems: brief lives and memorable mathematics/by George F. Simmons.
p.   cm.
Includes index.
ISBN 0-07-057566-5
1. Mathematics—History.   2. Mathematicians—Biography.
3. Calculus—History.   I. Title.
QA21.S54   1992
510—dc20   91-41213

CALCULUS GEMS
Brief Lives and Memorable Mathematics

ABOUT THE AUTHOR

George F. Simmons has the usual academic degrees (CalTech, Chicago, Yale), and taught at several colleges and universities before joining the faculty of Colorado College in 1962, where he is Professor of Mathematics. He is also the author of *Introduction to Topology and Modern Analysis* (McGraw-Hill, 1963), *Differential Equations with Applications and Historical Notes* (McGraw-Hill, 1972, 2d edition 1991), *Precalculus Mathematics in a Nutshell* (Janson Publications, 1981), and *Calculus with Analytic Geometry* (McGraw-Hill, 1985).

When not working or talking or eating or drinking or cooking, Professor Simmons is likely to be traveling (Western and Southern Europe, Turkey, Israel, Egypt, Russia, China, Southeast Asia), trout fishing (Rocky Mountain states), playing pocket billiards, or reading (literature, history, biography and autobiography, science, and enough thrillers to achieve enjoyment without guilt).

For Hope and Nancy,
my wife and daughter,
who *still*
make it all worthwhile

CONTENTS

The Mature Moderns

Part B   Memorable Mathematics

PREFACE

On coming to the end of my work on this book and thinking again about its nature and purpose, I am reminded of W. H. Fowler's Preface to his great *Modern English Usage*: "I think of it as it should have been, with its prolixities docked, its dullnesses enlivened, its fads eliminated, its truths multiplied." And also of W. H. Auden's rueful admission: "A poem is never finished, only abandoned."

Some readers will recognize that this book has been reconstructed out of two massive appendices in my 1985 calculus book, with many additions, rearrangements and minor adjustments. Its direct practical purpose is to provide auxiliary material for students taking calculus courses, or perhaps courses on the history of mathematics. There have been a number of requests that this material be made separately available, and I have been happy to take advantage of this occasion to fill in some gaps and reconsider my opinions. I had a friend who said to me once, "I should probably spend about an hour a week revising my opinions." I treasure the remark and value the opportunity to act upon it.[1]

My overall aims are bound up with the question, "What is mathematics for?" and with its inevitable answer, "To delight the mind and help us understand the world." I hold the naive but logically impeccable view that there are only two kinds of students in our colleges and universities: those who are attracted to mathematics; and those who are not yet attracted, but might be. My intended audience embraces both types.

Part A. This half of the book, entitled Brief Lives, amounts to a biographical history of mathematics from the earliest times to the late nineteenth century. It has two main purposes.

---

[1] The friend was George S. McCue, late of the Colorado College English Department.

First, I hope in this way to "humanize" the subject, to make it transparently clear that great human beings created it by great efforts of genius, and thereby to increase students' interest in what they are studying. Science—and in particular mathematics—is something that men and women do, and not merely a mass of observed data and abstract theory. The minds of most people turn away from problems—veer off, draw back, avoid contact, change the subject, think of something else at all costs. These people—the great majority of the human race—find solace and comfort in the known and the familiar, and avoid the unknown and unfamiliar as they would deserts and jungles. It is as hard for them to think steadily about a difficult problem as it is to hold together the north poles of two strong magnets. In contrast to this, a tiny minority of men and women are drawn irresistibly to problems: their minds embrace them lovingly and wrestle with them tirelessly until they yield their secrets. It is these who have taught the rest of us most of what we know and can do, from the wheel and the lever to metallurgy and the theory of relativity. I have written about some of these people from our past in the hope of encouraging a few in the next generation.

My second purpose is connected with the fact that many students from the humanities and social sciences are compelled against their will to study calculus as a means of satisfying academic requirements. The profound connections that join mathematics to the history of philosophy, and also to the broader intellectual and social history of Western civilization, are often capable of arousing the passionate interest of these otherwise indifferent students.

Part B. In teaching calculus over a period of many years, I have collected a considerable number of miscellaneous topics from number theory, geometry, science, etc., which I have used for the purpose of opening doors and forging links with other subjects . . . and also for breaking the routine and lifting the spirits. Many of my students have found these "nuggets" interesting and eye-opening. I have collected most of these topics in this part in the hope of making a few more converts to the view that mathematics, while sometimes rather dull and routine, can often be supremely interesting. The English mathematician G. H. Hardy said, "A mathematician, like a painter or poet, is a maker of patterns. If his patterns are more permanent than theirs, it is because they are made with ideas." Part B of this book contains a wide variety of these patterns, arranged in an order roughly corresponding to the order of the ideas in most calculus courses. Some of the sections even have a few problems, to give additional focus to the efforts of students who may read them: Sections A.14, B.1, B.2, B.16, B.21, B.25.

I repeat the fervent hope I have expressed in other books, that any readers who detect flaws or errors of fact or judgment will do me the great kindness of letting me know so that repairs can be made.

George F. Simmons

CALCULUS GEMS
Brief Lives and Memorable Mathematics

PART A

BRIEF LIVES

Biographical history, as taught in our public schools, is still largely a history of boneheads: ridiculous kings and queens, paranoid political leaders, compulsive voyagers, ignorant generals—the flotsam and jetsam of historical currents. The men who radically altered history, the great creative scientists and mathematicians, are seldom mentioned if at all.

Martin Gardner

In the index to the six hundred odd pages of Arnold Toynbee's *A Study of History*, abridged version, the names of Copernicus, Galileo, Descartes and Newton do not occur . . . yet their cosmic quest destroyed the mediaeval vision of an immutable social order in a walled-in universe and transformed the European landscape, society, culture, habits and general outlook, as thoroughly as if a new species had arisen on this planet.

Arthur Koestler

Gentlemen, anatomy [read: mathematics] may be likened to a harvest-field. First come the reapers, who, entering upon untrodden ground, cut down great store of corn from all sides of them. These are the early anatomists of modern Europe. Then come the gleaners, who gather up ears enough from the bare ridges to make a few loaves of bread. Such were the anatomists of the last century. Last of all

1

come the geese, who still continue to pick up a few grains scattered here and there among the stubble, and waddle home in the evening, poor things, cackling with joy because of their success. Gentlemen, we are the geese.

Dr. Barclay,
lecturer on anatomy at
Edinburgh University

A.1

THALES
(ca. 625–547 B.C.)

Truth is whatever survives the cleansing fires of skepticism after they have burned away error and superstition. The healthy growth of civilization depends on skepticism more than it does on faith.

Oliver Wendell Holmes

In all history, nothing is so astonishing as the sudden appearance of intellectual civilization in Greece. Many of the other components of civilization—art, religion, complex societies capable of organizing and carrying out enormous projects—had already existed for hundreds or thousands of years, in Egypt, in Mesopotamia, and in China. No one has ever surpassed the Egyptians or Babylonians in monumental stone sculpture, or the Chinese in the production of incomparable works of art in bronze and ceramics. And even today we stand in awe of the organized effort that built the Great Pyramid of Egypt and the Great Wall of China. But what the Greeks achieved in the realm of the intellect is greater still: they invented mathematics, science, and philosophy; they were the first to write genuine analytical history as opposed to mere descriptive annals; and they were the first to speculate freely about the nature of the world and the meaning of life, without being confined by the chains of any stultifying traditional orthodoxy.

It is a good saying that everything has happened at least twice in China, but nevertheless intellectual civilization was born only once, in Greece. It flickered out after several hundred years and was reborn in Western Europe in

3

the 17th century. This greatest creation of the Greek genius has been the powerhouse of Western civilization for more than two thousand years; it has set this civilization apart from all others and has spread over the whole earth, from China to Peru; and it started with Thales and his discovery of skepticism. What lies behind these wonderful events?

In the 7th and 6th centuries B.C. the static older world of Egypt and Babylonia was fading. Newer, more vital peoples, especially the Hebrews, Phoenicians and Greeks, were coming to the center of the stage. The Iron Age was displacing the Bronze Age, and brought with it revolutionary changes in the weapons of war and the tools of agriculture. The alphabet was invented by the Phoenicians, coins were introduced in Lydia, and trade expanded everywhere. The world was ready for a new kind of civilization. This new civilization first appeared in the small trading cities of Ionia on the western coast of Asia Minor, of which Thales' home town, Miletus, was the most important. It later continued its development on the Greek mainland and especially in the Greek colonies on Sicily and the coast of southern Italy.

Who was Thales, and what did he think, and why does it matter? He was the first and most interesting of the pre-Socratic philosophers of ancient Greece, that cluster of a half-dozen or so profoundly original minds that blossomed as if by a miracle in the small city-states of the eastern Aegean Sea during the two centuries before the time of Socrates. Even though very little is definitely known about him, he is the center of a vivid anecdotal tradition nourished by many ancient writers, including Herodotus, Aristophanes, Plato, Aristotle, and Plutarch.[1] He seems to have spent the early part of his life as a successful merchant, acquiring enough wealth to provide himself with security and comfort. He then put business aside and devoted the rest of his life to travel and study, as any man of good sense would do.[2]

He is said to have visited Egypt and the Near East, where he talked with the priests and learned much of their lore; and he astounded them in turn by using shadows and similar triangles to calculate the height of the Great Pyramid. He brought back with him to Miletus knowledge—primitive though it

---

[1] A good account of the various sources and their relations to one another is given by S. Bochner, *The Role of Mathematics in the Rise of Science,* Princeton University Press, 1966, pp. 364–68.

[2] His business acumen is illustrated by the famous story of the olive-presses. (Olive oil was as basic a commodity in Thales' time as wheat and potatoes are in ours.) Aristotle tells it well (*Politics* 1259a): "He was reproached for his poverty, which was supposed to show that philosophy was of no use. According to the story, he knew by his meteorological skill while it was still winter that there would be a great harvest of olives in the coming year; so, having a little money, he gave deposits for the use of all the olive-presses in Chios and Miletus, which he reserved at low prices because no one bid against him. When the harvest-time came, and many were needed all at once, he rented them out at any rate he pleased, and made a quantity of money. Thus he showed the world that philosophers can easily be rich if they like, but that their ambition is of another sort."

was—of Babylonian astronomy and Egyptian geometry. He then vitalized these fragmentary rules of thumb by placing them in the context of the emerging rationality of which he himself was probably the major primary source.

He was apparently the first Greek astronomer, and his studies began the process of establishing this subject as a legitimate science and disentangling it from its oriental associations with astrology. According to Herodotus, writing more than a century later, he astonished all Ionia by successfully predicting an eclipse of the sun that took place (as modern astronomers tell us) on May 28, 585 B.C.[3] Thales' involvement with astronomy gave rise to another famous anecdote that is probably the ancestor of all absent-minded professor stories.[4] This story goes as follows: One night Thales was looking at the stars, and fell into an irrigation ditch. A witty Thracian maidservant pulled him out and chided him by saying, "How do you expect to understand what's happening up in the sky, when you can't even see what's under your own feet?"

He is also said to have diverted the course of the river Halys so that King Croesus' army could pass more easily; to have advised the small Greek cities of Ionia that if they did not unite, they would be absorbed one by one by the Persian empire; to have offered some new practical rules for navigation by using the stars; and to have calculated the distance of a ship at sea by means of observations made at two points along the shore. And then there is the story that once he was in charge of some mules that were heavily loaded with sacks of salt. While crossing a river one of the animals slipped, with the result that the salt dissolved in the water and the load became instantly lighter. This clever beast deliberately rolled over at the next river-crossing to reduce its load, and continued doing so until Thales cured him of this troublesome trick by loading him with sponges instead of salt.

Thales endeared himself to all later generations of Greeks by his many memorable adventures, together with his uniquely personal combination of practical and theoretical wisdom. On being asked what was most difficult, he answered, "To know thyself." And then when asked what was most easy, he replied, "To give advice." To the question "What is God?" he answered, "That which has neither beginning nor end." When he was asked how men might live most virtuously and justly, he answered, "By never doing ourselves what we blame in others."[5] He was versatile, curious, intense, sensible,

[3] See Book I, Chapter 74, in the *History* of Herodotus. This eclipse was of interest to Herodotus because it stopped a battle between the Medes and the Lydians. For more on the eclipse, see pp. 137–38 in vol. I. of T. L. Heath, *A History of Greek Mathematics*, Oxford University Press, 1921.

[4] Plato, *Theaetetus* 174a.

[5] See pp. 37 and 39 in vol. I of Diogenes Laertius, *Lives of Eminent Philosophers*, Loeb Classical Library, Harvard University Press, 1966.

thoroughly normal, and very highly intelligent—altogether an ideal hero for the down-to-earth yet philosophy-loving Greeks.

In philosophy he is remembered mainly for a brief remark by Aristotle: "Thales of Miletus taught that 'All things are water.'"[6] This teaching may seem unpromising to people living in the late twentieth century, but it is really exceedingly remarkable, both in the assumption that underlies it and in the kind of answer it proposes. Never before had anyone attempted to answer the question, What are all things? In oriental civilizations before Thales, explanations of the natural world always took the form of animism, astrology, or mythology. Merely to formulate Thales' question required great insight and imagination, for the question assumes that everything forms a part of some single world of being and it raises the issue of what *being* is. And in trying to answer the question as he did, Thales assumed that there is enough order in the infinite variety of things in the world to make possible some sort of single answer. The fact that it is not true that all things are water is trivial compared with the importance of Thales' point of view, in which we find the first stirrings of genuine philosophy.

Furthermore, in order better to understand the weight of Thales' teaching, one should realize that at that time the Greek word for water (*hydor*) was not restricted to H_2O but more loosely meant "any kind of matter in a liquid state," as, for example, molten lava. We might also remember in this connection that even in the twentieth century it was thought only a few decades ago that everything is made of the nuclei of atoms of hydrogen—which after all is two-thirds of water—and many physicists still think that everything is made of quarks. Thales' teaching should therefore be regarded as an early scientific hypothesis, and by no means an absurd one.[7]

In mathematics he is the first man in history to whom specific discoveries are attributed. He is credited with the following theorems:[8]

1. A circle is bisected by any diameter.

2. The base angles of an isosceles triangle are equal.

3. The vertical angles formed by two intersecting straight lines are equal.

---

[6] *Metaphysics* 983b.

[7] Even with the restriction to H_2O the teaching is extraordinarily acute in its early recognition of the importance of water for living creatures. In the passage quoted above, Aristotle continues by speculating that Thales arrived at his idea "perhaps from seeing that the nutriment of all things is moist . . . and from the fact that the seeds of all things have a moist nature." See p. 89 of G. S. Kirk, J. E. Raven and M. Schofield, *The Presocratic Philosophers,* Cambridge University Press, 2nd ed., 1983. Cf. also the dictum of Sir Arthur Shipley: "Even a bishop is eighty percent water."

[8] Heath, *loc. cit.,* pp. 130–37.

4. A triangle is completely determined if two angles and the included side are given.

5. An angle inscribed in a semicircle is a right angle.

The first four of these theorems are so simple that intuition assures us at once of their truth. Their importance is therefore not to be judged by their content, but rather by the persistent tradition that Thales supported them by logical reasoning of some kind.

For example, let us consider theorem 3 and draw two intersecting straight lines as shown in Fig. A.1. The bald assertion that angle a "obviously" equals angle b contributes nothing to rational discussion. It is equally useless to perform the experiment of cutting out angle a with scissors and applying it to angle b, because geometric truths of this kind are not a matter of experiment. Of course no one knows what Thales actually thought about this, because he left no written works that have survived, but he might have reasoned somewhat as follows to show that angle a = angle b: angle a is the supplement of angle c; angle b is also the supplement of angle c; therefore, since things equal to the same thing are equal to one another, we conclude that angle a = angle b. In this way the desired result is obtained from a short chain of deductive reasoning that rests on even more basic ideas.

As we pointed out above, the first four theorems listed are so simple that it is difficult for us to doubt them enough to take seriously the problem of providing proofs. The fifth theorem is quite different, because it is not intuitively clear at all and most people are rightly unwilling to believe it unless convinced by a clear and compelling proof. Such a proof is easy to give if we allow ourselves to use the familiar theorem that the sum of the interior angles of any triangle equals two right angles, or equivalently—and this is the form we need—any exterior angle of a triangle equals the sum of the two opposite interior angles (Fig. A.2). Briefly, the argument for theorem 5 can now be given as follows: in Fig. A.3 we have $f = a + b$ and $b = a$, so $f = 2a$ and $a = \frac{1}{2}f$; similarly, $c = d + e$ and $e = d$, so $c = 2d$ and $d = \frac{1}{2}c$; therefore $a + d = \frac{1}{2}f + \frac{1}{2}c = \frac{1}{2}(f + c) = \frac{1}{2}(2 \text{ right angles}) = 1$ right angle. Was Thales aware of the interior angles theorem stated on the left in Fig. A.2? Again, no one knows, but the theorem is traditionally ascribed to Pythagoras, who lived two generations later and grew up only a few miles from Thales' home town. Figure A.4 gives an elegant proof of this theorem in the so-called

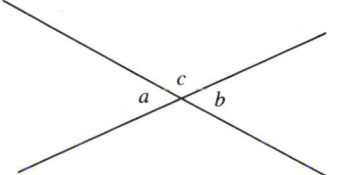

FIGURE A.1

$a + b + c = 2$ right angles $d = a + c$

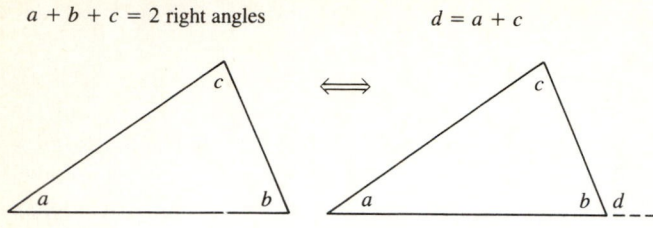

FIGURE A.2

"behold!"style—which means that the reader is invited to inspect the diagram and perceive the truth of the theorem at once by a flash of insight.

The educational experience most of us share makes us so familiar with the concept of a geometric theorem that it rarely occurs to us to wonder where this concept came from. And where did the idea of proving such theorems by reasoning come from? Both probably originated with Thales, as we shall now see.

Eudemus, a pupil of Aristotle who wrote a history of geometry in the late fourth century B.C., tells us that "Thales, who had traveled to Egypt, was the first to introduce this science to Greece." But Thales did not find a science of geometry in Egypt, or in Babylonia either. The distinguishing feature of Greek geometry is precisely its abstract and theoretical character anchored in demonstrable truth, whereas what Thales found was an oriental mishmash of 'recipes'—partly right and partly wrong—for calculating various areas and volumes, and these areas and volumes were always conceived concretely, as the area of a piece of land or the volume of a grain storage bin. This older mathematics had been mindlessly transmitted from one generation of priests to the next from time immemorial; it was dead tradition, mostly computational procedures amounting to approximate formulas, without any logical connections or foundation, and with the underlying train of thought—if any—long forgotten.

Thus, for example, according to the evidence of many cuneiform tablets the Babylonians would tell him that the area of a circle is $3r^2$, so for them $\pi = 3$, as it also was for the ancient Chinese; on the other hand, according to the famous Rhind Papyrus in the British Museum, the Egyptians would assert

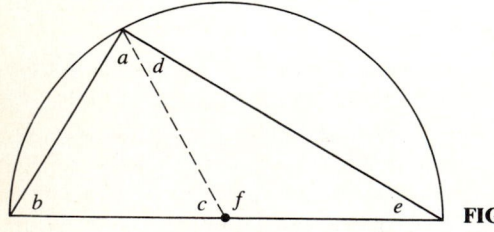

FIGURE A.3

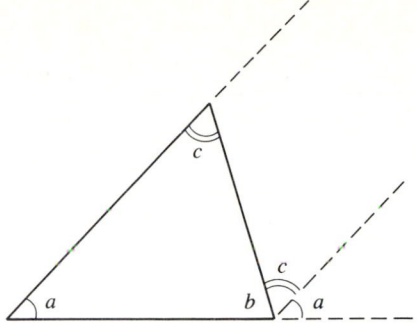

FIGURE A.4

that the area of a circle is the area of a square whose side is eight-ninths the diameter of the circle, that is, $(\frac{8}{9} \cdot 2r)^2 = \frac{256}{81}r^2 \cong (3.1605)r^2$, so for them $\pi = \frac{256}{81} \cong 3.1605$. This Egyptian formula was surely accurate enough for most practical purposes. But how was Thales—whose purposes in these matters were theoretical rather than practical—to distinguish the true from the false in geometry, the exactly correct formulas and the valid properties of figures from mere approximations and guesses? Only by making a clean sweep and doubting everything that rested on faith or authority or tradition, by formulating clear general statements, and by accepting only those that could be established by careful reasoning *capable of withstanding criticism.* In this way deliberate disbelief, deliberate skepticism that welcomes critical discussion and produces further reasoning and evidence, becomes a powerful instrument for the discovery of truth.

If this assessment is correct, it is likely that Thales himself originated abstract geometry as the study of the properties of idealized geometric figures, and also that he created the concepts of theorem and proof as natural modes of thought in dealing with this new subject. These ideas were then developed and elaborated by many other thinkers over the next several centuries, and systematically expounded by Euclid about 300 B.C.

The field of geometry was the perfect place for these modes of thought to germinate, and Thales' dawning awareness of the value of skepticism soon spread into other areas of intellectual life. Our discussion here is therefore not only about the origins of geometry, but also about the origins of Western philosophy and science, and with them Western civilization, which is the only civilization based upon science, though of course not upon science alone. What are the connections?

In all civilizations and at all times, faith and authority and tradition have had little to do with truth and have operated chiefly to maintain the stability of society, but with too much stability society tends to sicken and wither at the root. Even so, skepticism has always gotten short shrift in a world where cries of "Believe this or that!", "Have faith in my message!", and "Keep the faith!" ring out on all sides. The still, small voice of skepticism that says "Wait! Wait!

How do you know?" is usually drowned in the furor of enthusiastic and uncritical believers. The Will to Believe has always been strong among men and women, and the Will to Doubt will probably always be weak.[9]

Yet there is strong evidence that back at the earliest beginnings of Western civilization, in generation after generation, the Ionian philosophers laid upon their pupils the obligation to question and criticize their own teachings. And there is substantial evidence that Thales, the founder of the Ionian school, not only learned from his creation of geometry to welcome criticism, but even more, began this great tradition by teaching his pupils that for the sake of truth it is *necessary* to welcome criticism.[10]

This spark of skepticism, lighted by Thales and nurtured by most of his successors, was the crucial difference between the Greek thinkers and those of all other civilizations. These "other civilizations" are sharply described by Karl Popper as follows:

> In all or almost all civilizations we find something like religious and cosmological teaching, and in many societies we find schools. Now schools, especially primitive schools, all have, it appears, a characteristic structure and function. Far from being places of critical discussion they make it their task to impart a definite doctrine, and to preserve it, pure and unchanged. It is the task of a school to hand on the tradition, the doctrine of its founder, its first master, to the next generation, and to this end the most important thing is to keep the doctrine inviolate. A school of this kind never admits a new idea. New ideas are heresies, and lead to schisms; should a member of the school try to change the doctrine, then he is expelled as a heretic.
>
> There cannot, of course, be any rational discussion in a school of this kind. There may be arguments against dissenters and heretics, or against some competing schools. But in the main it is with assertion and dogma and condemnation rather than argument that the doctrine is defended.

Thus skepticism was at the heart of the Greek way of doing philosophy, which is unique in the history of thought; it led to the concept of mathematical proof—which arose nowhere else—because skepticism and demonstrative proof are opposite sides of the same coin; and this notion of proof is the

---

[9] However, the Will to Doubt has had its occasional eloquent spokesman, as in the lament of Bertrand Russell: "A habit of basing convictions upon evidence, and of giving to them only that degree of certainty which the evidence warrants, would, if it became general, cure most of the ills from which the world suffers." Cf. also the barb of Oliver Wendell Holmes, the greatest of American jurists: "Certitude is not the test of certainty. We have been cocksure of many things that are not so."

[10] See Karl Popper's essay "Back to the Presocratics," in his book *Conjectures and Refutations,* Basic Books, New York, 1962.

essence of genuine mathematics, which gave rise in turn to its companion, mathematicized science, both of which also appeared nowhere else.[11]

This is the way intellectual civilization was born in ancient Greece. The Greeks were a self-confident people with scarcely a trace of false modesty; and if they themselves could have surveyed the history of the world over the past 2500 years, up to our own time, they would certainly have had no hesitation in saying that other cultures emerged from barbarism in direct proportion to the degree to which they absorbed this civilization.

[11] The puzzling question of why these developments did not occur in China has been thoroughly studied by Joseph Needham in his great work, *Science and Civilisation in China,* Cambridge University Press, 1954–84 (15 volumes, occupying about 2 feet of shelf space). Briefly, he found that there was a rich flowering of skepticism at an early period, but that it led into humanistic studies rather than mathematics and science (vol. 2, pp. 365–71, 386–89). Also, and no doubt connected with this, there never arose any concept of rigorous proof; and even though the practical science and technology of the Chinese people was immensely rich—including printing, the magnetic compass, and gunpowder long before these inventions appeared in the West—they never developed the kind of abstract, conceptual science linked to mathematics that has been so important in Western civilization (vol. 3, pp. 150–51, 154–60, 166–68).

A.2

PYTHAGORAS
(ca. 580–500 B.C.)

. . . three-fifths of him genius and two-fifths sheer fudge.

J. R. Lowell

Western civilization is like a great river flowing through time, nourished and strengthened by many rich tributaries from other cultures. Let us project our imagination backward along its course a few thousand years, to its headwaters in ancient Greece. There, near the primary source of the stream, stands the hazy, half-mythical figure of Pythagoras. At the present time most people think of him as a mathematician, but to his contemporaries he was many things—a teacher of wisdom, a religious prophet, a saint, a magician, a charlatan, a political agitator, depending on the point of view. His fanatical followers spread his ideas throughout the Greek world, which ignored some and absorbed others. In terms of his enduring influence on mathematics, science, and philosophy in European civilization—and this influence was not always benign—Pythagoras was as important as anyone who has ever lived.

His master Thales created geometry as the contemplation of abstract patterns of lines and figures, and constructed the first proofs of the first theorems; but Pythagoras was the first person to see geometry as an organized system of thought held together by deductive proof, with one theorem depending on another in a tightly woven fabric of logic. He was even the first to use the word *mathematike* to mean mathematics; before him there was only *mathemata,* which meant knowledge or learning in general.

Thales and his great pupil Anaximander proposed bold theories about the nature of the world, leading to Anaximander's far-reaching idea that the earth is freely suspended in space; but Pythagoras performed the first deliberate scientific experiment and was the first person to conceive the supremely daring conjecture that the world is an ordered, understandable whole. He was the first to apply the word *kosmos*—which previously meant order or harmony—to this whole.

Western philosophy began with Thales and his discovery of skepticism; but Pythagoras deviated into belief in a form of mysticism that crystallized two centuries later into the core of Plato's metaphysical system—and subsequent philosophic thought in the West has often been described as a series of footnotes to Plato. Also, he seems to have originated the very word philosophy (*philosophia,* love of wisdom), in rejecting *sophia* (wisdom) as too weak to convey his own degree of commitment.

His germinal influence on Western thought therefore took several forms leading in several different directions. Let us see how they arose.

First, however, what can be said about his life? He was a contemporary of Confucius, Buddha, and Zoroaster. Like these other great figures from the childhood of the race, Pythagoras is known to us only through legends and traditions recorded hundreds of years after his death.

In broad outline, these traditions agree. He was born on the island of Samos, off the western coast of Asia Minor. He was studious as a youth (what sage was not?), and then traveled for about 30 years in Egypt, Babylonia, Phoenicia, Syria, and perhaps even Persia and India. In the course of his journeys he acquired a little astronomy and primitive empirical mathematics, and a good deal of ridiculous nonsense in the form of Oriental mysticism—the so-called "immemorial wisdom of the East." On returning at last to Samos, he disliked what he found there—an efficient but unsympathetic tyrant—and at the age of about 50 migrated to the Greek colony of Crotona in southern Italy.

Here his public life began. He established himself as a teacher and founded the famous Pythagorean school, which was a semisecret association of several hundred disciples with some claim to the honor of being the world's first university. In the beginning, this school seems to have been as much a religious brotherhood aiming at the moral reformation of society as it was a focus of intellectual activity. However, society does not always welcome moral reformation, and outsiders came to regard the Pythagoreans as an offensively puritanical political party. Eventually, their increasing political activities aroused the ire of the citizens to such an extent that they were violently stamped out and their buildings sacked and burned. Pythagoras himself fled to the nearby colony of Metapontum, where he died at an advanced age. The other surviving Pythagoreans, though scattered throughout the Mediterranean world, kept the faith and carried on as an active philosophical school for more than a century.

What was this faith? Its starting point was Pythagoras' theory of the soul as an objective entity, which no doubt evolved from his experiences in Egypt and Asia. He believed in the doctrine of metempsychosis, or the transmigration of the individual soul at death from one body into another, either human or animal.[1] Each soul continues this process of reincarnation indefinitely, moving up or down to higher or lower animals according to merit or demerit. The only way to escape from this "wheel of birth" and attain unity with the Divine is through purification, both of the body and of the mind. These ideas, though fantastic to modern minds, were widespread in antiquity, and played formative roles in many of the world's religions.

The Pythagorean community was bound together by vows of loyalty to one another and obedience to the Master, and purification was sought in a variety of ways. They practiced a communistic sharing of material things. They dressed plainly, behaved modestly, and did not laugh or swear. It was forbidden to eat beans or meat. The prohibition of beans was probably an echo of some primitive taboo, and vegetarianism was a natural precaution against the abomination of eating an ancestor. Also, the drinking of water instead of wine was recommended—advice of doubtful wisdom in southern Italy today.

It appears that Pythagoras himself surpassed all his students in the perfection of his life according to these standards. His moral and intellectual authority were so great that the phrase *autos epha*—"he himself has said it"—became their formula for a final decision on any issue. Also, it was customary to attribute all ideas and discoveries to the Master, which makes it almost impossible for us to distinguish his own achievements from those that may have originated among his disciples.

As suggested above, the Pythagoreans sought purification of the body through austerity, abstinence, and moderation. This was common then, and is common now in many lands of the East. The uniqueness of Pythagoras lies in his plan for attaining purification of the mind: through the active study of mathematics and science. This is diametrically opposed to the passive "meditation" urged by most mystical cults, which an unsympathetic observer might describe as little more than presiding over a vacuum. This plan of Pythagoras is the source of his lasting influence on Western civilization, and accounts in part for the distinctive character of that civilization as it has developed through the past 2500 years.

The course of study required by Pythagoras consisted of four subjects: geometry, arithmetic, music, and astronomy. In the Middle Ages this group of subjects came to be known as the *quadrivium* (or "fourfold way"), and it was then enlarged by adding the *trivium* of grammar, rhetoric, and logic. These

---

[1] The skeptic Xenophanes (ca. 570–478 B.C.) mocked this Pythagorean doctrine as follows: "Do not hit him," Pythagoras is said to have ordered a man who was beating a dog, "for in this dog lives the soul of a friend of mine; I recognize him by his voice."

were the seven liberal arts that came to be regarded as essential parts of the education of any cultured person.

Greek geometry is certainly one of the half-dozen supreme intellectual achievements of human history. Thales started it all, not for the practical reasons of the Egyptian surveyors, but perhaps for the personal satisfaction to be found in knowing something beyond doubt. However, Pythagoras brought it to the center of the stage as a mental discipline which is capable of lifting the mind to higher levels of order and clarity. He seems to have created the pattern of interrelated definitions, axioms, theorems, and proofs, according to which the intricate structure of geometry is produced from a small number of explicitly stated assumptions by the action of rigorous deductive reasoning. Tradition tells us that Pythagoras himself discovered many theorems, most notably, the fact that the sum of the angles in any triangle equals two right angles, and the famous Pythagorean theorem about the square of the hypotenuse of a right triangle (see Section B.1). According to one source, his joy on discovering this magnificent theorem was so great that he sacrificed an ox in thanksgiving, but this is an unlikely story, since such an action would have been a shocking violation of Pythagorean beliefs. The brotherhood also knew many properties of parallel lines and similar triangles, and arranged all this material into a coherent logical system roughly equivalent to the first two books of Euclid's *Elements* (ca. 300 B.C.). That is to say, starting almost from the beginning, they discovered for themselves about as much geometry as a student learns today in the first half of a high school course.

The Pythagoreans were also entranced by arithmetic—not in the sense of useful computational skills, but rather as the abstract theory of numbers. They seem to have been the first to classify numbers into even and odd, prime and factorable, etc. Their favorites were the figurate numbers, which arise by arranging dots or points in regular geometric patterns. We mention the triangular numbers $1, 3, 6, 10, \ldots$, which are the numbers of dots in the following triangular arrays:

These are evidently numbers of the form $1 + 2 + 3 + \cdots + n$. There are also the square numbers $1, 4, 9, 16, \ldots$:

As indicated, each square number can be obtained from its predecessor by adding an L-shaped border called a *gnomon*, meaning a carpenter's square. The Pythagoreans established many interesting facts about figurate numbers by merely inspecting pictures. For example, since the successive gnomons are the successive odd numbers, it is immediately clear from the square arrays that the sum of the first n odd numbers equals n^2:

$$1 + 3 + 5 + \cdots + (2n - 1) = n^2.$$

In much the same way, the formula

$$1 + 2 + 3 + \cdots + n = \tfrac{1}{2}n(n + 1)$$

for the nth triangular number can be made to appear visibly true by writing it in the form

$$2 + 4 + 6 + \cdots + 2n = n(n + 1),$$

for the left side of this is the sum of the first n even numbers, and the equality is seen at once when this sum is expressed in the form of a rectangular array with n dots along one side and $n + 1$ along the other, as follows:

The tremendous idea that mathematics is the key to the correct interpretation of nature originated with the Pythagoreans, and probably with Pythagoras himself. The discovery that suggested this idea arose from a simple experiment with music. Pythagoras stretched a lyre string between two pegs on a board. When this taut string was plucked, it emitted a certain note. He found that if the string is stopped at its midpoint by a movable wedge inserted between the string and the board, so that the vibrating part is reduced to $\frac{1}{2}$ its original length, then it emits a note one octave above the first; if it is reduced to $\frac{2}{3}$ its original length, it emits a note which is a "fifth" above the first; and if it is reduced to $\frac{3}{4}$ its original length, it emits a note which is a "fourth" above the first. The octave, the fifth, and the fourth were already well-known melodic concepts. The Pythagoreans were deeply impressed by this remarkable relation between the simple fractions $\frac{1}{2}$, $\frac{2}{3}$, and $\frac{3}{4}$ and musical intervals whose recognized significance was based on purely aesthetic considerations. Further, in what seemed to them a natural next step, they held that every body moving in space produces a sound whose pitch is proportional to its speed. It followed from this that the planets moving at different speeds in their various orbits around the earth produce a celestial harmony, which they called the "music of the spheres". As an additional contribution to astronomy, Pythagoras also asserted that the earth is spherical—probably for the simple reason that the sphere is the most beautiful solid body—and he was apparently the first person to do so.

The law of musical intervals described here was the first quantitative fact ever discovered about the natural world. Along with its "philosophically obvious" extension to the planets, this led the Pythagoreans to the conviction that numbers—which to them meant positive whole numbers and fractions—

are the essence of all that is intelligible in the universe.[2] "Everything is Number" became their motto, apparently in the sense that the only basic and timeless aspects of any object or idea lie in the numerical attributes it possesses.

Almost at once, this doctrine tripped over geometry and fell flat on its face. Because everything is number—meaning rational number, since no others existed for them—it was evident that the length of any line segment must be a rational multiple of the length of any other line segment. Unfortunately this is false, as they soon discovered, for the Pythagorean theorem guarantees that a square of side 1 has diagonal of length $\sqrt{2}$, and according to tradition, Pythagoras himself proved that there is no rational number whose square is 2. This fatal discovery confronted the brotherhood with two alternatives, one unthinkable and the other intolerable: either the diagonal of a square of side 1 has no length, or it is not true that everything is number. To them the crumbling of their simple generalization reducing the universe to rational numbers shattered the necessary foundations of thought, and one legend states that they even went so far as to drown a renegade Pythagorean who revealed their unspeakable secret to the outside world. To us, however, it represents the discovery of irrational numbers (see Section B.2), which was one of the finest achievements of early Greek mathematics. It is often seen in the history of ideas that one generation's disaster is an opportunity for the next.

In spite of this setback, Pythagoras and his followers kept their faith in Number as they conceived it. If Number contradicts reality, so much the worse for reality. In the ecstasy of their enthusiasm they abandoned all further interest in learning about the world by combining observation, experiment and thought, and instead sought their own "higher reality" by plunging joyously into the quagmire of number mysticism.

Like the tenets of any religion, the numerological beliefs of the Pythagoreans are difficult to make plausible to the uninitiated. The central concept of their system seems to have been the sacred *tetractys,* consisting of the numbers 1, 2, 3, 4, whose sum is the holy number 10—holy because 1 is the point, 2 is the line, 3 is the surface, and 4 is the solid, and therefore $1 + 2 + 3 + 4 = 10$ is everything, the number of the universe. It was doubtless a great day for them when they learned that the fractions $\frac{1}{2}, \frac{2}{3}, \frac{3}{4}$, which are the successive ratios of the numbers 1, 2, 3, 4, are closely linked to musical harmony; and it should be satisfying to us to realize that our own decimal system has a more rational foundation than the accidental fact that human

---

[2] As Aristotle wrote in his *Metaphysics* (Book I, Chapter 5, ca. 330 B.C.), "The so-called Pythagoreans, who were the first to take up mathematics, not only advanced this subject, but, saturated with it, they fancied that the principles of mathematics were the principles of all things."

beings have 10 fingers. Their next basic article of faith lies much deeper, so deep indeed that modern people can scarcely hope to comprehend it: odd numbers (except 1) are male, and even numbers are female. In addition to such general principles as these, they believed that each number has its own individual significance and personality. Thus, 1 is the generator of all numbers, the omnipotent One; 2 is diversity, the first female number; $3 = 1 + 2$ is the first male number, being composed of unity and diversity; $4 = 2 + 2 = 2 \cdot 2$ is the number of justice, being evenly balanced; $5 = 3 + 2$ is the number of marriage, since it is the union of the first male and female numbers; $6 = 1 + 2 + 3$ is perfect, because it is the sum of its proper divisors, and these are unity, diversity, and the sacred trinity, whose meaning expanded considerably in early Christian numerology.[3] And so on, and so on.

The importance for us of this chaotic mass of fanciful mumbo-jumbo is that it passed into the mind of Plato (428–348 B.C.) and emerged in an altered form as part of a powerful torrent of belief that swept almost unabated through the early Christian era, the Middle Ages, and the Renaissance, and is still a potent influence in our own time.

Plato, of course, is one of the titans of world literature. His half-dozen greatest dialogues have held their place in the affection and respect of mankind mainly because of their dramatic and poetic qualities and the personality of their chief character, Socrates. The Socratic element in Plato's thought is primarily concerned with human affairs—with morality, politics, and the problem of how to live a good life. In addition to his love and admiration for Socrates, Plato was enthralled by mathematics, especially in its role as a body of knowledge that appears to be independent of the evidence of the senses. In his middle years he spent considerable time in southern Italy and came into personal contact with the Pythagorean communities there, whose philosophy was mathematical but whose driving forces were religious and mystical.[4] This decisive experience imparted a Pythagorean tinge to much of the rest of Plato's thought, which can be characterized as a mixture of a few gold nuggets and a

---

[3] As St. Augustine said in *The City of God* (ca. A.D. 420): "Six is a number perfect in itself, and not because God created the world in six days; rather the contrary is true; God created the world in six days because this number is perfect." (Perfect numbers are discussed as mathematics rather than theology in Section B.2). In a very different mood, this remarkable man also uttered the following unforgettable prayer as he approached the end of his licentious youth and looked forward to a monastic old age: "O Lord, give me chastity, but not quite yet."

[4] See F. M. Cornford, *Before and After Socrates,* Cambridge University Press, 1932, especially Chapter III.

substantial amount of gravel, with the gravel gilded by his exalted literary style. Unfortunately, gold-plated gravel is still basically gravel.[5]

The quintessence of Plato's Pythagoreanism is found in his mystical doctrine of Ideas or Forms. This doctrine asserts a view of reality consisting of two worlds: first, the everyday world perceived by our senses, the world of change, appearance, and imperfect knowledge; and second, the world of Ideas perceived by the reason, the world of permanence, reality, and true knowledge. Thus Justice is an Idea imperfectly reflected in human efforts to be just, and Two is an Idea participated in by every pair of material objects. Each of Plato's Ideas was for him an objective reality lying outside of space and time; they can be approached by thought but are not created by it; they are the timeless and perfect patterns of Being, whose blurred and shadowy copies constitute the deceptive phenomena of the world around us.

These stable absolutes imagined by Plato appeal to all people battered by change and hungry for permanence. Aristotle tried to dilute them with a little common sense; but like oil and water, Platonism and common sense do not mix easily, and his efforts failed.[6] This Pythagorean apotheosis of abstract concepts is now called "Platonic realism." It has had a long and controversial history in Western thought, and was still alive and well in the early twentieth century.[7]

We hope these remarks have clarified the assertion made at the beginning, that Pythagoras was much more than merely an ancient mathematician: he is entitled to be recognized as one of the principal founders of

---

[5] The following are a few of the gamiest items in Plato's numerology: his Geometrical Number, $60^4 = 12,960,000$, "which has control over the good and evil of births" (*Republic* 546); his apparent rejection of the well-known length of the year—approximately $365\frac{1}{4}$ days—in favor of $364\frac{1}{2}$ days, because this is 729 or 9^3 days and nights, and 9^3 is "the interval by which the tyrant is parted from the king" (*Republic* 588); his number 5040 ($= 1 \cdot 2 \cdot 3 \cdot 4 \cdot 5 \cdot 6 \cdot 7$), which he concludes is the exact number of citizens suitable for his ideal city (*Laws* 738, 741, 747, 771, 878); and the whole of the *Timaeus,* in which the structure of the universe and the nature of life are given thoroughgoing Pythagorean explanations in terms of triangles. For more details on the mysterious Geometrical Number, see T. L. Heath, *A History of Greek Mathematics,* Oxford University Press, 1921, vol. I, pp. 305–307.

[6] "The Forms [Ideas] we can dispense with, for they are mere sound without sense."—*Posterior Analytics* 83a. See also *Metaphysics* 990b, 991a, 1079a, 1079b, 1090a, etc. Compared with Plato, Aristotle suffered the serious disadvantage of having about as much charm as an old shoe.

[7] Consider the statement of the English astronomer Sir Arthur Eddington (1935): "I believe that all the laws of nature that are usually classed as fundamental can be foreseen wholly from epistemological considerations." Also that of the English mathematician G. H. Hardy (1940): "I believe that mathematical reality lies outside us, that our function is to discover or observe it, and that the theorems which we prove, and which we describe grandiloquently as our 'creations,' are simply the notes of our observations."

mathematics, science, and philosophy in European civilization. He was also the first to open up the enduring gulf of incomprehension between the scientific spirit, which hopes that the universe is ultimately understandable, and the mystical spirit, which hopes—perhaps unconsciously—that it is not.[8]

---

[8] The view expressed here, that in the core of his thinking Plato was a deep-dyed mystic, is rejected with such indignation by many modern philosophers that perhaps it requires some independent support. For example, in his *Mysticism and Logic,* Bertrand Russell writes: "In Plato, the same twofold impulse [toward mysticism and science] exists, though the mystic impulse is distinctly the stronger of the two, and secures ultimate victory whenever the conflict is sharp." Two other eminent contemporary philosophers who have no difficulty in seeing Plato for what he was—a very great writer but unfortunately a mystic and would-be tyrant— are Karl Popper (*The Open Society and its Enemies,* 1945) and Gilbert Ryle (review of Popper's book in *Mind,* 1947). And in a letter to John Adams (July 5, 1814), Thomas Jefferson—at the age of 71—writes: "I amused myself with reading seriously Plato's republic [in the original Greek, of course]. I am wrong however in calling it amusement, for it was the heaviest task-work I ever went through. I had occasionally before taken up some of his other works, but scarcely ever had the patience to go through a whole dialogue. While wading thro' the whimsies, the puerilities, and unintelligible jargon of this work, I laid it down often to ask myself how it could have been that the world should have so long consented to give reputation to such nonsense as this? . . . But fashion and authority apart, and bringing Plato to the test of reason, take from him his sophisms, futilities, and incomprehensibilities, and what remains?" In other letters Jefferson expressed himself on this subject with much less restraint.

A.3

DEMOCRITUS
(ca. 460–370 B.C.)

While the philosophy of Plato and Aristotle was noised and celebrated in the schools amid the din and pomp of professors, this of Democritus was held in great honour with the wiser sort.

Francis Bacon

Democritus, long famous as one of the founders of the atomic theory of matter and the greatest philosopher of physical science among the ancient Greeks, came into his own as a mathematician with the discovery in 1906 of the lost treatise of Archimedes known as the *Method*—which has the form of a letter to his friend Eratosthenes in Alexandria. In the chatty introduction to this treatise–letter, Archimedes tells us that Democritus was the discoverer of the wonderful theorems of solid geometry that the volume of a cone is one-third the volume of a cylinder having the same base and equal height (Fig. A.5), and also that the volume of a pyramid is one-third the volume of a prism having the same base and equal height.

Democritus was a native of Abdera in Thrace, on the northern coast of the Aegean Sea. He inherited considerable wealth, which enabled him to

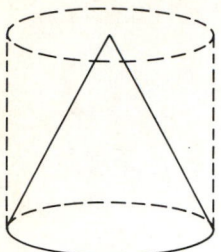

FIGURE A.5

travel widely in search of knowledge. He says:

> I have traveled more extensively than any man of my time, and have made the most exhaustive inquiries; I have seen the most climates and countries, and have listened to the greatest number of learned men.[1]

His travels lasted for many years and are said to have taken him to Egypt, Babylon, and Persia, where he talked with the priests and magi; and some writers say that he even went as far as Ethiopia and India. Democritus himself says that he spent five years in Egypt alone.

According to the list of titles given by Diogenes Laertius, he wrote at least 75 works on almost every conceivable subject, from physics and mathematics to logic, ethics, magnets, fevers, diets, agriculture, law, "the sacred writings in Babylon," "the right use of history," and even the growth of animal horns, spiders and their webs, and the eyes of owls.[2] The Roman orator–essayist Cicero tells us that his writings had a purity of style not inferior to that of Plato. In the breadth of his knowledge and the importance of his influence on both ancient and modern thinkers, Democritus was the Aristotle of the 5th century; and in metaphysics the sanity of his views has led many to consider him the equal, and perhaps the superior, of Plato.

Plato apparently hated him and was jealous of him, and is said to have wished to burn all the writings of Democritus he could lay his hands on; but his companions prevented him, saying that it would do no good, for his books were already widely circulated. Diogenes Laertius adds the following comment:

> There is clear evidence for this in the fact that Plato, who mentions almost all the early philosophers, never once alludes to Democritus, not even where it would be

---

[1] See Democritus fragment 299 in Kathleen Freeman, *Ancilla to the Pre-Socratic philosophers* (a complete translation of the fragments in Diels, *Fragmente der Vorsokratiker*), Harvard University Press, 1948.

[2] *Lives of Eminent Philosophers*, vol. II, pp. 455–63, Harvard University Press, 1966.

necessary to controvert him, obviously because he knew that he would have to match himself against the prince of philosophers.[3]

Aristotle, on the other hand, discusses his ideas in many places and pays judicious tribute to his genius, remarking, for example, that on the subject of growth and change no one except Democritus had observed anything other than trivialities, whereas Democritus seemed to have thought of everything.[4] Even though Plato was frustrated in his desire for a book-burning, the works of Democritus perished anyway, except for several hundred so-called fragments, that is, isolated sentences and brief paragraphs of his culled from the writings of others. Here is a sampling:[5]

It is hard to be governed by one's inferior. (49)

It is better to examine one's own faults than those of others. (60)

Many very learned men have no intelligence. (64)

I would rather discover one cause than gain the kingdom of Persia. (118)

Pigs revel in mud. (147)

A life without festivity is a long road without an inn. (230)

To a wise man the whole earth is his home. (247)

The rearing of children is full of pitfalls. Success is attended by strife and care, failure means grief beyond all others. (275)

As we see, these observations on the world as he knew it can also be thought of as pointed comments about our own modern political life, academic life, and family life. And finally, a poignant autobiographical fragment:

I came to Athens and no one knew me. (116)

This is sometimes interpreted as a lament that his merits were not sufficiently recognized in his own time. If so, he would have been pleased to enter Athens today from the direction of Thrace, and to pass the imposing Democritus Nuclear Research Laboratory. And on reaching central Athens, he would find

---

[3] *Op. cit.*, p. 451.

[4] *De Generatione et Corruptione*, I.2.315ª33, Oxford University Press, 1982.

[5] See Freeman, *loc. cit.*

that one of the most necessary objects of daily life—a Greek 100-drachma note—carries an elegant engraved portrait of himself.

In physical science, atoms and the void (= empty space) are the basis of Democritus' system for explaining the universe—solid, corporeal atoms scattered throughout infinite space. "Nothing exists but atoms and the void," he said. According to his system, as deduced partly from his own words but mostly from the commentaries of his successors, matter consists of tiny separate particles that are physically indivisible (*atomos* means "unable to be cut"); these atoms are eternal and invisible, differ only in size and shape, move about in empty space, and coagulate here and there into ordinary matter; and all change is produced by the direct contact of moving atoms with one another. Our earth and everything in it and on it, like everything everywhere, consists of atoms and space, and the universe has no other constituents of any kind. Everything that exists or happens can be explained within this mechanical system, and in particular there is no need or place for a providence or gods or any form of supreme being with its own inscrutable purposes.

It is surprising for us to realize that this familiar atomic theory was then little more than wild but logical speculation based on the slimmest evidence. Democritus worked his ideas out in considerable detail, and some of the reasoning is quite interesting.[6] However, ways in which the concepts of atoms and the void might have occurred to him are even more interesting. Each atom, Democritus said, is impenetrable and indivisible because it contains no void. The existence of empty space inside apparently solid matter is suggested by the fact that when we cut an apple with a knife, the blade has to find empty space between the atoms where it can penetrate; for if there were no void in the apple, it would be physically impenetrable and no knife could cut it. An attractive line of thought suggesting the existence of atoms might have stemmed from the discovery by Empedocles (ca. 490–430 B.C.) of air as a corporeal though invisible substance. The essence of Empedocles' experiment is very simple and can be carried out in any kitchen sink: invert an empty glass over a bowl of water and press it down into the water; the water in the bowl visibly does not enter the glass because its entry is blocked by an invisible material substance called 'air'; and if the glass is tilted slightly so that some of the air escapes in the form of bubbles, then an equivalent amount of water is able to enter the glass.[7] In this way Empedocles discovered that *the invisible really exists,* and we can easily imagine Democritus hearing of this and saying to himself, "Aha! So that's what the world is made of."

For Democritus and his followers, the gods were invented by priests and

---

[6] See pp. 413–33 of G. S. Kirk, J. E. Raven, and M. Schofield, *The Presocratic Philosophers*, 2nd ed., Cambridge University Press, 1983.

[7] See Kirk, Raven, and Schofield, *op. cit.*, pp. 359–60.

all the religions of his time were evil.[8] He was himself a complete materialist: life and thought were physical processes and the universe operated by mechanical laws, without meaning or purpose or any special concern for human beings. It was his atomism that furnished the concepts needed for materialism to appear as a coherent and powerful system of philosophy. This philosophy has been attacked by every other throughout the ages, but has survived them all and sprouts up here and there in each new generation.

In both atomism and materialism, Democritus was followed by Epicurus (341–270 B.C.), who was more interested in ethics than logic; and Epicurus was followed by Lucretius (ca. 95–51 B.C.), the greatest of philosophical poets, whose unfinished masterpiece *De Rerum Natura* (*On the Nature of Things*) gave passionate literary expression to these ideas. Needless to say, these views of life and the world were totally incompatible with Christianity, and they disappeared from Western civilization almost completely until their revival at the time of the French Enlightenment in the 18th century. In 1841 the young Karl Marx (1818–83) began his career as a critic of society with a dreary, inflated doctoral dissertation for the University of Jena entitled *Differenz der demokritischen und epikureischen Naturphilosophie* (*Differences between the natural philosophy of Democritus and Epicurus*). Of course, atomism in its 20th century manifestation in physics and chemistry is the most basic fact—no longer a theory—of modern physical science.

We turn at last to the mathematics of Democritus, about which very little is known. There were six mathematical titles among the many listed by Diogenes Laertius:

On the contact of a circle and a sphere;
On geometry;
Geometrica;
Numbers;
On irrational lines and solids;
Projections.

Nothing whatever is known about the content of any of these works, so our knowledge of specific mathematical ideas is restricted to the two great theorems attributed to him by Archimedes as described at the beginning of this

---

[8] Democritus may have been influenced by the skeptic Xenophanes (ca. 570–478 B.C.): "Mortals believe that the gods are born as they are, and have clothes and speech and bodies like theirs ... But if cattle and horses had hands, and could draw and create works of art as men do, horses would draw pictures of gods like horses, and cattle of gods like cattle, and make the bodies of their gods in the image of their own kinds ... The Ethiopians make their gods black and snub-nosed, and the Thracians have gods with blue eyes and red hair." *Ibid.*, p. 169.

section. Nothing definite is known about how Democritus discovered or proved these theorems, but most mathematicians believe that the only plausible line of thought is the one described in the following paragraphs.

In Fig. A.6 we consider the cone shown in Fig. A.5. Our purpose is to establish the volume formula $V = \frac{1}{3}Bh$, and in particular to understand where the factor $\frac{1}{3}$ comes from. In the base of this cone we inscribe a regular polygon with n sides, where n is some large number (in the figure, $n = 8$). Using this polygon as a base, we construct a pyramid whose vertex is the vertex of the cone. As n increases, the volume of the cone is the limiting value approached by the volume of the pyramid and the area of the base of the cone is the limiting value approached by the area of the base of the pyramid. To prove the volume formula for the cone, it therefore suffices to show that the volume of the pyramid is one-third the area of its base times its height. Since the pyramid can be divided into n congruent pyramids of the type shown in Fig. A.7, it suffices to show that the volume formula is valid for these special pyramids. This we now do.

On the left in Fig. A.8 we show the pyramid in Fig. A.7 in a slightly different position. On the base OPQ we construct a prism with height h and base area B (Fig. A.8, center). This prism can be divided into three pyramids as shown on the right in the figure. Pyramids I and II have height h and triangular bases OPQ and RST of equal area, so they have equal volumes. Pyramids II and III have the same height (the distance from R to the plane $PQST$) and triangular bases PST and PQT of equal area, so they also have equal volumes. This reasoning shows that all three pyramids have the same volume, so the volume of each is one-third the volume Bh of the prism. This establishes the volume formula in Fig. A.7, and with it the volume formula $V = \frac{1}{3}Bh$ for the cone.

There is a logical gap in this argument, where we assume without proof that two pyramids have equal volumes if they have the same height and

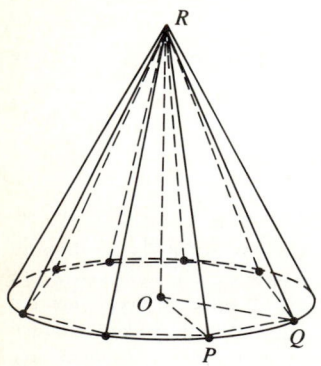

FIGURE A.6

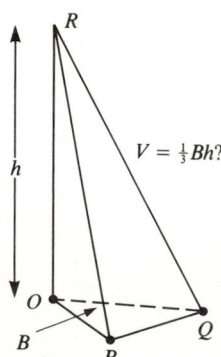

$V = \frac{1}{3}Bh$?

FIGURE A.7

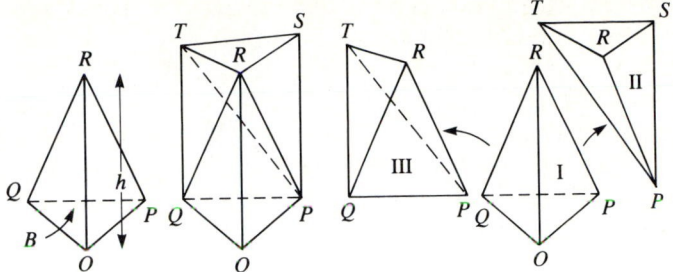

FIGURE A.8

triangular bases with equal areas. This is plausible but not easy to prove. Its rigorous proof depends on a kind of geometrical atomism that we describe later in our account of the 17th century Italian mathematician Bonaventura Cavalieri.

Democritus was right: It is much better to have conceived these beautiful ideas about cones and pyramids than to be King of Persia.

A.4

EUCLID
(flourished ca. 300 B.C.)

Euclid's *Elements* is certainly one of the greatest books ever written.

Bertrand Russell

It is also one of the dullest, and by any educational standards whatever, it ought to have been a student's and teacher's nightmare throughout the twenty-three centuries of its existence.

The *Elements* purports to begin at the very beginning of geometry, with nothing required of the reader in the way of previous knowledge or experience. Yet it offers no preliminary explanations, and nowhere does it provide illuminating remarks of any kind. It has no direct scientific content and doesn't even hint at a single application. It makes no attempt to place its subject in any mathematical or historical context, and nowhere is the name of any person mentioned. Its stony impersonality stuns the mind. The Bible also starts impersonally enough—"In the beginning God created the heaven and the earth"—but even this compresses the greatest possible action into the fewest possible words, and after several sentences living creatures make their appearance. The *Elements* begins with a definition—"A point is that which has no part"—and marches with inhuman, undeviating monotony through 13 Books and 465 Propositions, none of which are discussed or motivated in any way. It conveys the impression of simply existing, like a rock, indifferent to human concerns. Such are the outward qualities of a book that has gone through more than a thousand editions since the invention of printing and has often been described as having had an influence on the human mind greater than that of any other work except the Bible.

In view of its tone and style, the most surprising thing about the *Elements* is that it seems to have had an author. Who was this Euclid, whose name was almost synonymous with geometry until the twentieth century? Only three facts are known about him, and two slender anecdotes.

The facts are these: He was younger than Plato (b. 428 B.C.), he was older than Archimedes (b. 287 B.C.), and he taught in Alexandria. When Alexander the Great died in 323 B.C., his African empire was inherited by Ptolemy, his favorite Macedonian general, who ruled as king from 305 to 285 B.C. It is surmised that Ptolemy brought Euclid from Athens to Alexandria to join the staff of the great center of Hellenistic learning—known as the Museum, with its famous Library—that he founded there.[1]

And the anecdotes are these:

> Ptolemy once asked Euclid if there was any shorter way to a knowledge of geometry than that of the *Elements,* and he replied that there is no royal road to geometry.

> Someone who had begun to read geometry with Euclid, when he had learned the first proposition, asked him, "What shall I get by learning these things?" Euclid called his slave and said, "Give this person a penny, since he must make a profit out of what he learns."

The reliability of these smug little stories can be judged from the fact that their authors (Proclus and Stobaeus) lived in the fifth century A.D., more than 700 years after the time of Euclid.

In addition to its systematic account of elementary geometry, the *Elements* also contains all that was known at the time about the theory of numbers. Euclid's role as the author was mainly that of an organizer and arranger of the scattered discoveries of his predecessors. It is possible that he contributed some of the ideas and proofs himself, and in the absence of evidence to the contrary, several important theorems are traditionally credited to him.

---

[1] The Alexandrian Museum was not a museum at all in the modern sense of this word. The Greek word from which the term museum is derived meant a home for the Muses, the nine daughters of Zeus, who came to be associated with the arts and sciences most valued by the Greeks. It implied an environment suited to literary and scientific studies, that is, what we would call a great university. For more information about the origin and history of the Alexandrian Museum and Library, see two articles by David E. H. Jones in *Smithsonian Magazine,* December 1971 and January 1972.

Book I of the *Elements* begins with 23 definitions (point, straight line, circle, etc.), 5 postulates, and 5 axioms or "common notions." Among the Greek philosophers, axioms were thought of as general truths common to all fields of study ("The whole is greater than the part"), while postulates were considered to be assumptions that have meaning only in the specific subject under discussion ("It is possible to draw a straight line from any point to any other point"). This distinction has been dropped in modern mathematics, and the words *axiom* and *postulate* are now used interchangeably. Broadly speaking, Books I to VI deal with plane geometry, Books VII to IX with number theory, Book X with irrationals, and Books XI to XIII with solid geometry. The 47th proposition in Book I (usually designated as I 47) is the Pythagorean theorem.[2] The following are a few other individual items of particular interest: VII 1 and VII 2 give the Euclidean algorithm, a process for finding the greatest common divisor of two positive integers; VII 30 is Euclid's lemma, which asserts that a prime number that divides the product of two positive integers necessarily divides one of the factors; IX 20 is Euclid's theorem on the infinity of primes (Theorem 2 in our Section B.16); IX 36 is Euclid's theorem on perfect numbers (Theorem 1 in Section B.2); and XII 10 gives the volume of a cone.

Students will recall from their study of geometry that a regular polygon with n sides (also known as a regular n-gon) has all of its n sides equal and all of its n angles equal. Figure A.9 shows a regular 3-gon, 4-gon, 5-gon, and 6-gon, which of course are usually called an equilateral triangle, a square, a regular pentagon, and a regular hexagon. Book IV of the *Elements* gives the classical constructions, using only a straightedge and compass, of the regular

  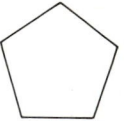 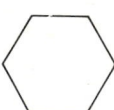

FIGURE A.9
Regular polygons.

[2] This theorem was the stimulus for the English philosopher Thomas Hobbes's first encounter with geometry at the age of 40. He happened to be in a friend's library and saw a copy of the *Elements* lying open at I 47. As Aubrey tells it in his *Brief Lives*: "He read the proposition. 'By G–,' sayd he, (he would now and then sweare, by way of emphasis), 'this is impossible!' So he reads the demonstration of it, which referred him back to such a proposition; which proposition he read. That referred him back to another, which he also read. *Et sic dienceps* [and so in succession] back to the first, so that at last he was demonstratively convinced of that trueth. This made him in love with geometry." We add Bertrand Russell's remark about this incident: "No one can doubt that this was for him a voluptuous moment, unsullied by the thought of the utility of geometry in measuring fields."

polygons with 3, 4, 5, 6, and 15 sides. These constructions were known to the Pythagoreans long before Euclid's time, and it was Plato and his pupils who insisted that the so-called Euclidean tools—a straightedge and compass—are the only ones "philosophically suitable" for use in geometry.[3] By means of angle bisections, it is easy to construct from a given regular polygon with n sides another regular polygon with $2n$ sides. The Greeks were therefore able to construct regular n-gons where n has the following values:

$$3, 6, 12, 24, \ldots,$$
$$4, 8, 16, 32, \ldots,$$
$$5, 10, 20, 40, \ldots,$$
$$15, 30, 60, 120, \ldots.$$

The next obvious step was to seek Euclidean constructions for regular polygons with 7, 9, 11, 13, . . . sides. Many tried, but all such efforts failed, and there the matter rested for about 2100 years, until March 30, 1796.

On that day one of the greatest geniuses of recorded history, a young German named Carl Friedrich Gauss, proved the constructibility of the regular polygon with 17 sides. Gauss was 18 years old at the time, and his discovery pleased him so much that he decided to adopt mathematics as a career in preference to philology. He continued his investigations and quickly solved the constructibility problem completely. He proved by rather recondite methods involving algebra and number theory that a regular n-gon is constructible if and only if n is the product of a power of 2 (including $2^0 = 1$) and distinct prime numbers of the form $p_k = 2^{2^k} + 1$. In particular, when $k = 0, 1, 2, 3$, we see that each of the corresponding numbers $p_k = 3, 5, 17, 257$ is prime, so regular polygons with these numbers of sides are constructible. The prime number 7 is not of the stated form, so the regular polygon with 7 sides is not constructible.[4]

Book XIII of the *Elements* is devoted to the construction of the regular polyhedra, which are less familiar to most people than the regular polygons. A *polyhedron* is simply a solid whose surface consists of a number of polygonal faces; it is said to be *regular* if its faces are congruent regular polygons and if

---

[3] Plato's Academy, which was the second genuine university in the Western world (after the school of Pythagoras) and which lasted over a thousand years, placed great importance on the study of mathematics. It is supposed to have had a sign over its entrance saying, "Let no one ignorant of geometry enter here." In all probability, Euclid was a member of the Academy before moving to Alexandria. The name "Academy" comes from the place where it was located, in a grove named after a certain hero, Hecademus (Diogenes Laertius, Loeb edition, I, p. 283).

[4] Additional details on Euclidean constructions and constructibility are given very clearly and briefly in Chapter 3 of R. Courant and H. Robbins, *What Is Mathematics?*, Oxford University Press, 1941; and also in Chapter IX of H. Tietze, *Famous Problems of Mathematics*, Graylock Press, 1965. Gauss's prodigious creative life continued for another 60 years, and he is now recognized as the greatest of all mathematicians.

the solid angles at all its vertices are congruent. There are clearly an infinite number of regular polygons, but there are only five regular polyhedra. They are named for the numbers of faces they possess (Fig. A.10): the tetrahedron (4 triangular faces), the hexahedron or cube (6 square faces), the octahedron (8 triangular faces), the dodecahedron (12 pentagonal faces), and the icosahedron (20 triangular faces). Plato and his followers studied these polyhedra so persistently that they have come to be known as the "Platonic solids." In his fantastic dialogue *Timaeus*, he associates the tetrahedron, octahedron, cube, and icosahedron with the four classical "elements" of fire, air, earth, and water (in this order), while in some mystical sense he makes the dodecahedron symbolize the entire universe. The first three regular polyhedra actually occur in nature as crystals, and the last two as the skeletons of microscopic sea animals called radiolarians.[5] However, it is the symmetry and beauty of these figures, and not their applications, that have fascinated people through the centuries. The construction of the regular polyhedra provides a superb climax to Euclid's geometry, and some have conjectured that this was the primary purpose for which the *Elements* was assembled—to glorify the Platonic solids.

It is so easy to prove that there are only five regular polyhedra that we venture to give the argument here. Let m be the number of sides of each regular polygonal face, and n the number of polygons meeting at each vertex. The size (in degrees) of each angle in each face is $180 - (360/m)$. Also, the

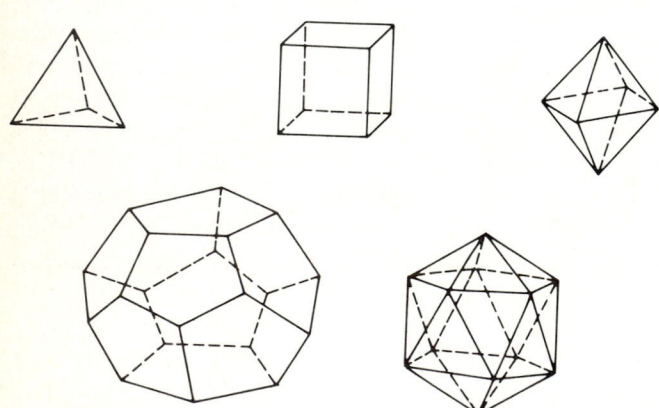

FIGURE A.10
Regular polyhedra.

---

[5] See the illustrations on p. 75 of H. Weyl, *Symmetry*, Princeton University Press, 1952.

sum of the angles at each vertex of the polyhedron is less than 360 degrees, so

$$n\left(180 - \frac{360}{m}\right) < 360$$

or

$$n\left(1 - \frac{2}{m}\right) < 2,$$

which is easily shown to be equivalent to

$$(m - 2)(n - 2) < 4.$$

But m and n are both greater than 2. Therefore if $m = 3$, n can only be 3, 4, or 5; if $m = 4$, n can only be 3; and if $m = 5$, n can only be 3, which gives five possibilities in all.

To mathematicians, some of the theorems in the *Elements* are important, some are interesting, and a few are both. However, the source of the immense influence of this book on all subsequent thought lies elsewhere, not so much in the exposition of particular facts as in the methodology of it all. It is clear that one of Euclid's main aims was to give a connected logical development of geometry in such a way that every theorem is rigorously deduced from the "self-evident truths" which are explicitly stated at the beginning. This pattern of thought was conceived by Pythagoras, but it was Euclid who worked it out in such stupefying detail that for more than 2000 years no one was capable of doubting that his success had been complete and final. It is true that from time to time critics pointed out definitions that do not define and proofs that do not prove.[6] Nevertheless, these flaws were considered relatively minor and easy to mend. All thinking people continued to believe that the Euclidean system of geometry was *true*, in the sense that it described correctly the geometry of the actual world in which we live, and *necessary*, in the sense that it could be deduced by unassailable reasoning from axioms whose self-evident character is apparent to all.

This happy state of affairs in geometry led to the hope that in a similar way the remotest truths of science and society could be discovered and proved by simply pointing out those things that are self-evident and then reasoning from these foundations. No more attractive or tenacious idea has ever appeared in the intellectual history of the Western world. The prestige of

---

[6] Recall the definition of a point quoted earlier. Also: "A *line* is breadthless length"; "A *straight line* is a line which lies evenly with the points on itself"; "A *unit* is that by virtue of which each of the things that exist is called one"; "A *number* is a multitude composed of units." The defects in the proofs often consist of the use of additional assumptions that are not explicitly recognized.

geometry was so great, especially in the seventeenth and eighteenth centuries, that True Knowledge in any field almost required the Euclidean deductive form as a seal of legitimacy. The more disorderly branches of knowledge, which evaded this pattern, were considered to be somehow less respectable, a stage or two beneath the aristocratic disciplines.

Thus Spinoza's *Ethics,* whose subjects are God and the human passions, consists of definitions, axioms, and propositions which he attempts to support by means of proofs in the Euclidean manner.[7] Kant taught that the axioms of Euclidean geometry are imposed on our minds *a priori,* and are therefore necessary modes of perceiving space; and he built his entire system of philosophy on this foundation. Newton's *Principia,* with its empirical content centered on the laws of motion and the astronomy of the solar system, is wholly dominated by the Euclidean scheme of definitions, axioms, lemmas, propositions, corollaries, and proofs, with a liberal sprinkling of Q.E.D.'s. The seventeenth century doctrine of natural rights proclaimed by Locke was an attempt to deduce the laws of politics and government from axioms of a Euclidean type.[8] Even the American Declaration of Independence, in saying "We hold these truths to be self-evident," was seeking clarity and credibility by emulating the Euclidean model.

Unfortunately, self-evident truths are much scarcer now than they used to be. Since the advent of general relativity and cosmology, it has been known that Euclidean geometry is not an adequate mathematical framework for the universe at large, and in this sense is no longer "true." Since the advent of non-Euclidean geometries, it has been known that Euclid's axioms are not self-evident at all: on the contrary, they can be replaced by others which contradict them and have just as good a claim to acceptance from the point of view of logic. Axioms in government and human behavior are now recognized as hopes and expressions of preference rather than immutable truths.

In spite of these lost illusions, the axiomatic method first elaborated by Euclid is still widely used in the more abstract parts of higher mathematics as a convenient way of clearly delineating the mathematical system to be investigated. It is no exaggeration to say that modern abstract mathematics could hardly exist without this method.

All in all, for more than 2000 years the intellectual architecture of the *Elements* has rivaled the Parthenon as a symbol of the Greek genius. Both have deteriorated somewhat in recent centuries, but perhaps the book has sustained less damage than the building.

---

[7] "I shall consider human actions and desires just as if I were studying lines, planes, and solid bodies."—*Ethics,* Part III, Introduction.

[8] "To understand political power right, and derive it from its original, we must consider what state all men are naturally in, and that is, a state of perfect freedom to order their actions, and dispose of their possessions and persons, as they think fit."—*Second Treatise of Government,* Section 4.

A.5

ARCHIMEDES
(ca. 287–212 B.C.)

There was more imagination in the head of Archimedes than in that of Homer.

Voltaire

Archimedes will be remembered when Aeschylus is forgotten, because languages die and mathematical ideas do not.

G. H. Hardy

Archimedes, who was certainly the greatest mathematician, physicist, and inventor of the ancient world, was one of the supreme intellects of Western civilization. Another genius of comparable power and creativity did not appear until Isaac Newton in the seventeenth century.

Archimedes was born in the Greek city of Syracuse on the island of Sicily. He was on intimate terms with the royal family, and was probably a relative of King Hieron II. In his youth he studied in the great intellectual center of Alexandria. It was perhaps during this period that he met his friend Eratosthenes, later director of the Alexandrian Library, to whom he communicated many of his discoveries. On his return to his native city, he settled down and devoted the rest of his life to mathematicial research. At the age of 75 he was killed by a Roman soldier when Syracuse was conquered by the army of Marcellus during the Second Punic War.

Archimedes was famous throughout the Greek world during his lifetime, and has been a legendary figure ever since, not because of his profound mathematical discoveries, but rather because of his vivid and memorable achievements, his many ingenious inventions, and the manner of his death. Few solid facts are known, but traditional accounts of his activities are found in the writings of numerous Roman, Greek, Byzantine, and Arabic authors down through the centuries. He put the stamp of his personality on the world, and the world has not forgotten.[1]

Perhaps the most famous of the traditional stories concerns the occasion when he was asked by King Hieron to determine whether a newly made crown was made of pure gold, as specified, or whether the goldsmith had cheated him by substituting silver for some of the gold. Archimedes was perplexed until one day he noticed the overflow of water as he stepped into a public bath. He suddenly realized that since gold is denser than silver, a given weight of gold occupies a smaller volume than an equal weight of silver, and will therefore displace less water than an equal weight of silver. He was so overjoyed with his discovery that he forgot he was naked and ran home through the streets of the city without his clothes, shouting "Eureka, eureka!" which means "I have found it, I have found it!" He quickly determined that Hieron's new crown displaced more water than an equal weight of gold, thereby convicting the goldsmith of fraud. This story is often associated with his discovery of the basic law of hydrostatics, which states that a floating body displaces its own weight of liquid. From this beginning he created the science of hydrostatics, and proved many theorems about equilibrium positions of floating bodies of various shapes.[2] Further, one of his best-known inventions was a spiral-shaped water pump called the "screw of Archimedes." This device is still used along the Nile for raising water from the river to the adjoining fields.[3]

In mechanics, he discovered the principle of the lever, originated the concept of center of gravity, and found the centers of gravity of many plane and solid figures.[4] According to one writer, it was his study of levers that led him to utter his famous saying, "Give me a place to stand on, and I can move the earth." Plutarch gives another version:

Archimedes one day asserted to King Hieron, whose friend and kinsman he was, that with a given force he could move any given weight whatever; and he further claimed that if he were given another earth, he could cross over to it and move this one. When Hieron, full of wonder, begged him to give a demonstration of

---

[1] For information about the sources, see the introductory chapters (with references) of E. J. Dijksterhuis, *Archimedes*, Humanities Press, 1957; or T. L. Heath, *The Works of Archimedes*, Dover, n.d. The most detailed account is that given by Plutarch in his *Life of Marcellus*.

[2] See the treatise *On Floating Bodies, Works*, pp. 252–300.

[3] The present writer has actually seen this device being used on the banks of the Nile.

[4] See the treatise *On the Equilibrium of Planes, Works*, pp. 189–220.

some great weight moved by a small force, Archimedes caused one of the king's galleys to be drawn up on shore by many men and great labor; and loading her with many passengers and much freight, he placed himself at a distance, and without visible effort, only moving with his hand the end of a machine, which consisted of a variety of ropes and pulleys, he drew the ship to him as smoothly and safely as if she were moving through the water.

Hieron was so astounded at this feat that he declared, "From this day forth Archimedes is to be believed in everything he says."

Archimedes' greatest fame in antiquity came from the many stories of the engines of war he devised to defend Syracuse against the army and navy of the Roman general Marcellus. Plutarch devotes several vivid pages of description to the attacks of the Romans and the shattering effect of Archimedes' defensive machines. There were catapults of adjustable range for hurling enormous stones, huge movable beams for projecting suddenly over the walls and dropping heavy weights on enemy galleys that approached too close, and giant grappling cranes that caught hold of ships by the prow, lifted them up, and plunged them to the bottom of the sea. There were even burning mirrors that set the enemy ships on fire at a distance.[5] As Plutarch writes:

> The Romans, being infinitely distressed by an invisible enemy, began to think they were fighting against the gods. Marcellus escaped unhurt, and deriding his own engineers said, "We must give up fighting with this geometrical Briareus [a hundred-armed mythological monster], who sitting on the shore and acting as if it were only a game, plays pitch-and-toss with our ships, and striking us in a moment with such a multitude of bolts, outdoes even the hundred-handed giants of mythology." At last the Romans were so terrified, that if they saw only a rope or a stick put over the walls, they cried out that Archimedes was leveling some engine at them, and turned their backs and fled.

Thus Marcellus abandoned his intention of assaulting the city and put his hopes on a siege. This siege of Syracuse lasted three years and ended in 212 B.C. with the fall of the city.

According to all accounts Archimedes died in a manner consistent with his life, absorbed in mathematical contemplation. In the general confusion and slaughter that followed the fall of the city, he was found concentrating on some diagrams he had drawn in the sand, and was killed by a marauding soldier who did not know who he was. In one version of the story he said to the intruder, who came too close, "Do not disturb my circles," whereupon the enraged soldier ran a sword through his body. Marcellus was greatly saddened by this, since he had given strict orders to his men to spare the house and person of

---

[5] For the details of a modern experiment by the Greek navy showing that this use of solar power in warfare is quite feasible, see *Newsweek,* Nov. 26, 1973, p. 64.

Archimedes. He mourned the death of his formidable antagonist, befriended his surviving relatives, and saw to it that he had an honorable burial. The eminent modern philosopher A. N. Whitehead found a larger meaning in this event than the death of a single man:

> The death of Archimedes at the hands of a Roman soldier is symbolic of a world change of the first magnitude. The Romans were a great race, but they were cursed by the sterility which waits upon practicality. They were not dreamers enough to arrive at new points of view, which could give more fundamental control over the forces of nature. No Roman lost his life because he was absorbed in the contemplation of a mathematical diagram.

Archimedes is said to have asked his friends to place on his tombstone a representation of a cylinder circumscribing a sphere, and in memory of his greatest mathematical achievement to inscribe the proportion 3/2 which the containing solid bears to the contained. This was done at the order of Marcellus. The Roman orator Cicero, when he was quaestor in Sicily in 75 B.C., searched out this monument, discovered it neglected and overgrown with brambles, and had it cleaned and restored out of respect for the great mathematician.[6]

Cicero also saw and described an invention of Archimedes that made such a deep impression on the ancient world that it is mentioned by many classical authors. This device was apparently a small planetarium, an open revolving sphere of bronze and glass with internal machinery driven by water, in which during one revolution the sun, moon, and five planets moved in the same way relative to the sphere of fixed stars as they do in the sky in one day, and in which one could also observe the phases and eclipses of the moon. Closed spheres like modern terrestrial globes that revolved uniformly and imitated the daily motion of the fixed stars had long been known, but that Archimedes was able to represent by one mechanism the independent and very different motions of the sun, moon, and planets, together with the revolution of the fixed stars, seemed to his contemporaries to be evidence of superhuman abilities. Cicero, having seen this planetarium himself, writes:

> When Gallus set the sphere in motion, one could actually see the moon rise above the earth's horizon after the sun, just as occurs in the sky every day; and then one saw how the sun disappeared and how the moon entered the shadow of the earth with the sun on the opposite side.[7]

---

[6] See Cicero's *Tusculan Disputations,* Loeb Classical Library, p. 491. The Romans were so uninterested in mathematics that Cicero's act of respect in cleaning up Archimedes' grave was perhaps the most memorable contribution of any Roman to the history of mathematics.

[7] See Cicero's *De Re Publica,* Loeb Classical Library, p. 43.

This hydraulic mechanism was captured by the Romans as part of their booty in the sack of Syracuse, and it was evidently treasured by them for more than a hundred years afterward as one of the wonders of the world.

Archimedes was clearly a very ingenious and successful inventor, but Plutarch says that his inventions were only "the diversions of geometry at play." In a famous passage he purports to tell us about Archimedes' attitude toward practical life in general and his own inventions in particular:

> Archimedes possessed such a lofty spirit, so profound a soul, and such a wealth of scientific knowledge, that although his inventions had won him renown for superhuman sagacity, he would not deign to leave behind him any written work on these subjects; but regarding as ignoble and vulgar the construction of instruments and every art concerned with use and profit, he devoted his whole ambition and effort to those studies whose beauty and subtlety have no relation to the practical needs of life.

Though eloquent, the truth of what Plutarch says here is more than doubtful, because Archimedes is known to have written a treatise that is now lost (*On Sphere-making*) which probably dealt with the detailed techniques required for the construction of his planetarium. Plutarch was thoroughly infected, as Archimedes certainly was not, by the Platonic contempt for scientific instruments and measurement that was one of the many foolish legacies of the philosopher Plato to his worshipful posterity.

Nevertheless, it is very clear that in pure mathematics Archimedes was able to satisfy to the full the deepest desires of his nature. Plutarch is more convincing when he tells us that few people have ever lived who were so preoccupied with mathematics as he was:

> We are not, therefore, to reject as incredible, what is commonly told of him, that being perpetually charmed by his familiar Siren, that is, by his geometry, he neglected to eat and drink and took no care of his person; that he was often carried by force to the baths, and when there he would trace geometrical figures in the ashes of the fire, and with his finger draw lines upon his body when it was anointed with oil, being in a state of great ecstasy and divinely possessed by his science.

What, precisely, *were* his achievements in mathematics? Most of his wonderful writings still survive, and even on cursory inspection are obvious works of genius. Virtually all of the subject matter of his nine treatises is completely original, and consists of entirely new discoveries of his own. Though he treated a wide range of subjects, including plane and solid geometry, arithmetic, astronomy, hydrostatics, and mechanics, he was no compiler of earlier discoveries like Euclid, no mere writer of textbooks. His aim was always to provide some new contribution to knowledge. As to the overall impression conveyed by his works, Heath says:

> The treatises are, without exception, monuments of mathematical exposition; the gradual revelation of the plan of attack, the masterly ordering of the propositions, the stern elimination of everything not immediately relevant to the purpose, the finish of the whole, are so impressive in their perfection as to create a feeling akin to awe in the mind of the reader. As Plutarch said [with understandable exaggeration], "It is not possible to find in geometry more difficult and troublesome questions or proofs set out in simpler and clearer propositions." There is at the same time a certain mystery veiling the way in which he arrived at his results. For it is clear that they were not *discovered* by the steps which lead up to them in the finished treatises.[8]

Thus, from one point of view his writings present an aspect of austere architectural perfection. On the other hand, in most of the mathematical treatises (though not in those concerned with physics) there is a personal preface in which he addresses his friends, explains his purposes, and generally sets the stage for the intellectual drama to come. Compared with Euclid, his writings are throbbing with life.

The range and importance of the mathematical work of Archimedes can best be understood from a brief account of his six geometrical treatises, three dealing with plane geometry, two with solid geometry, and one with his method of making discoveries.

1. Quadrature of the parabola. This treatise of 24 propositions contains two proofs of his theorem that the area of a segment of a parabola, that is, the region cut from a parabola by any transverse line, is $\frac{4}{3}$ the area of the triangle with the same base and height. A complete proof of this theorem is provided in Section B.3. In his last two propositions Archimedes sums the infinite geometric series $1 + \frac{1}{4} + (\frac{1}{4})^2 + \cdots$ in such a way as to demonstrate that he is fully aware of the subtlety of the concept of limit. This did not become clear to other mathematicians until the early nineteenth century.

2. On spirals. The subject of this treatise of 28 propositions is the curve now known as the *spiral of Archimedes*. He defines it as follows:

> If a straight line of which one extremity remains fixed be made to revolve at a uniform rate in a plane until it returns to the position from which it started, and if, at the same time as the straight line revolves, a point moves at a uniform rate along the straight line, starting from the fixed extremity, the point will describe a spiral in the plane.

---

[8] T. L. Heath, *A History of Greek Mathematics*, Oxford University Press, 1921, vol. II, p. 20.

His main achievements were to determine the tangent at any point (Prop. 20) and to find the area enclosed by the first turn (Prop. 24), the latter being $\frac{1}{3}$ the area of the circle whose radius is the distance the moving point travels along the moving line. Later mathematicians used this spiral as an auxiliary curve for trisecting an angle and squaring a circle. Also, it has been conjectured that he discovered how to determine the tangent at a point by methods verging on those of differential calculus.[9]

3. Measurement of a circle. In this short work of three propositions he demonstrates with full rigor, as no one had done before him, that the area of a circle is equal to that of a triangle with base equal to its circumference and height equal to its radius, $A = \frac{1}{2}cr$; and since $c = 2\pi r$ by the definition of π, we have the familiar formula $A = \pi r^2$. He also establishes the inequalities

$$3\tfrac{10}{71} < \pi < 3\tfrac{1}{7},$$

by an elaborate calculation of the perimeters of regular polygons of 96 sides inscribed in and circumscribed about a given circle.

4. On the sphere and cylinder. This is the profoundest of the treatises, for it contains rigorous proofs of his great discoveries of the volume and surface area of a sphere (Props. 33 and 34), as well as much else besides. As to how he made these discoveries, see item 6 below.

5. On conoids and spheroids. By these terms Archimedes means solids of revolution generated by revolving parabolas, hyperbolas, and ellipses about their axes. He calculates the volumes of segments of these solids, and incidentally proves and uses the formulas

$$1 + 2 + \cdots + n = \frac{n(n+1)}{2}$$

and

$$1^2 + 2^2 + \cdots + n^2 = \frac{n(n+1)(2n+1)}{6}$$

for the sums of the first n integers and squares (see pp. 162 and 105–109 of the *Works*). Also, he proves the formula πab for the area of an ellipse with semiaxes a and b (Prop. 4).

6. Method. This, the most interesting of all the treatises, takes the form of a letter to Eratosthenes in which Archimedes explains his method of making

---

[9] See the Appendix to vol. II of Heath's *History*.

discoveries in geometry and illustrates his ideas with 15 propositions. This work was accidentally discovered on a palimpsest parchment in Constantinople in 1906, after having been lost for nearly a thousand years. As Heath says:

> The *Method,* so happily recovered, is of the greatest interest for the following reason. Nothing is more characteristic of the classical works of the great geometers of Greece, or more tantalizing, than the absence of any indication of the steps by which they worked their way to the discovery of their great theorems. As they have come down to us, these theorems are finished masterpieces which leave no traces of any rough-hewn stage, no hint of the method by which they were evolved. We cannot but suppose that the Greeks had some method or methods of analysis hardly less powerful than those of modern analysis; yet, in general, they seem to have taken pains to clear away all traces of the machinery used and all the litter, so to speak, resulting from tentative efforts, before they permitted themselves to publish, in sequence carefully thought out, and with definitive and rigorously scientific proofs, the results obtained. A partial exception is now furnished by the *Method*; for here we have a sort of lifting of the veil, a glimpse of the interior of Archimedes' workshop.

In one of Archimedes' illustrations of his method, he shows us how he discovered his favorite theorem about the volume of a sphere. The details of his ideas are given in Section B.6; and, as we point out there, his way of thinking is essentially equivalent to the basic process of integral calculus. He then goes on to tell us how he was led by this to discover the surface area of a sphere, by thinking of a sphere as if it were a cone wrapped around its vertex:

> From this theorem, to the effect that a sphere is four times as great as the cone with a great circle of the sphere as base and height equal to the radius of the sphere, I conceived the notion that the surface of any sphere is four times as great as a great circle in it; for, judging from the fact that any circle is equal to a triangle with base equal to the circumference and height equal to the radius of the circle, I apprehended that, in like manner, a sphere is equal to a cone with base equal to the surface of the sphere and height equal to the radius.

Among other discoveries included in this treatise are two standard examples (or problems) often used in modern calculus textbooks, about the location of the center of gravity of a solid hemisphere (Prop. 6), and the volume common to two equal cylinders whose axes intersect at right angles (Prop. 15, also Preface).

In addition to these six geometrical treatises, and the two dealing with physics, there is one more that ought to be mentioned. This is concerned with arithmetic and astronomy and is called *The Sand-Reckoner.* In it he constructs a system of notation for designating very large numbers, a system that enables him (without our decimal system or exponent notation, which came two thousand years later) to express numbers as large as N^{N^N} where N is 10^8. He then applies his ideas to find an upper bound for the number of grains of sand that would fill a sphere whose radius is the distance from the sun to what Aristarchus called "the sphere of the fixed stars," and this turns out to be a mere 10^{63}. It is here, among interesting related comments about astronomy, that we learn that Aristarchus had propounded the Copernican theory of the solar system a few decades earlier; this was nearly two thousand years before the time of Copernicus.

Finally, if we consider what all other men accomplished in mathematics and physics, on every continent and in every civilization, from the beginning of time down to the seventeenth century in Western Europe, the achievement of Archimedes outweighs it all. He was a great civilization all by himself.

APPENDIX:
THE TEXT OF ARCHIMEDES

Students of literature, philosophy and the history of science are not always aware of how complex and difficult a problem it often is to determine what the great writers of the past actually wrote. In the case of ancient writers who worked many centuries before the invention of printing, this problem is particularly daunting: their handwritten manuscripts, copies of these manuscripts, and copies of these copies containing bogus interpolations and the errors of scribes at almost every stage, had to survive fires, floods, theft, wars, and general decay, to the point where it seems almost a miracle that anything is left at all.[10]

For example, when Aristotle died in 322 B.C. all of his books and writings came into the possession of his pupil and successor Theophrastus. When Theophrastus died 35 years later, the whole of his library, including the material from Aristotle, went to a certain Neleus of Scepsis, a town near the site of ancient Troy on the other (eastern) side of the Aegean Sea, and in Scepsis the manuscript treatises of Aristotle lay neglected and rotting in a cellar for the next 200 years. These manuscript treatises were then sold to a book-collector named Apellicon of Teos (near Ephesus), who had some of them copied, with many erroneous restorations of worm-eaten passages of the original text; were then seized as booty by the victorious Roman general Sulla, who took them home to Rome about 84 B.C.; and there became the basis of the edition of Aristotle's works published by Andronicus of Rhodes between

---

[10] See Gilbert Highet's essay "The Written Past: Loss, Destruction, and Survival," in his book *Explorations*, Oxford University Press, 1971.

43 and 20 B.C. Our knowledge of Aristotle's writings and ideas from that day to this has rested almost wholly on this fragile sequence of events.[11]

With these ideas in mind, the reader may find it interesting to follow the equally fragile sequence of events that enables us to know what Archimedes wrote, as put together by T. L. Heath in his *History* (vol. II, pp. 25–27):

> Heron, Pappus and Theon all cite works of Archimedes which no longer survive, a fact which shows that such works were still extant at Alexandria as late as the third and fourth centuries A.D. But it is evident that attention came to be concentrated on two works only, the *Measurement of a Circle* and *On the Sphere and Cylinder*. Eutocius (*fl.* about A.D. 500) only wrote commentaries on these works and on the *Plane Equilibriums,* and he does not seem even to have been acquainted with the *Quadrature of the Parabola* or the work *On Spirals,* although these have survived. Isidorus of Miletus revised the commentaries of Eutocius on the *Measurement of a Circle* and the two Books *On the Sphere and Cylinder,* and it would seem to have been in the school of Isidorus that these treatises were turned from their original Doric into the ordinary language, with alterations designed to make them more intelligible to elementary pupils. But neither in Isidorus's time nor earlier was there any collected edition of Archimedes's works, so that those which were less read tended to disappear.
>
> In the ninth century Leon, who restored the University of Constantinople, collected together all the works that he could find at Constantinople, and had the manuscript written (the archetype, Heiberg's A) which, through its derivatives, was, up to the discovery of the Constantinople manuscript (C) containing *The Method,* the only source for the Greek text. Leon's manuscript came, in the twelfth century, to the Norman court at Palermo, and thence passed to the House of Hohenstaufen. Then, with all the library of Manfred, it was given to the Pope by Charles of Anjou after the battle of Benevento in 1266. It was in the Papal Library in the years 1269 and 1311, but, some time after 1368, passed into private hands. In 1491 it belonged to Georgius Valla, who translated from it the portions published in his posthumous work *De expetendis et fugiendis rebus* (1501), and intended to publish the whole of Archimedes with Eutocius's commentaries. On Valla's death in 1500 it was bought by Albertus Pius, Prince of Carpi, passing in 1530 to his nephew, Rodolphus Pius, in whose possession it remained till 1544. At some time between 1544 and 1564 it disappeared, leaving no trace.
>
> The greater part of A was translated into Latin in 1269 by William of Moerbeke at the Papal Court at Viterbo. This translation, in William's own hand, exists at Rome (Cod. Ottobon. lat. 1850, Heiberg's B), and is one of our prime sources, for, although the translation was hastily done and the translator sometimes misunderstood the Greek, he followed its wording so closely that his version is, for purposes of collation, as good as a Greek manuscript. William used also, for his translation, another manuscript from the same library which

---

[11] All this on the authority of the Greek geographer–historian Strabo, lived in Rome at the end of the first century B.C.

contained works not included in A. This manuscript was a collection of works on mechanics and optics; William translated from it the two Books *On Floating Bodies,* and it also contained the *Plane Equilibriums* and the *Quadrature of the Parabola,* for which books William used both manuscripts.

The four most important extant Greek manuscripts (except C, the Constantinople manuscript discovered in 1906) were copied from A. The earliest is E, the Venice manuscript (Marcianus 305), which was written between the years 1449 and 1472. The next is D, the Florence manuscript (Laurent. XXVIII. 4), which was copied in 1491 for Angelo Poliziano, permission having been obtained with some difficulty in consequence of the jealousy with which Valla guarded his treasure. The other two are G (Paris. 2360) copied from A after it had passed to Albertus Pius, and H (Paris. 2361) copied in 1544 by Christopherus Auverus for Georges d'Armagnac, Bishop of Rodez. These four manuscripts, with the translation of William of Moerbeke (B), enable the readings of A to be inferred.

A Latin translation was made at the instance of Pope Nicholas V about the year 1450 by Jacobus Cremonensis. It was made from A, which was therefore accessible to Pope Nicholas though it does not seem to have belonged to him. Regiomontanus made a copy of this translation about 1468 and revised it with the help of E (the Venice manuscript of the Greek text) and a copy of the same translation belonging to Cardinal Bessarion, as well as another 'old copy' which seems to have been B.

The *editio princeps* was published at Basel (*apud Hervagium*) by Thomas Gechauff Venatorius in 1544. The Greek text was based on a Nürnberg MS. (Norimberg. Cent. V, app. 12) which was copied in the sixteenth century from A but with interpolations derived from B; the Latin translation was Regiomontanus's revision of Jacobus Cremonensis (Norimb. Cent. V, 15).

A translation by F. Commandinus published at Venice in 1558 contained the *Measurement of a Circle, On Spirals,* the *Quadrature of the Parabola, On Conoids and Spheroids,* and the *Sand-reckoner.* This translation was based on the Basel edition, but Commandinus also consulted E and other Greek manuscripts.

Torelli's edition (Oxford, 1792) also followed the *editio princeps* in the main, but Torelli also collated E. The book was brought out after Torelli's death by Abram Robertson, who also collated five more manuscripts, including D, G and H. The collation, however, was not well done, and the edition was not properly corrected when in the press.

The second edition of Heiberg's text of all the works of Archimedes with Eutocius's commentaries, Latin translation, apparatus criticus, &c., is now available (1910–15) and, of course, supersedes the first edition (1880–1) and all others. It naturally includes *The Method,* the fragment of the *Stomachion,* and so much of the Greek text of the two Books *On Floating Bodies* as could be restored from the newly discovered Constantinople manuscript.

The Constantinople manuscript, which is referred to as C in this account, is the sole source for the text of Archimedes' *Method.* It was discovered in 1906 by the Danish scholar Heiberg in what was certainly one of the greatest adventures of scholarship in the 20th century—or any other century. Here is

Heath's description of the circumstances of this discovery (see his edition of the *Works,* pp. 5–6 of the Introductory Note to the *Method*):

> From the point of view of the student of Greek mathematics there has been, in recent years, no event comparable in interest with the discovery by Heiberg in 1906 of a Greek MS. containing, among other works of Archimedes, substantially the whole of a treatise which was formerly thought to be irretrievably lost.
>
> The full description of the MS. as given in the preface to Vol. 1 (1910) of the new edition of Heiberg's text of Archimedes now in course of publication is—
>
> Codex rescriptus Metochii Constantinopolitani S. Sepulchri monasterii Hierosolymitani 355, 4to.
>
> Heiberg has told the story of his discovery of this MS. and given a full description of it. His attention having been called to a notice in Vol. IV (1899) of the Ἱεροσολυμιτικὴ βιβλιοθήκη of Papadopulos Kerameus relating to a palimpsest of mathematical content, he at once inferred from a few specimen lines which were quoted that the MS. must contain something by Archimedes. As the result of inspection, at Constantinople, of the MS. itself, and by means of a photograph taken of it, he was able to see what it contained and to decipher much of the contents. This was in the year 1906, and he inspected the MS. once more in 1908. With the exception of the last leaves, 178 to 185, which are of paper of the 16th century, the MS. is of parchment and contains writings of Archimedes copied in a good hand of the 10th century, in two columns. An attempt was made (fortunately with only partial success) to wash out the old writing, and then the parchment was used again, for the purpose of writing a Euchologion thereon, in the 12th–13th or 13th–14th centuries. The earlier writing appears with more or less clearness on most of the 177 leaves; only 29 leaves are destitute of any trace of such writing; from 9 more it was hopelessly washed off; on a few more leaves only a few words can be made out; and again some 14 leaves have old writing upon them in a different hand and with no division into columns. All the rest is tolerably legible with the aid of a magnifying glass. Of the treatises of Archimedes which are found in other MSS., the new MS. contains, in great part, the books *On the Sphere and Cylinder,* almost the whole of the work *On Spirals,* and some parts of the *Measurement of a Circle* and of the books *On the Equilibrium of Planes.* But the important fact is that it contains (1) a considerable proportion of the work *On Floating Bodies* which was formerly supposed to be lost so far as the Greek text is concerned and only to have survived in the translation by Wilhelm von Mörbeke, and (2), most precious of all, the greater part of the book called, according to its own heading, Ἐφοδοζ and elsewhere, alternatively, Ἐφόδιον or Ἐφοδικόν, meaning *Method.* The portion of this latter work contained in the MS. has already been published by Heiberg (1) in Greek, and (2) in a German translation with commentary by Zeuthen. The treatise was formerly only known by an allusion to it in Suidas, who says that Theodosius wrote a commentary upon it; but the *Metrica* of Heron, newly discovered by R. Schöne and published in 1903, quotes three propositions from it, including the two main propositions enunciated by Archimedes at the beginning as theorems novel in character which the method furnished a means of investigating.

A.6

APOLLONIUS
(ca. 262–190 B.C.)

He who understands Archimedes and Apollonius will admire less the achievements of the foremost men of later times.

Leibniz

Apollonius was born about 25 years later than Archimedes and thought of
himself as a rival of the great Silician mathematician, and for this he deserves
the sympathy we extend to all able people who stand in the shadows of the
very great. His most important work was his immense treatise entitled *Conics*,
on the geometry of the conic sections, that is, the ellipse, parabola and hyperbola.
Without this treatise, Kepler would never have been able to discover his laws
of planetary motion, one of which states that the orbit of each planet around
the sun is an ellipse with the sun at one focus (1609); without Kepler's laws,
Newton would probably have been unable to formulate his theory of universal
gravitation and his laws of motion (1687), which mark the real beginning of
modern physics and astronomy; and for this crucial influence on our own
civilization Apollonius deserves our respect and attention.

The little that is known of Apollonius' life is mostly what can be inferred
from the personal letters that form the prefaces to the various books of the
Conics. He was born at Perga, which is now a small but well-preserved ruin
near the modern Turkish city of Antalya on the southern coast of Asia Minor.
As a young man he went to Alexandria, where he studied with the successors
of Euclid and remained for many years. He visited the kingdom of Pergamum

on the western coast of Asia Minor, where he became acquainted with a certain Eudemus, to whom he dedicated the first two books of the *Conics*. During this visit he probably also met King Attalus I (230–197 B.C.), to whom he dedicated the fourth and following books of his treatise.

In his *Conics* Apollonius seems to have originated the genre of an exhaustive monograph on a specific, limited topic in mathematics. The work was prepared in eight "books," which means papyrus scrolls. The author says in his General Preface to Book I that "Volume 1," consisting of Books I–IV, deals with the elementary properties of conic sections—which presumably everyone interested in geometry ought to know—while "Volume 2," consisting of Books V–VIII, is intended more for specialists than for beginning students. Perhaps because of these statements by the author himself, Volume 2 was less widely distributed and many libraries did not acquire it, with the result that it did not survive in the original Greek: Book VIII, the last "chapter" in Volume 2, is completely lost, whereas Books V–VII survived only in Arabic translation.[1]

The *Conics* is very long (387 Propositions) and quite difficult for a modern student or mathematician to read, because of the extreme generality of the treatment, the elaborate structure of the whole, and the cumbersome verbal statements of many complicated theorems made necessary by the minimal use of symbols.[2]

The conic sections had been studied for about a century and a half before the time of Apollonius. However, he organized, generalized, and greatly extended this earlier work. For instance, these curves had previously been defined by means of three types of right circular cones, according as the vertex angle of the cone is less than, equal to, or greater than a right angle; by cutting such cones by a plane perpendicular to a generator (or element), one obtains an ellipse, a parabola, and a hyperbola, respectively. Apparently for the first time, Apollonius showed that it is not necessary to take sections perpendicular to a generator, and that all three curves can be obtained from a single cone simply by varying the inclination of the cutting plane (Fig. A.11); and incidentally, he introduced the names *ellipse, parabola, hyperbola*. Also, he replaced the previous single-napped cone (resembling a single ice cream cone) by a double-napped cone (resembling two ice cream cones placed end to

---

[1] The first printed edition of the Greek text of Books I–IV, with a Latin translation of Books V–VII, was published in 1710 by Newton's friend Edmund Halley (of Halley's Comet). Halley—a very remarkable and interesting man—learned Arabic for the specific purpose of making this translation.

[2] Surprisingly, Books I–IV are included in the well-known set of books entitled *Great Books of the Western World* (1952). The editors (Robert M. Hutchins and Mortimer J. Adler) to appear to believe that Apollonius is suitable leisure-time cultural reading for liberally educated citizens of 20th century America.

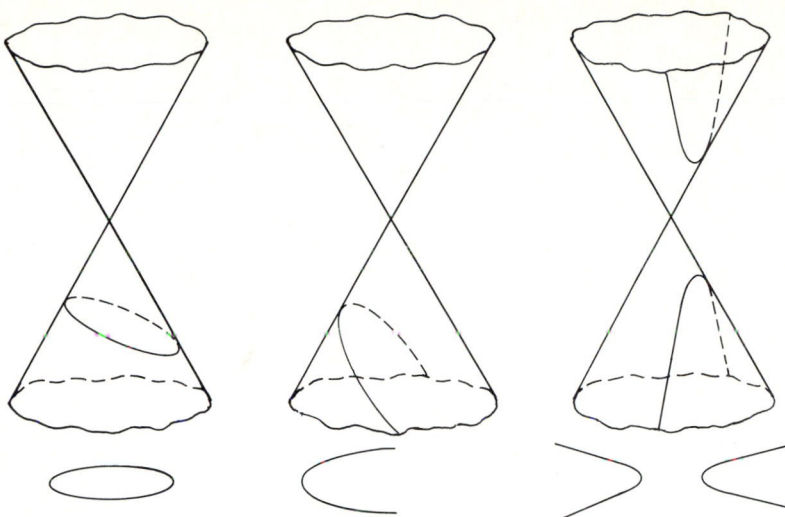

FIGURE A.11

end), thereby considering the two branches of a hyperbola as two parts of a single curve rather than two separate curves. Further, he showed that it is not necessary to use *right* circular cones; *oblique* circular cones will do just as well.

Perhaps his most original and far-reaching contribution is in Book V, where he discusses normals to a conic section as minimum and maximum straight lines drawn from a given point to the curve, and tangents as lines perpendicular to normals. It is impossible to study the orbits of the planets without reference to the tangents to an ellipse, and Kepler made indispensable use of this part of the *Conics*.

Most of Apollonius' other treatises are lost, but various remains and traces have survived in the form of partial Arabic translations and fairly full descriptions by other ancient writers. These remains and traces were used as the basis for several 'restorations' in the 18th century.

One of the most interesting of these lost works was called *Tangencies*, and dealt with a set of tangency problems that have intrigued geometers ever since. These problems require the construction of a circle tangent to each of three given circles, where the given circles are allowed to degenerate independently into points (by shrinking) or lines (by expanding infinitely). The two easiest cases, involving three points or three lines, had appeared in Euclid's *Elements* in connection with constructing circumscribed and inscribed circles for a triangle. The most difficult case, and the most interesting historically, is that of constructing a circle tangent to each of three nondegenerate circles. This is now known as the *problem of Apollonius*. It attracted the attention of many later mathematicians, including Vieta and Newton.[3]

Another lost work was called *Comparison of the Dodecahedron With the Icosahedron*. It contained the remarkable theorem that if a dodecahedron and an icosahedron are inscribed in the same sphere, then the ratio of the volumes of the two solids equals the ratio of their surface areas. This was established as an easy consequence of the theorem that the plane pentagonal faces of the dodecahedron are the same distance from the center of the sphere as are the plane triangular faces of the icosahedron (think of the formula for the volume of a pyramid).

It is abundantly clear that the Greeks studied geometry for the sheer pleasure of it, as a form of play, without any thought of possible applications to science or practical life. For them, geometry—and perhaps also philosophy and drama—were serious games, in much the same way as chariot races and gladiatorial combats were for the Romans. If we consider the Greeks, and the Romans, and perhaps ourselves as well from this point of view, it begins to appear that nothing reveals the quality of a culture so clearly as the games it plays.

APPENDIX:

APOLLONIUS' GENERAL PREFACE TO HIS TREATISE[4]

Apollonius to Eudemus, greeting.

If you are in good health and things are in other respects as you wish, it is well; with me too things are moderately well. During the time I spent with you at Pergamum I observed your eagerness to become acquainted with my work in conics; I am therefore sending you the first book, which I have corrected, and I will forward the remaining books when I have finished them to my satisfaction. I dare say you have not forgotten my telling you that I undertook the investigation of this subject at the request of Naucrates the geometer, at the time when he came to Alexandria and stayed with me, and, when I had worked it out in eight books, I gave them to him at once, too hurriedly, because he was on the point of sailing; they had therefore not been thoroughly revised, indeed I had put down everything just as it occurred to me,

---

[3] The problem of Apollonius has eight solutions, which are ingeniously illustrated on p. 2002 (vol. 3) of *The World of Mathematics,* ed. James R. Newman, Simon and Schuster, 1956.

[4] T. L. Heath, *A History of Greek Mathematics,* Oxford University Press, 1921, vol. II, pp. 128–30, with a few adjustments to smooth out archaic terminology.

postponing revision till the end. Accordingly I now publish, as opportunities serve from time to time, instalments of the work as they are corrected. In the meantime it has happened that some other persons also, among those whom I have met, have got the first and second books before they were corrected; do not be surprised therefore if you come across them in a different shape.

Now of the eight books the first four form an elementary introduction. The first contains the modes of producing the three sections and the opposite branches of the hyperbola, and the fundamental properties subsisting in them, worked out more fully and generally than in the writings of others. The second book contains the properties of the diameters and the axes of the sections as well as the asymptotes, with other things needed for determining limits of possibility; and what I mean by diameters and axes you will learn from this book. The third book contains many remarkable theorems useful for the construction of solid loci and for limits of possibility; the largest number and most beautiful of these theorems are new, and it was their discovery which made me aware that Euclid did not work out the construction of the locus with respect to three and four lines, but only a chance portion of it and that not successfully; for it was not possible for the said construction to be completed without the aid of the additional theorems discovered by me. The fourth book shows in how many ways the sections of cones can meet one another and the circumference of a circle; it contains other things in addition, none of which have been discussed by earlier writers, namely the questions in how many points a section of a cone or a circumference of a circle can meet a double-branch hyperbola, or two double-branch hyperbolas can meet one another.

The rest of the books are more specialized: one of them deals somewhat fully with *minima* and *maxima,* another with equal and similar sections of cones, another with theorems of the nature of determinations of limits, and the last with determinate conic problems. But of course, when all of them are published, it will be open to all who read them to form their own judgement about them, according to their own individual tastes. Farewell.

A.7

HERON
(first century A.D.)

The most outstanding trait of his authorship is its clarity. A man who is always able to present his subject in such a way that it is readily understood, is a man who understands it himself. The literary cast of mind, which values ambiguity for its own sake, was alien to him.

A. G. Drachmann

Heron of Alexandria was an extremely ingenious mathematician, inventor and scientific engineer. He is best known in the history of mathematics for the formula, bearing his name, for the area of a triangle in terms of its sides:

$$A = \sqrt{s(s - a)(s - b)(s - c)},$$

where a, b, c are the sides and $s = \frac{1}{2}(a + b + c)$ is the semiperimeter. We know from Arabic sources in the Middle Ages that "Heron's formula" is due to Archimedes, but the proof given in Heron's *Metrica* is the earliest that survives (see Section B.1).

Heron is also well known to students of mathematics and physics for *Heron's principle of reflection* in elementary optics. The ancient Greeks were familiar with mirrors and burning glasses, but their theories of optics, especially those of the Pythagoreans and Platonists, were metaphysical fantasy rather than science. They knew very well as an empirical fact that the angles of incidence and reflection are equal when a ray of light traveling from A to B is reflected by a mirror ($\alpha = \beta$ in Fig. A.12). But it was Heron in his *Catoptrica*

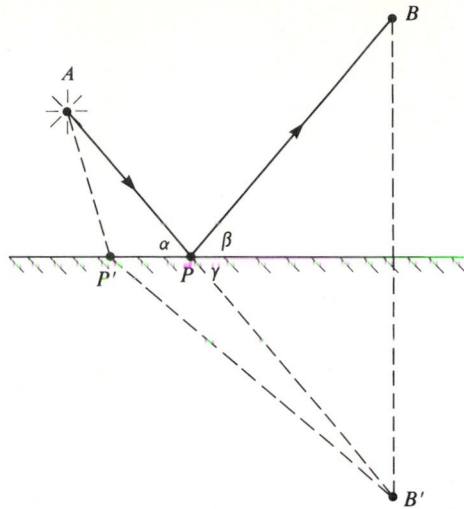

FIGURE A.12

(theory of mirrors) who first gave an elegant proof that this equality of angles can be derived from the principle that the ray of light must follow the shortest path in its journey from A to B by way of the mirror.[1] This was the first appearance of the far-reaching principle of least action that now dominates not only geometrical optics but also dynamics and most of mathematical physics as well.[2]

Nothing whatever is known about Heron's personal life apart from the association with Alexandria implied by his name—for he has always been called Heron "of Alexandria"—and the century in which he lived.[3] In fact, he is only a name attached to about a dozen works that were widely used in the Roman, Byzantine and Islamic Empires and were still familiar to many Renaissance scientists, in particular Galileo.

His writings suggest that he was probably an Egyptian with a Greek

---

[1] *Proof*: Let B' in Fig. A.12 be the mirror image of B, so that the surface of the mirror is the perpendicular bisector of the segment BB'. Then the total length of the path is $AP + PB = AP + PB'$; and if this length is required to be a minimum, then P must be chosen to lie on the straight line joining A and B', since for any other position P' the total length $AP' + P'B'$ is evidently greater because the sum of two sides of a triangle is greater than the third side. This implies that $\alpha = \gamma$, and since $\gamma = \beta$, we conclude that $\alpha = \beta$.

[2] See pp. 528–29 of George F. Simmons, *Differential Equations*, 2nd ed., McGraw-Hill, 1991.

[3] It was known for a very long time that Heron "flourished" during the first century. Unfortunately there were two first centuries, and the question of whether his activities belonged to B.C. or A.D. was not settled until 1938. In that year the eminent scholar Otto Neugebauer noticed that an eclipse of the moon described by Heron in such a way that he must have seen it with his own eyes, corresponds to an eclipse in A.D. 62 and to none other during the centuries in question.

education who taught a variety of subjects at the great Museum (essentially a university) in Alexandria. Some of his books appear to be workbooks for students of mathematics, with extensive lists of computational procedures, and occasional proofs, for finding areas and volumes of many geometric figures. Others are collections of instructions for making and using instruments and machines for surveyors and architects, including an odometer and theodolite invented by himself. For instance, his *Mechanica* is a treatise on theoretical and practical mechanics with an exposition of the theory of what he called the "five powers"—the wheel-and-axle, lever, pulley, wedge, and screw—and many examples of their practical uses, including screw-presses and several kinds of cranes.[4] He also left detailed instructions in his *Pneumatica* for making many kinds of moving toys and small mechanical contrivances powered by water, steam, and descending weights.

Among these curiosities was the first jet engine or primitive steam reaction turbine (Fig. A.13). This consisted of a hollow globe pivoted so that it could rotate freely on a pair of hollow tubes through which steam was supplied from a covered pot of water with a fire beneath. The steam escaped from the globe through two bent tubes facing in opposite directions at the ends of a diameter perpendicular to the axis, and the globe rotated by reaction to the escaping steam. Another device was his automatic temple door-opener (Fig. A.14), which was activated by lighting a fire on the altar and can be thought of

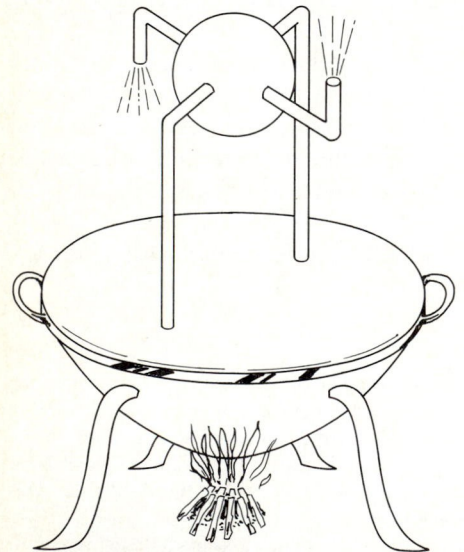

FIGURE A.13
Heron's steam turbine.

[4] See A. G. Drachmann, *The Mechanical Technology of Greek and Roman Antiquity,* University of Wisconsin Press, 1963; or A. P. Usher, *A History of Mechanical Inventions,* Harvard University Press, 1954.

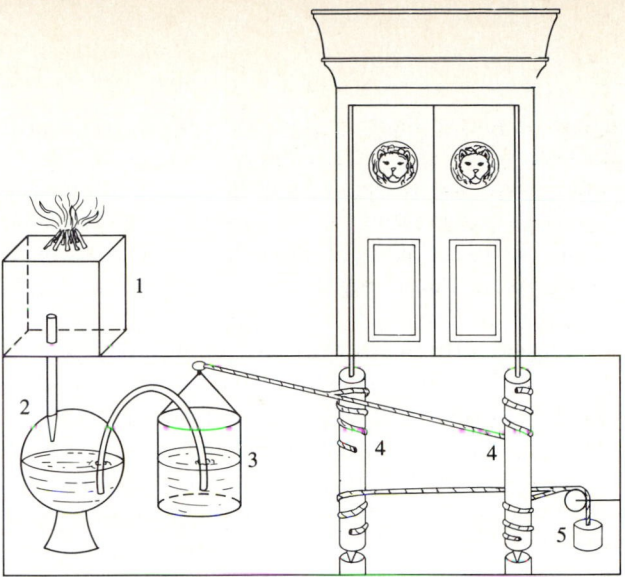

FIGURE A.14
Heron's automatic temple door-opener. A fire lit on an airtight hollow altar (1) raised the air pressure inside, forcing some of the water stored in the globe (2) through a siphon into the bucket (3). As the bucket descended, it pulled ropes attached to the pivots (4), which swung open the temple doors. When the fire went out and the air pressure in the altar decreased, water returned from 3 to 2 and a falling counterweight (5) shut the doors.

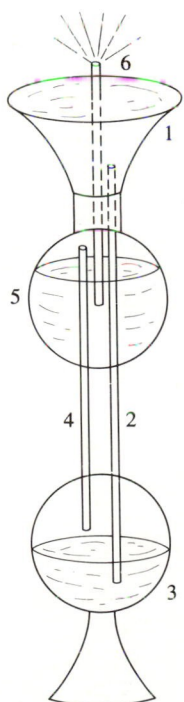

FIGURE A.15
Heron's Fountain. Water poured into the basin 1 flows down through tube 2 into globe 3, expelling air through tube 4 into globe 5, where the air pressure then forces water up through tube 6 into a jet.

as a primitive pressure engine. And yet another was the famous toy called "Heron's Fountain," which is shown in Fig. A.15. There were also siphons, water clocks, a fire engine with piston and valves, a wind organ, a trick jar that gave wine and water separately or mixed from the same spout, penny-in-the-slot machines, toys using steam to make a mechanical bird sing or cause a statue to blow a horn, and many other ingenious proofs of a playful imagination.

Heron's interest in such topics represents a radical departure from the more theoretical interests of the earlier Greek mathematicians—except Archimedes, of course, who would have loved it all. This was characteristic of the period, and some historians see a linkage between this movement toward applications in mathematics and science and the loss of vigor in philosophy and religion which led many people of the Hellenistic Age to sink into mysticism and cults.

A.8

PAPPUS
(fourth century A.D.)

Pappus of Alexandria was an able, enthusiastic, and elegant mathematician who had many good ideas. However, he had the bad luck to be born when the great age of Greek mathematics—which had lasted roughly 900 years, from the time of Thales and Pythagoras—was drawing its last breath.

His main work, the *Mathematical Collection,* is a combined encyclopedia, commentary, and guidebook for Greek geometry as it existed in his time, enriched by many new theorems, extensions and new proofs of old ones, and valuable historical remarks. Unfortunately, the *Collection* turned out to be the requiem of Greek mathematics instead of a breath of new life, for after Pappus, mathematics shriveled and almost disappeared, and had to wait 1300 years for a rebirth in the early seventeenth century.

He is best known for his beautiful geometric theorems connecting centers of gravity with solids and surfaces of revolution. The first of these theorems states (Fig. A.16) that the volume generated by the complete revolution of a region bounded by a closed plane curve that lies entirely on one side of the axis of revolution equals the product of the area of the region and the distance traveled by the center of gravity around its circular path. Pappus was rightly proud of the generality of his theorems, for, as he says, "they include any

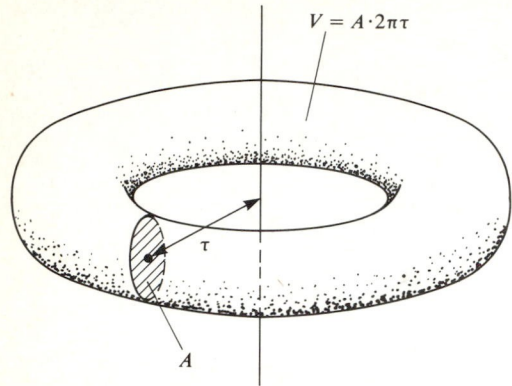

$$V = A \cdot 2\pi\tau$$

FIGURE A.16

number of theorems of all sorts about curves, surfaces, and solids, all of which are proved at once by one demonstration."[1]

He gave the first statement and proof of the focus–directrix–eccentricity characterization of the three conic sections (see the Appendix). Since he was scrupulous about crediting the sources of his material, and no source is given for this, it is reasonable to infer that it was his own discovery.

He gave the following interesting extension of the theorem of Pythagoras (see Fig. A.17): Let ABC be any triangle and $ACDE$ and $BCFG$ any parallelograms constructed externally on the sides AC and BC; if DE and FG intersect at H, and AJ and BI are equal and parallel to HC, then the area of the parallelogram $ABIJ$ equals the sum of the areas of the parallelograms $ACDE$ and $BCFG$ (proof: $ACDE = ACHR = ATUJ$ and $BCFG = BCHS = BIUT$). It is not difficult to see that this statement really does yield the theorem of Pythagoras as a special case, when the angle C is a right angle and the constructed parallelograms are squares.

Finally, we mention the important result of projective geometry known as Pappus' Theorem: If the vertices of a hexagon lie alternately on a pair of intersecting lines (Fig. A.18), then the three points of intersection of the opposite sides of the hexagon are collinear. (The "opposite" sides can be recognized from the numbered vertices of the schematic hexagon shown in the figure.) The full significance of this classic theorem was at last revealed only in 1899, by the great German mathematician David Hilbert, as part of his program to clarify the foundations of geometry.[2]

---

[1] T. L. Heath, *A History of Greek Mathematics,* Oxford University Press, 1921, vol. II, p. 403.

[2] For mathematicians: Hilbert discovered that the validity of Pappus' Theorem in a Desarguesian projective plane is equivalent to the commutativity of the field of coefficients. See pp. 82–86 of A. Seidenberg, *Lectures in Projective Geometry,* Van Nostrand, 1962; or E. Artin, *Geometric Algebra,* Interscience, 1957.

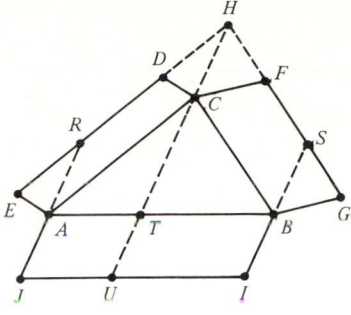

FIGURE A.17

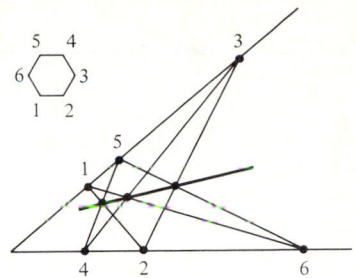

FIGURE A.18
Pappus' theorem.

APPENDIX:
THE FOCUS–DIRECTRIX–ECCENTRICITY DEFINITIONS OF THE CONIC SECTIONS

There are several distinct but equivalent ways of defining the conic sections, each with its own merits. In Section A.6 we gave the Apollonius definition, by means of a given right circular cone and a slicing plane that cuts through the cone more or less steeply, yielding our three curves by varying the degree of steepness. This three-dimensional approach is vivid and geometric, and provides a clear visual impression of what the curves look like (Fig. A.11). Our purpose in this appendix is to use this Apollonius concept to obtain an equivalent two-dimensional definition that refers only to the plane in which the curves lie. Thus, let there be given a line d (called the *directrix*) in this plane, a point F (called the *focus*) not on the line, and a positive number e (called the *eccentricity*). We shall see that a *conic section* can be defined as the path (or locus) of a point P that moves in such way that the ratio of the distances PF/FD (see Fig. A.19) is constant and equal to e, and that the ellipse, parabola and hyperbola correspond to the cases $e < 1$, $e = 1$ and $e > 1$.[3]

Our discussion is based on Fig. A.20, which shows a cone with vertex angle α and a slicing plane with tilting angle β. This tilting angle can be defined as the angle between the axis of the cone and a normal line to the plane, but it plays its main role in our argument as the indicated acute angle of the right triangle PQD. The figure is drawn to illustrate the case of an ellipse, but the argument is valid for the other cases as well.

---

[3] For reasons that will soon be clear, circles must be excluded from this discussion, because the necessary geometric constructions are not possible when the slicing plane is perpendicular to the axis of the cone.

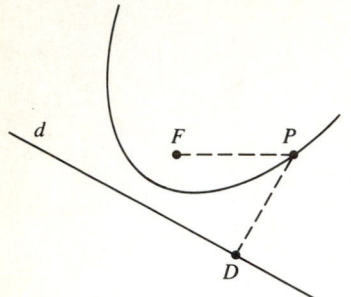

FIGURE A.19

We begin at the beginning. Let there be inscribed in the cone a sphere which is tangent to the slicing plane at a point F, and tangent to the cone along a circle C. If d is the line in which the slicing plane intersects the plane of the circle C, we shall prove that the conic section has F as its focus and d as its directrix, and the facts about the eccentricity will emerge in the course of our discussion.

To this end, let P be a point on the conic section, let Q be the point where the line through P and parallel to the axis of the cone intersects the plane of C, let R be the point where the generator through P intersects C, and let D be the foot of the perpendicular from P to the line d. Then PR and PF are two segments which are tangent to the sphere from the same point P, and therefore have the same length,

$$PR = PF. \tag{1}$$

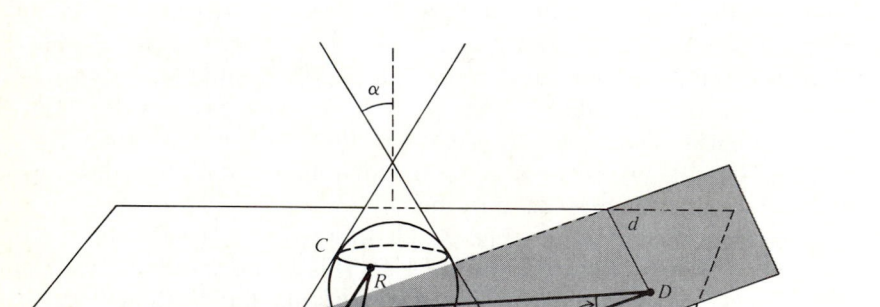

FIGURE A.20

Also, from the right triangle PQR we have

$$PQ = PR \cos \alpha;$$

and from the right triangle PQD we have

$$PQ = PD \sin \beta.$$

It follows that

$$PR \cos \alpha = PD \sin \beta,$$

so

$$\frac{PR}{PD} = \frac{\sin \beta}{\cos \alpha}.$$

In view of (1) this means that

$$\frac{PF}{PD} = \frac{\sin \beta}{\cos \alpha}.$$

This can be written in the slightly more convenient form

$$\frac{PF}{PD} = \frac{\cos \gamma}{\cos \alpha}, \tag{2}$$

where γ is the other acute angle in the right triangle PQD. If we now define the eccentricity e by

$$e = \frac{\cos \gamma}{\cos \alpha},$$

then this number is constant for a given cone and a given slicing plane, and (2) becomes

$$\frac{PF}{PD} = e \begin{cases} <1 & \text{for an ellipse} \\ =1 & \text{for a parabola} \\ >1 & \text{for a hyperbola,} \end{cases}$$

where the statements on the right are easily verified by inspecting the figure. Thus, for a parabola we see that PD is parallel to a generator of the cone, so $\gamma = \alpha$ and $e = 1$; for an ellipse, we have $\gamma > \alpha$, so $\cos \gamma < \cos \alpha$ and $e < 1$; and for a hyperbola, we have $\gamma < \alpha$, so $\cos \gamma > \cos \alpha$ and $e > 1$.

The words "parabola," "ellipse," and "hyperbola" come from three Greek words meaning "a comparison," "a deficiency," and "an excess," referring to the fact that for the corresponding curves we have $e = 1$, $e < 1$, and $e > 1$. One should also compare these words with the words "parable," "ellipsis," and "hyperbole" in modern English.

A.9

HYPATIA
(A.D. 370?–415)

In an era in which the domains of intellect and politics were almost exclusively male, Theon [her father] was an unusually liberated person who taught an unusually gifted daughter and encouraged her to achieve things that, as far as we know, no woman before her did or perhaps even dreamed of doing.

Ian Mueller

The mathematical life of the ancient world came to a sudden and violent end one March day in the year 415. On that day Hypatia, the first woman mathematician in history—beautiful, eloquent, and brilliant—was dragged from her carriage on a street in Alexandria and brutally slaughtered by a howling Christian mob.

Hypatia was the daughter of the mathematician Theon, who was the last man known to have been a professor in the Museum. This famous Alexandrian institution of higher learning had been founded seven hundred years earlier by the first Ptolemies, the rulers of Egypt after the death of Alexander the Great in 323 B.C. Its associated Library was intended to be the central repository for the world's memory—all the books in the world gathered and safeguarded in a single place for an elite of intellectuals. As Canfora writes:

Ptolemy I composed a letter "to all the sovereigns and governors on earth," imploring them "not to hesitate to send him" works by authors of every kind:

"poets and prose-writers, rhetoricians and sophists, doctors and soothsayers, historians, and all the others too." He gave orders that any books on board ships calling at Alexandria were to be copied: the originals were to be kept, and the copies given to their owners.[1]

It is not known how far the original ideal was realized, of collecting together not only all of Greek literature but also translations into Greek from the other languages of the Mediterranean and the Middle East. However, it is known that the earliest important translation of the Bible was carried out here—the Septuagint, the first five books of the Old Testament—from Hebrew into Greek; this was accomplished in the middle of the 3rd century B.C., at about the same time as the young Archimedes was a student at the Museum.

The Museum and Library survived for many centuries, but were reaching the end of their long life in Theon's time. This was a period of violent Christian–pagan confrontations. To the Christians the Museum was a nest of hated pagans, and during one of these riots in the year 391 they sacked and burned the Temple of Serapis, which then housed the main library of the Museum. This Temple was "second in splendour only to the Campidoglio at Rome. The marble, alabaster and priceless ivory of its furnishings had been smashed in fragments, and the parchment of its books burned splendidly."[2] After this orgy of destruction, according to contemporary testimony this district of the once great city was "now a desert."[3] Such was the state of affairs in Alexandria around the year 400, when Hypatia wrote and lectured on mathematics and Neoplatonic philosophy.

The scientific writings of Hypatia are completely lost, but we are told that she wrote learned commentaries on Diophantus and the *Conics* of Apollonius. The commentary was a form of explanatory exposition popular at that time, whose purpose was to help students grapple with difficult mathematical classics. We need not linger over Apollonius, whose work we have already discussed in Section A.6, but perhaps a few remarks about Diophantus are in order.

Of Diophantus himself nothing whatever is known, apart from his association with Alexandria and the Museum, and the speculation that he may have lived in the 3rd century A.D.[4] However, his treatise entitled *Arithmetica* had a substantial impact on modern mathematics through its influence in the 17th and 18th centuries on the great mathematicians Fermat and Euler, the

---

[1] See p. 20 of Luciano Canfora, *The Vanished Library,* Hutchinson Radius, London, 1989.

[2] Canfora, *op. cit.,* p. 91.

[3] Canfora, *op. cit.,* p. 195.

[4] See J. D. Swift, "Diophantus of Alexandria," *American Mathematical Monthly,* vol. 63 (1956), pp. 163–70.

founders of modern number theory. This treatise was one of the main glories of the mathematics section of the Museum in the time of Hypatia, and was certainly the focus for much of her teaching, writing and research.

To give a brief indication of the kinds of problems dealt with in this part of number theory, we point out that in modern terminology a *Diophantine equation* is an equation in which there is more than one unknown and in which positive integer solutions are sought. As an example we mention the classical problem of finding all right triangles with integral sides. By the theorem of Pythagoras this amounts to seeking positive integer solutions of the equation $x^2 + y^2 = z^2$. It is easy to see by inspection that 3, 4, 5 and 5, 12, 13 are solutions. But are there any others, and if so, what are they? A simple calculation shows that

$$x = p^2 - q^2, \qquad y = 2pq, \qquad z = p^2 + q^2 \qquad (1)$$

is always a solution if $p > q$, and the specific solutions just mentioned are obtained by taking $p = 2$, $q = 1$ and $p = 3$, $q = 2$. Diophantus knew that multiples of (1) comprise all possible solutions:

$$x = d(p^2 - q^2), \qquad y = d(2pq), \qquad z = d(p^2 + q^2).$$

The proof is not difficult, and is given in the Appendix for the convenience of readers who prefer to skip the details.

Unfortunately we know nothing at all about the particular topics Hypatia discussed in her lectures and commentaries on number theory and the conic sections. But whatever the content of her writings may have been, the fact that they existed establishes her as the first woman known to have written on mathematical subjects—a unique distinction, and one worthy of being remembered and celebrated.

Along with mathematics, Hypatia also cultivated philosophy with great success. She is said to have lectured on Plato, Aristotle and other philosophers, and drew to her lectures large audiences of enthusiastic pupils from many cities in Africa, Asia and Europe. The late 5th century church historian Socrates (*not* the Athenian philosopher of the 5th century B.C.) wrote that she "far surpassed all the philosophers of her time" and became the leader of the Neoplatonic school in Alexandria. Socrates reports that her home, as well as her lecture room, was crowded with the brightest young people of the day and was, along with the Museum and Library, one of the liveliest intellectual centers in the city.[5]

---

[5] The main sources for information about Hypatia, both ancient and modern, are given in the good account by Ian Mueller, pp. 74–79 of *Women of Mathematics*, Greenwood Press, Westport, Connecticut, 1987. It should also be mentioned that the 19th century English novelist Charles Kingsley, who thought of himself as competing on equal terms with Dickens, wrote a historical romance about Hypatia's life entitled *Hypatia: New Foes With an Old Face* (1853). Kingsley's Preface is quite interesting; but as for the novel itself, this is one of those books that the present writer would rather mention than actually read.

Hypatia is described in the *Suidas Lexicon* (a late 10th century Byzantine encyclopedia) as a beautiful and well-proportioned woman. It is generally thought that she never married, though students often fell in love with her. Suidas transmits a story that she once brought a desperately amorous would-be lover to his senses in a drastic and shocking manner.[6]

Hypatia's most famous pupil was Synesius of Cyrene, who later became the wealthy and influential Bishop of Ptolemais, a provincial city on the coast of what is now Libya. His many published letters include a number to Hypatia. They are full of affection and admiration and are a major source of information about her work and personality. He addresses her as "mother, sister, reverend teacher, and benefactress," and often asks for scientific advice. He wrote her, for example, to ask just how to construct an astrolabe and hydrometer of her own devising, and he discusses an astronomical instrument called a planisphere which they seem to have designed together.

She usually dressed in the well-aged cloak favored by many philosophers of antiquity, and like them often engaged in philosophical discussions while walking in the streets of her city. "Such was her confidence and ease of manner," says Socrates, "arising from the refinement and cultivation of her mind, that she often appeared before the leading citizens of Alexandria without ever losing in the company of men that dignified modesty of deportment for which she was conspicuous, and which gained for her universal respect and admiration."

But the admiration was not quite universal, and this led to the final events of her life that so much resemble the catastrophic conclusion of a classic Greek tragedy. Although she was on friendly terms with many Christians, she held to the views and traditions of pagan Hellenism. She was not only a brilliant and beautiful unbeliever, but also an intimate friend of Orestes, the prefect of the city who ruled its secular affairs on the authority of the Emperor Theodosius II in Constantinople. When Archbishop Cyril egged on his monastic followers to expel the Jews from Alexandria by force and violence, Orestes sent a damningly impartial report to the Emperor, thereby incurring Cyril's mortal enmity. It was believed by the Archbishop and his followers that Hypatia was a major force inciting Orestes against Cyril. Let the great historian Gibbon tell the rest of the story in a vivid paragraph from his *Decline and Fall of the Roman Empire*:[7]

> Hypatia, the daughter of Theon the mathematician, was initiated in her father's studies; her learned comments have elucidated the geometry of Apollonius and Diophantus; and she publicly taught, both at Athens and Alexandria, the

---

[6] J. B. Bury, *History of the Later Roman Empire*, vol. I, p. 217.

[7] Abridgement by D. M. Low, 1960, p. 601.

philosophy of Plato and Aristotle. In the bloom of beauty, and in the maturity of wisdom, the modest maid refused her lovers and instructed her disciples; the persons most illustrious for their rank or merit were impatient to visit the female philosopher; and Cyril beheld with a jealous eye the gorgeous train of horses and slaves who crowded the door of her academy. A rumor was spread among the Christians that the daughter of Theon was the only obstacle to the reconciliation of the prefect and the archbishop; and that obstacle was speedily removed. On a fatal day, in the holy season of Lent, Hypatia was torn from her chariot, stripped naked, dragged to the church, and inhumanly butchered by the hands of Peter the reader [a minor clerk on Cyril's staff] and a troop of savage and merciless fanatics: her flesh was scraped from her bones with sharp oyster-shells, and her quivering limbs were delivered to the flames. The just progress of inquiry and punishment was stopped by seasonable gifts; but the murder of Hypatia has imprinted an indelible stain on the character and religion of Cyril of Alexandria.

After these events the prefect Orestes disappeared from history, but Archbishop Cyril continued to be active in church politics and originated the theological position that was condemned as the Monophysite Heresy by the Council of Chalcedon in 451.[8]

APPENDIX: A PROOF OF DIOPHANTUS' THEOREM ON PYTHAGOREAN TRIPLES

We continue the discussion started in Section A.9 by noticing that if x, y, z are positive integers that satisfy the equation

$$x^2 + y^2 = z^2, \tag{1}$$

and if d is a common factor of x, y, z, then d^2 can be canceled out of the equation. In examining the form of solutions x, y, z, we may therefore assume without loss of generality that these positive integers are relatively prime, that is, that they have no common factors greater than 1. Under this assumption we shall prove that x, y, z are given by the formulas

$$x = p^2 - q^2, \qquad y = 2pq, \qquad z = p^2 + q^2$$

for certain positive integers p and q, where $p > q$, p and q are relatively prime, and one of the numbers p and q is even and the other is odd. For ease of understanding, we divide the argument into several distinct steps.

---

[8] See the article "Saint Cyril of Alexandria" in the *Encyclopaedia Britannica*.

1. First, we observe that the positive integers x and y are not both odd. For if both are odd, say $x = 2a + 1$ and $y = 2b + 1$, then we have $z^2 = x^2 + y^2 = 4a^2 + 4a + 1 + 4b^2 + 4b + 1 = 4(a^2 + a + b^2 + b) + 2 = 4n + 2$, and this contradicts the fact that any square must have the form $4n$ or $4n + 1$.[9]

2. Next, the positive integers x and y are relatively prime and are therefore not both even. For if they have a common factor $k > 1$, so that $x = ku$ and $y = kv$, then $z^2 = x^2 + y^2 = k^2(u^2 + v^2)$, so k^2 is a factor of z^2 and therefore k is a factor of z.[10] This contradicts our assumption that x, y, z are relatively prime, so x and y must be relatively prime. In just the same way, since $y^2 = z^2 - x^2$ and $x^2 = z^2 - y^2$, we know that x, z and y, z are also relatively prime.[11]

3. Since one of the numbers x, y is even and the other is odd, and equation (1) is symmetric in x and y, we may assume that x is odd and y is even. Then x^2 is odd and y^2 is even, so $z^2 = x^2 + y^2$ is odd and therefore z is odd. We can write

$$y^2 = z^2 - x^2 = (z + x)(z - x), \tag{2}$$

and since x and z are both odd, $z + x$ and $z - x$ are both even and we have

$$z + x = 2r, \qquad z - x = 2s, \tag{3}$$

where r, s are positive integers [(1) implies that $x < z$] and $r > s$. By adding and subtracting equations (3), we obtain

$$z = r + s, \qquad x = r - s, \tag{4}$$

where r and s are relatively prime because x and z are. Therefore r and s cannot both be even. Also, r and s cannot both be odd, for if they were, by (4) x and z would both be even, whereas we know that both are odd.

4. Substituting (3) in (2) gives

$$y^2 = 4rs. \tag{5}$$

---

[9] Every integer has precisely one of the forms $4n$, $4n + 1$, $4n + 2$, $4n + 3$. But even and odd numbers $2m$ and $2m + 1$ have squares

$$(2m)^2 = 4m^2 = 4(m^2) = 4n$$

and

$$(2m + 1)^2 = 4m^2 + 4m + 1 = 4(m^2 + m) + 1 = 4n + 1.$$

[10] The last inference follows from the prime factorization theorem, namely, that every integer > 1 can be expressed uniquely as a product of primes.

[11] It is worth noticing that in general it can easily happen that three numbers x, y, z are relatively prime and that no two of them are relatively prime, for example, $x = 6$, $y = 10$, $z = 15$. The difference is that in our situation the numbers we are considering satisfy equation (1).

Since y is even, $y = 2t$ and $y^2 = 4t^2$, so by (5)

$$t^2 = rs.$$

Since r and s are relatively prime, it follows from the unique factorization theorem [footnote (2)] that r and s are both squares,

$$r = p^2, \qquad s = q^2,$$

where $p > q$ because $r > s$. We see that p and q are relatively prime, because any common factor >1 would be a common factor of r and s, and they are relatively prime. Further, one of the numbers p and q is even and the other is odd, because this is true of r and s.

5. Finally, by (4) and (5) we have

$$x = p^2 - q^2, \qquad y = 2pq, \qquad z = p^2 + q^2, \tag{6}$$

where $p > q$, p and q are relatively prime, and one of the numbers p and q is even and the other is odd—as promised at the beginning.

Triples of positive integers x, y, z that satisfy the Pythagorean equation (1) are often called *Pythagorean triples*. The effect of the ideas we have developed here is to prove that multiples of the solutions (6) comprise all possible Pythagorean triples.

For readers who wish to pursue these matters further, with liberal servings of the fascinating history that surrounds early number theory, we recommend Chapter 14 of H. Rademacher and O. Toeplitz, *The Enjoyment of Mathematics*, Princeton University Press, 1957; and Chapter I of André Weil, *Number Theory: An Approach Through History*, Birkhäuser, Boston, 1984.

A.10

KEPLER
(1571–1630)

The story of his persistence in spite of persecution and domestic tragedies that would have broken an ordinary man is one of the most heroic in science.

E. T. Bell

. . . an unanalyzable compound of rationality and irrationality.

Salomon Bochner

Johannes Kepler is one of the great watershed figures in the history of science: half his mind churned with medieval fantasies and the other half was pregnant with the beginnings of the mathematicized science that formed the modern world.

Besides being talented and immensely hard-working—neither of which is of much use without the other—he was friendly, trusting, bursting with imagination, and filled with enthusiasm for new ideas. These are not qualities normally associated with pious German Lutherans, but tenacity is such a quality. Though bullied by his mentor, the Danish astronomer Tycho Brahe, ignored by Galileo, who of all people in the world should have welcomed his discoveries, and kept as a lapdog court astrologer by several successive German emperors, Kepler held fast to his vision throughout his life. He sought with fanatical fervor for the simple mathematical laws that govern the harmony of the world. And he found them, or at least enough of them to earn permanent and well-deserved fame as the father of modern dynamical astronomy. He was also the effective creator of the basic geometric idea that underlies integral calculus, which arose in his work on planetary orbits and again in connection with his calculations of the volume of wine barrels. We shall briefly consider the precise nature of these epoch-making achievements against the background of his turbulent life.

Kepler was born in a village near Stuttgart in southern Germany. His father was a shiftless, drunken mercenary soldier in the service of the Duke of Württemberg, and his mother an ignorant, superstitious woman of violent temper. From his earliest years his family was mired in poverty, and he himself was a sickly child whose health was permanently damaged by several serious illnesses. However, the Duke had created an excellent educational system to produce the stalwart clergy he needed for his struggle against Catholicism, and Kepler was given a good education and received his master's degree in 1591 at the great Protestant University of Tübingen.

His original purpose at the University was to become a Lutheran minister, and most of his early studies were in theology. Fortunately, however, he was diverted from this purpose by his professor of mathematics and astronomy, who awakened his interest in these subjects and gave him private lessons in the heliocentric theory of Copernicus. This radical theory, which maintained that the earth is merely a planet like the others and that all of them travel at constant speeds in circular orbits around the sun, was then less than 50 years old, and Kepler's teacher was one of the very few people who took it seriously. The standard teaching of the churches and schools was the earth-centered theory worked out in staggering detail by Ptolemy of Alexandria in the 2nd century A.D., in which the sun, the moon, and the other five planets were thought to travel in complicated wandering paths around the fixed earth against the frozen background of the fixed stars (the word "planet" comes from a Greek word meaning wanderer). Almost all intellectuals of the time believed what they were taught and rejected the theory of Copernicus, either because they failed to understand it or because they were unable to cope with revolutionary ideas. For them, displacing the earth and its human inhabitants from the favored position at the center of the universe which they had occupied for so many centuries was unthinkable blasphemy. And anyway, everyone can see that the earth is at rest and the sun and moon move through the sky around it—what could be clearer than that? The idea that the solid earth beneath our feet is spinning on its axis and hurtling through empty space at many miles per second was too absurd to be worth the attention of serious people. But for the young Kepler it was obviously true, and he resolved to devote the rest of his life to unraveling the mysteries of the planetary system whose simpler features had been described by Copernicus. His first substantial effort in this direction is memorable because it was so wonderfully imaginative and yet wholly erroneous, as we shall see.

Before we get into the details, however, it should be pointed out that the growing freedom of Kepler's thinking made him clearly unfit for a career in the

Church, where rigid orthodoxy prevailed. In 1594 he was happy to obtain a position as lecturer in mathematics and astronomy at the Protestant seminary in Graz, in southern Austria. His teaching career was a dismal failure: in the first year his classes were attended by only a dozen students, and in the second year by none at all. As he himself said in a fascinating self-analysis, he was a poor teacher not because he was boring, but because he was so eager and enthusiastic that his lectures were "confusing and not very intelligible"; he often got so excited that he "burst into speech without taking the time to consider what he was saying," and because of the turmoil of his ideas was "constantly drawn into digressions and thought of new words and new subjects, and new ways of expressing or proving his point."

In spite of these pedagogical problems, his superiors had a high opinion of both his character and his intellect. In their report on the new teacher they said that the lack of students should not be blamed on him, "because the study of mathematics is not every man's cup of tea." To prevent his salary from being a complete waste of money, they assigned him the task of compiling almanacs containing the usual weather forecasts and astrological predictions, and to prepare himself for this duty he applied himself with great diligence to the study of astrology. No blame should be attached to Kepler for this astrological work—ridiculous as it seems to modern eyes—because at the end of the 16th century no one had any idea of what causes the movement of the planets across the sky, and it was widely believed, even by educated people, that planetary movements and positions influence the course of day-to-day events on earth. When Kepler's first predictions—numbing cold weather, invasion by the Turks, and peasant uprisings—all came true, he gained a solid reputation as an astrologer.[1]

However, by this time his desire to understand the reasons why the planetary system is constructed in just the way it is had become the ruling passion of his life. From the period of his earliest researches, Kepler was inspired by his belief that God had created the world according to some simple

---

[1] He reported as follows to his teacher back in Tübingen: "By the way, the almanac's predictions are proving correct. There is unheard-of cold in our land. In the Alpine farms people die of the cold. It is reliably reported that when they arrive home and blow their noses, the noses fall off . . . As for the Turks, on January 1 they devastated the whole country from Vienna to Neustadt, setting everything on fire and carrying off men and plunder." As for his own attitude toward astrology, sometimes he thought there might be a grain of truth in it, but mostly he did it for a living with his tongue in his cheek and despised himself for having to support his family in this way. At one point he wrote: "A mind accustomed to mathematical deduction, when confronted with the faulty foundations of astrology, resists a long, long time, like an obstinate mule, until compelled by beating and curses to put its foot into that dirty puddle." See pp. 242–46 of Arthur Koestler's *The Sleepwalkers*, Macmillan, 1959. The middle 300 pages of this book provide a superb account of Kepler's life and work—without, however, any reference to his substantial place in the history of mathematics.

and harmonious plan, and that clues to this plan might be found in the number of the planets and the sizes of their orbits. After the long night of the Middle Ages, the Greek view that the truths of nature lie in mathematical laws—in numbers and geometry—was reasserting its grip on the mind of Europe.

While Kepler was teaching in Graz, he made his first attempts to discover the hidden mathematical regularities underlying the structure of the solar system. At that time there were six known planets—Mercury, Venus, Earth, Mars, Jupiter, Saturn—which were known to be at successively greater distances from the sun. In probing many possibilities for the pattern God might have had in mind when constructing this system, Kepler hit on the idea that between the "spheres" of the six planets could be fitted the five regular polyhedra of Greek geometry: the octahedron, icosahedron, dodecahedron, tetrahedron, and cube (Fig. A.10), in this order, corresponding to the order of the planets out from the sun. Thus, for example, he began by noticing that the radii of the spheres inscribed in and circumscribed about a cube seem to be in the same proportion as the distances of Jupiter and Saturn from the sun; also, the tetrahedron can be accurately interpolated between the successive orbital spheres of Mars and Jupiter, and similarly for the other planetary intervals and regular polyhedra (Fig. A.21).

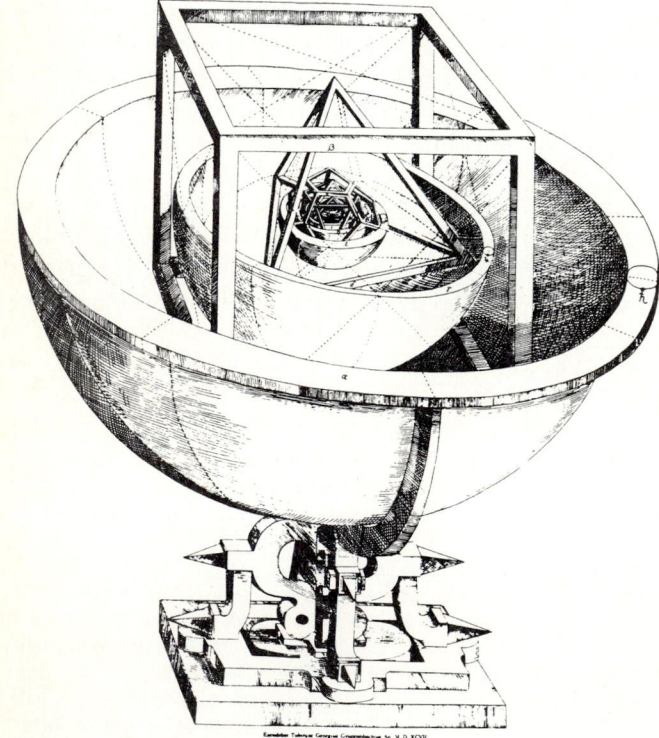

FIGURE A.21
Kepler's model of his interlaced nested "orbital spheres" and regular polyhedra.

These results satisfied Kepler that he had discovered one of the fundamental secrets of the universe. Furthermore, the existence of six planets, with five intervals between their orbital spheres, seemed to be connected with the existence of five regular polyhedra. Of course, as a mathematician Kepler knew very well that there are *only five* regular polyhedra (a proof is given in Section A.4), so his theory seemed to him to provide good reasons why God had created six planets and no more. He was dismayed to learn a few years later that Galileo claimed to have discovered four "new planets" with his newly invented 32-power telescope, and was greatly relieved when they turned out to be only satellites of Jupiter. (The word "satellite" was introduced by Kepler.) Fortunately for Kepler's faith in his theory, the seventh planet—Uranus—was not discovered until almost two centuries later, by Herschel in 1781.

Kepler published his ideas in 1596, in a work entitled *Mysterium Cosmographicum* ("The Secret of the Universe"). Of course, the discovery of additional planets in later centuries, as well as the improved accuracy of astronomical measurements, did not support his ingenious theory, and no scientific significance is now attached to it. Nevertheless, it gave clear evidence of his powerful imagination, contained the seeds of most of his later discoveries, and brought him to the attention of the eminent astronomers Tycho Brahe in Denmark and Galileo in Italy, to whom he sent copies of his book. Also, this book furnished the earliest enthusiastic support for Copernican ideas that appeared in print.[2]

Graz was a Catholic city, and one day in 1598, with animosity increasing between the Christian sects, all Protestant teachers and clergy were abruptly ordered to leave town before sundown, regardless of families, homes or property. Kepler was allowed to return a month later, but the message of the handwriting on the wall was clear, and he began looking for a new position in a safer environment. Tycho Brahe had recently moved to Prague as Imperial Mathematician to Emperor Rudolph II, and in 1600 he hired Kepler as his assistant. Relations between the two men were often difficult, because Tycho arrogantly pushed his own personal version of Ptolemaic astronomy. However, Tycho died a year later and Kepler was appointed as his successor, though at a reduced salary. This reduced salary was often unpaid, and to survive, Kepler was forced into the profitable business of casting horoscopes for members of the imperial household. He said, with wry humor, "Mother Astronomy would certainly starve if daughter Astrology did not earn the bread of both."[3]

Tycho is unforgettable on two counts, entirely apart from his importance as an astronomer. First, his nose had been sliced off in a youthful duel, and he

---

[2] There is an interesting modern translation by A. M. Duncan, Abaris Books, New York, 1981.

[3] Astrology could be dangerous as well as profitable. In 1570 the Italian mathematician Cardano was imprisoned for heresy because he published the horoscope of Jesus!

wore a golden nose to replace it, held on with a strap. He often disconcerted his opponents in the midst of fierce arguments by taking off his nose and polishing it with a special ointment that he carried for this purpose. And second, he died of a burst bladder caused by drinking too much at a long dinner party, because he refused to violate etiquette by leaving the table before his guests did.

Before moving to Prague, Tycho had spent years in the great observatory in Denmark that he had built for his own use, making many thousands of precise observations of the movements of the planets. When he died, all this data passed to Kepler, and proved to be a storehouse of information of enormous value for the future of astronomy. From the beginning Kepler paid particular attention to the observations of Mars, whose movements seemed impossible to account for by means of the circular orbits required by the Copernican system. Kepler combined a love of general principles with a deep respect for facts and the habit of attending to details; indeed, his favorite saying was, "God lives in the details." He undertook many long years of painstaking analysis of the data, testing one hypothesis after another with endless patience and never hesitating to discard hypotheses that did not fit the facts of Tycho's observations. At last his work led him to take the bold step that broke with the tradition of more than 2000 years and revolutionized astronomy. He discovered that the shape of a planetary orbit is not circular at all, but elliptical, and that the sun is not at the center but at one focus of the ellipse.[4]

This discovery was *Kepler's first law of planetary motion*—the first of three. It was published in 1609 in a momentous book called *Astronomia Nova* ("The New Astronomy"). The importance of Kepler's work was not immediately recognized, even by Galileo, to whom he sent a copy of his book. In fact, there is no evidence that Galileo ever examined the book; he did not reply with even a polite letter of thanks, and it probably stood unopened on his shelves until his own death in 1642. He was so wedded to the circular orbits of Copernicus that he made no mention of Kepler's laws at any time during his life, and probably was unaware of them. This is not very surprising, because the *Astronomia Nova* was an enormous tome in which the important new ideas were unfortunately buried almost out of sight among long accounts of Kepler's many fruitless hypotheses and blind alleys, which he took peculiar pleasure in describing. As he wrote in his Preface:

> What matters to me is not merely to impart to the reader what I have to say, but above all to convey to him the reasons, subterfuges, and lucky hazards which led

---

[4] The word "focus" was introduced into geometry by Kepler. It means "fireplace" or "hearth" in Latin, which is wonderfully appropriate in view of the fact that the sun is the central warming body of the solar system.

me to my discoveries. When Christopher Columbus, Magellan and the Portuguese relate how they went astray on their journeys, we not only forgive them, but would regret to miss their narration because without it the whole grand entertainment would be lost. Hence I shall not be blamed if, prompted by the same affection for the reader, I follow the same method.

An admirable sentiment, but most of his would-be readers did not appreciate it.

The second basic element of Copernican theory was the idea that planets must travel around their orbits at constant speeds, and this too Kepler found to be false according to the evidence. The facts are best understood in terms of the diagrams in Fig. A.22. If *AB, CD, EF* are the distances traversed by a planet along its elliptical path in equal intervals of time, then according to the doctrine of constant speed, *AB, CD*, and *EF* would all have to be equal. But by analyzing the orbit of Mars, Kepler found that these distances are not equal, and that in fact a planet travels faster when it is closer to the sun, *S*. Instead, he calculated the areas of the shaded sectors *SAB, SCD, SEF* and found these areas to be equal. He accomplished this by a method of approximation that used simple geometry and foreshadowed the ideas of integral calculus. He subdivided sector *SAB* into many very thin sectors, as shown on the right in the figure, and similarly for the sectors *SCD* and *SEF*. In each of the thin sectors there is very little difference between the small arc of the ellipse and its chord, so for the purpose of his approximation he took the area of the thin sector to be essentially equal to the area of the inscribed triangle, which he then calculated by the formula $\frac{1}{2}hb$ of elementary geometry. In this way, with much labor extending over several years, he arrived at his *second law of planetary motion*: the focal radius joining the sun to a planet sweeps out equal areas in equal intervals of time.

Besides making public his first two laws of planetary motion, Kepler's treatise of 1609 contained a discussion of celestial forces that developed ideas first suggested 13 years earlier in his *Mysterium Cosmographicum*. He believed that the planets are moved in their orbits by some kind of influence exerted by

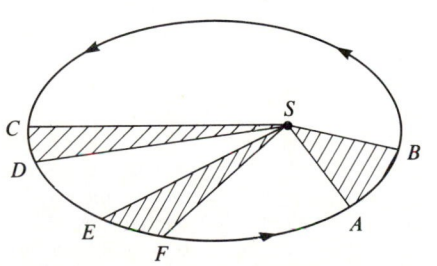
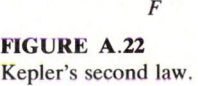

FIGURE A.22
Kepler's second law.

the sun. Even though the physical concept of force did not exist at that time, Kepler's vague ideas were the origin of the line of thinking that culminated about 60 years later in Newton's Force of Universal Gravitation. To emphasize this aspect of his purpose, Kepler's fuller title of his book was *Astronomia Nova Aitiologetos* (he ordered this last word to be printed in Greek capital letters for additional impact), which means "The New Astronomy Based on Causes." The idea that astronomical phenomena might be caused by physical mechanisms of the same kind that operate on earth was wholly new to science.[5]

Before turning to a discussion of Kepler's mathematics, we complete our description of his great astronomical discoveries by stating his *third law of planetary motion*: the squares of the periodic times of the planets are proportional to the cubes of their mean distances from the sun. This law is usually expressed by saying that the ratio T^2/a^3 is constant—where T is the planet's periodic time, that is, the time it takes to make one complete circuit of its orbit, and a is the planet's mean distance from the sun, that is, the average of its greatest and least distances from the sun. This can be seen most clearly from the Table A.1, in which the earth's periodic time and its mean distance

TABLE A.1
Data for Kepler's third law

Planet	Periodic time, T (in years)	Mean distance, a	T^2	a^3
Earth	1	1	1	1
Mercury	0.241	0.387	0.058	0.058
Venus	0.615	0.723	0.378	0.378
Mars	1.881	1.524	3.538	3.540
Jupiter	11.86	5.203	140.66	140.85
Saturn	29.46	9.539	867.89	867.98

---

[5] Some years later he published an explanation of his thinking intended for the general public. This appeared in parts from 1618 to 1621 under the title *Epitome Astronomiae Copernicanae* ("Epitome of Copernican Astronomy"). A partial translation is included in Volume 16 of *Great Books of the Western World,* 1952, eds. Robert M. Hutchins and Mortimer J. Adler. See especially pages 888–900. In his introductory note To The Reader, Kepler writes: "It has been ten years since I published my *Astronomia Nova.* As only a few copies of the book were printed, and as it has so to speak hidden the teaching about celestial causes in thickets of calculations and the rest of the astronomical apparatus, and since the more delicate readers were frightened away by the price of the book too; it seemed to my friends that I should be doing right and fulfilling my responsibilities, if I should write an epitome, wherein a summary of both the physical and astronomical teaching concerning the heavens would be set forth in plain and simple speech and with the boredom of the demonstrations alleviated." This book was accessible to ordinary educated people; it therefore had the honor of being at once suppressed and placed on the infamous Index of books condemned by the Church and forbidden to be read by believers.

are both taken as units, so that the constant ratio equals 1.[6] If we write this law in the form $T^2 = a^3$ or $T = a^{3/2}$, then the following (Kepler's own example) provides a simple illustration of the way it works. With the earth's mean distance from the sun as our unit of measurement, the mean distance a of Saturn is a bit more than 9, its square root is slightly greater than 3, and therefore the periodic time T, being the cube of this, is a bit more than 27. This means that the Saturn year is a bit more than 27 earth years; in fact, as the table shows, it is 29.46 earth years.

Kepler had struggled for more than 20 years to find this connection between a planet's *distance* from the sun and the *time* required to complete its orbit. He published his discovery in 1619 in a work entitled *Harmonices Mundi* ("The Harmonies of the World").[7] He was so filled with joy and pride over his achievement that he prefaced his book with the following extravagant—but characteristic—outburst:

> Since the first light dawned eight months ago, since broad daylight came three months ago, and since the sun of my wonderful discovery shone fully only a very few days ago, nothing holds me back . . . The die is cast, and I am writing the book—whether to be read by my contemporaries or by posterity matters not. It may be that my book will wait a hundred years for a reader; but has not God waited six thousand years for a proper observer of His handiwork?

At the present time we announce scientific discoveries in more pedestrian language, but Kepler's full-blown enthusiasm is one of his most engaging traits.

The *Harmonices Mundi* contains a few brief pages of great science and an overwhelming, luxuriant growth of mystical fantasy. Kepler thought he had found an analogy between the angular velocities of the planets and the frequencies of musical notes. As a planet's angular velocity varies in the course of its periodic journey around the sun, so the corresponding musical note changes, returning to its initial frequency when the planet returns to its starting point. For Kepler each planet sang its own song as it moved around its orbit, and together they harmonized. He wrote all this down at great length in musical notation—his own version of the Pythagorean "music of the spheres." The valid ideas in his books were accompanied by so much extraneous matter that no one until Isaac Newton more than half a century later was able to separate the wheat from all the chaff. Kepler was certainly the most fertile

---

[6] The data in this table are taken from Fred C. Whipple's authoritative book on the planets, *Orbiting the Sun,* Harvard University Press, 1981.

[7] The third law is stated on p. 1020 of the partial translation included in Volume 16 of *Great Books of the Western World* (see footnote 5).

scientific mind of his time, and he gave his readers everything he thought, without stint, the nonsense along with the sense.[8]

One of the difficulties with the acceptance of Kepler's laws was that no one could understand why they might be true. For Kepler's contemporaries they were purely empirical; the third law in particular seemed to be no more than just another of the many numerical curiosities he dredged up from his years of work on the data. However interesting these laws might be, there was no conceptual framework or causal mechanism to help one understand why they *must* be true. This was provided by Newton many years later: his Laws of Motion enabled him to *prove* Kepler's laws as mathematical consequences of his inverse square Law of Universal Gravitation; and this, of course, was one of the greatest of all intellectual achievements, for by doing this, Newton virtually created physics and astronomy as mathematical sciences.[9]

Kepler's contributions to mathematics were no less seminal than his contributions to astronomy. The Newtonian advances sketched above would have been quite impossible without the differential and integral calculus that Newton himself created in the 1660s, but the seeds of integral calculus seem to have originated with Kepler.

This development started with Kepler's area calculations described above in connection with his second law of planetary motion. He noticed that the same idea can be used in a slightly different way to find the area of a circle, by dividing the circle into a very large number of very thin sectors as shown in Fig. A.23. These sectors are so thin that we can think of them as triangles

---

[8] In this connection, the following fragment of conversation deserves to be more widely known. It took place between Linus Pauling, probably the greatest chemist of the 20th century—the only person ever to win two unshared Nobel prizes—and a former student of his named David Harker.

Harker: "Dr Pauling, how do you have so many good ideas?"
Pauling: "Have lots of ideas and throw away the bad ones. You aren't going to have good ideas unless you have lots of ideas and some sort of principle of selection."

Kepler certainly had a great many ideas, but he seems to have published them all and left it to posterity to choose.

[9] The details of the proofs are given in Section B.25. The history of Kepler's laws up to the time of Newton's use of them in the 1660s is very obscure. As we pointed out above, Galileo never mentioned them. Descartes had never heard of them when he died in 1650, nor had his friend Mersenne, who knew everybody and read everything. What caused this strange gap of knowledge? Was it the bizarre books in which Kepler published his laws? And if so, what impelled Newton to fix his attention on them with such momentous effect? It is possible that the answer to this question is related to the fact that the only eminent Western European who ever met Kepler face to face was the great English poet John Donne, who made a point of visiting him during a diplomatic trip to Germany in 1619. Donne may well have learned about the three laws of planetary motion from Kepler's own lips, and brought them back to England, where they somehow made their way to Newton half a century later.

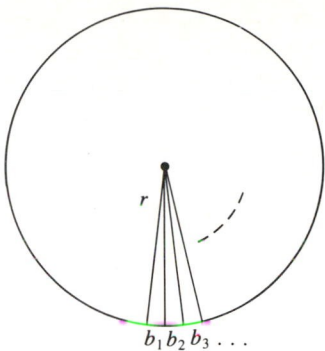

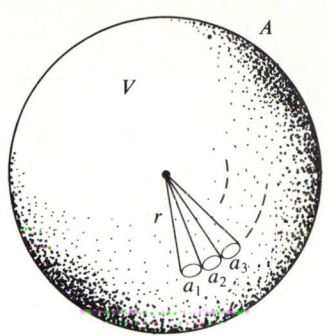

FIGURE A.23   **FIGURE A.24**

whose heights are all equal to the radius r of the circle. If we now denote the very small bases of these triangles by b_1, b_2, b_3, ... , then the area A of the circle, which is the sum of the areas of all the triangles, is found by repeatedly applying the formula $\frac{1}{2}hb$:

$$A = \tfrac{1}{2}rb_1 + \tfrac{1}{2}rb_2 + \tfrac{1}{2}rb_3 + \cdots$$
$$= \tfrac{1}{2}r(b_1 + b_2 + b_3 + \cdots).$$

But the sum of the b's is the circumference c of the circle, and $c = 2\pi r$ by the definition of π (this definition is that $\pi = c/2r$, the circumference divided by the diameter), so

$$A = \tfrac{1}{2}rc = \tfrac{1}{2}r(2\pi r) = \pi r^2.$$

This familiar formula was well known to the ancient Greeks, and Archimedes gave a fully rigorous formal proof by a complicated method called double reduction to absurdity.[10] Kepler's argument as given here is much more informal and intuitive, and this was the style he preferred. The proofs of Archimedes, he said, were completely rigorous, "*absolutae et omnibus numeris perfectae*" ("absolute and in every respect perfect") but he willingly left them to those people who demand logical perfection.

By yet another application of the same idea, Kepler was also able to show that the volume of a sphere is one-third the product of the radius and the surface area. To see this, we imagine the sphere to be dissected into a large number of small cones as shown in Fig. A.24, each cone having its vertex at the center and its base on the surface of the sphere. Of course, the bases of these "cones" are actually curved, but they are so small that they can be

---

[10] In this proof each of the hypotheses $A < \pi r^2$ and $A > \pi r^2$ is shown to imply a contradiction, leaving $A = \pi r^2$ as the only possibility.

thought of as flat. Let a_1, a_2, a_3, $\ldots$ be the areas of these small bases. Then the sphere is the sum of all the cones, and by repeatedly using Democritus' formula $\frac{1}{3}hB$ for the volume of a cone with height h and area of base B, the volume V of the sphere is

$$V = \tfrac{1}{3}ra_1 + \tfrac{1}{3}ra_2 + \tfrac{1}{3}ra_3 + \cdots$$
$$= \tfrac{1}{3}r(a_1 + a_2 + a_3 + \cdots).$$

But the sum of the a's is the surface area A of the sphere, so in this way Kepler obtained the formula

$$V = \tfrac{1}{3}rA,$$

as stated above. Archimedes had given well-known rigorous proofs of the formulas

$$V = \tfrac{4}{3}\pi r^3, \qquad A = 4\pi r^2$$

for the volume and surface area of a sphere, and Kepler's equation enabled him to pass easily from either one of these formulas to the other.[11]

The point of these discussions is not the results themselves, but rather Kepler's method of thinking, in which he breaks down larger geometric objects (circle and sphere) into smaller ones (triangles and cones) that are simpler and easier to work with. The atomic theory of matter was founded by Democritus in the 5th century B.C., and we here have the beginnings of a Democritus-like atomic theory of geometry. In each of the above discussions we use a kind of "geometric atom," first a tiny triangle, then a tiny cone. We think of areas and volumes as composed of large numbers of these atoms—an "infinite number" of "infinitely small" pieces—and our principle of calculation is to find the whole by summing the atoms of which that whole consists.

The year 1612 was a bumper year for wine, and Kepler began thinking about the crude methods used at that time for estimating the contents of wine barrels. When he bought wine he noticed that the merchants determined the contents of the partially full barrels by inserting a measuring rod into the bunghole all the way to the bottom and observing how much of the rod was wetted, without regard to the curvature of the sides. A wine barrel can be thought of as a solid of revolution, generated by revolving a longitudinal section about its axis. Kepler's plan for finding the true volume of such a solid was to divide it into a large number of elementary parts and then to sum the volumes of these parts. Thus, for example, he visualized a barrel as made up of many thin cylindrical layers or disks like that in Fig. A.25. The volume of each

---

[11] Archimedes was already familiar with these ideas of Kepler about circles and spheres, but this knowledge had no influence on the development of mathematics because it was locked in a lost manuscript that was not rediscovered until 1906.

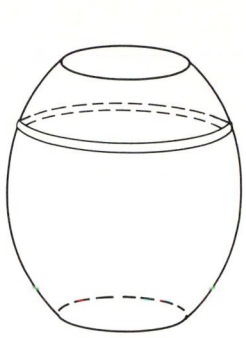

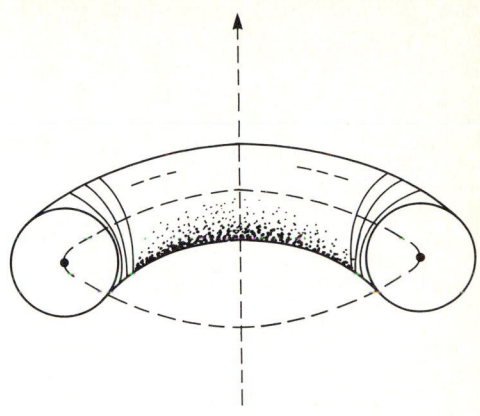

FIGURE A.25   **FIGURE A.26**

disk was then found by using the elementary formula $\pi r^2 h$ for the volume of a cylinder. The unstoppable Kepler measured the circumference c of the barrel at each level, calculated the radius r of the disk from the formula $c = 2\pi r$, and with the small thickness h known, calculated the volume. He also applied this idea to establishing Democritus' formula for the volume of a cone.

Another instructive example of Kepler's method is to the problem of finding the volume of a torus (Fig. A.26). He divided the torus into a large number of thin disks by planes through the axis. These disks are not uniformly thick, but are thinner in the parts nearer the axis and thicker on the opposite sides. According to Kepler these inequalities cancel each other out, and the volume of the torus equals that of a cylinder whose base equals the cross-section of the torus and whose height equals the circumference of the circle described by the center of this cross-section as it revolves about the axis. This reasoning may be a bit shaky, but the conclusion turns out to be exactly correct.

He presented these ideas in his 1615 treatise entitled *Nova Stereometria Doliorum Vinariorum* ("New Volume Measurements of Wine Barrels"), devoted to the computation of the volumes of solids obtained by revolving a conic section about a line in its plane. He placed his axis of revolution in different positions (parallel to the principal axis, the other axis, a given diameter, a given tangent) and passing through the center, intersecting the conic, not intersecting it, or tangent to it. In this way he generated 92 solids. Some of these (conoids and spheroids) had already been discussed by Archimedes, but many others were new, and he gave these his own names, such as apple, lemon, pear, nut, etc. For example, he revolved a segment of a circle about a chord and called the result an apple if the segment was more than a semicircle (Fig. A.27) and a lemon if it was less than a semicircle.

In the phrase we used at the beginning of this section, Kepler was here

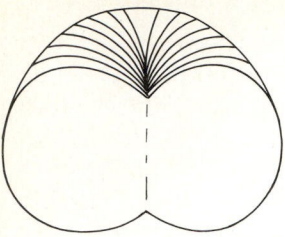

FIGURE A.27

exploiting on a broad front the "basic geometric idea that underlies integral calculus"—namely, to calculate something big, divide it into a large number of convenient small pieces and add up the pieces. No one paid much attention to his book until the 1630s, when his ideas were systematically developed into the *method of indivisibles* by the Italian mathematician Cavalieri, a disciple of Galileo. Others in Italy and France then became interested in this method as well, and applied it in many ingenious ways.

From the writings we have mentioned as well as several others, it is evident that Kepler was intimately familiar with the geometry of the conic sections. His astronomical discoveries furnished the first major scientific applications of what had previously been a part of pure mathematics as devoid of applications as the theory of prime numbers; also, they guaranteed that every decently educated mathematician from that day to this has needed to know something about this subject. He himself was the foremost expert of his time on the *Conics* of Apollonius.[12] Among his own contributions to the subject was the important idea of the continuity of the several types of conic sections with one another, as suggested in Fig. A.11, wherein we can pass in an unbroken transition from the circle through the ellipse, parabola, and hyperbola. Thus, when one focus of an ellipse is held fixed and the other moves off to infinity, the ellipse becomes a parabola. We mentioned earlier that he introduced the concept of the focus, which plays a central role in modern approaches to the subject.

Kepler contributed so much to the awakening of the European mind that the misery, poverty and misfortunes of his personal life are especially sad to contemplate. At the tender age of four he contracted smallpox, which nearly killed him and left him with impaired eyesight and crippled hands; as a small boy in elementary school he was often kept away from his classes and sent to work as a laborer in the fields; he had a joyless youth; his marriage was a constant source of despair; he was summarily fired from his lectureship in Graz

---

[12] There is an interesting historical puzzle here. It appears from the way he quotes Apollonius that he was thoroughly familiar with the content of Book V of the *Conics,* but this was not translated from the Arabic until 1661, long after Kepler was dead.

and expelled from the city in one of the preliminary upheavals of the Thirty Years' War; his favorite son died of smallpox and his epileptic wife went mad and died a few days later; his elderly mother was arrested and imprisoned for witchcraft under the imminent threat of torture and burning; he struggled to save her—in the end successfully—and narrowly escaped condemnation for heresy himself; his second marriage was even more unfortunate than the first, for he carefully analyzed the merits and faults of eleven women before confidently choosing the wrong one; and in 1630 he died of a fever in a strange bed while on a journey trying to collect some of his back salary—his condition when he left home being so bad that his son-in-law later wrote: "His widow, his children and friends expected to see the Last Judgment sooner than his return."

As to his scientific life, Kepler was a mathematician because he enjoyed it; but he was driven to astronomical research as a way of revealing the purposes of God—for much the same reasons as Saint Paul was driven to undertake his missionary journeys—by an irresistible force inside himself.

A.11

DESCARTES
(1596–1650)

Descartes commanded the future from his study more than Napoleon from his throne.

Oliver Wendell Holmes

Modern philosophy was born in the year 1637, in a short book by Descartes entitled *Discourse on Method*. In this work he rejected the sterile scholasticism prevailing at the time and set himself the task of rebuilding knowledge from the ground up, on a foundation of reason and science instead of authority and faith. He provided the fresh points of view needed for the vigorous development of the Scientific Revolution, whose influence has been the dominant fact of modern history. Further, in an appendix to the *Discourse* on his ideas about geometry, he foreshadowed the new forms of mathematics—analytic geometry and calculus—without which this Revolution would have died in infancy.

The seventeenth century, like the twentieth, was an age of religious hatred, oppression, and plundering warfare, presided over, for the most part, by squalid rulers with the political ethics of cattle rustlers and the personal morality of gigolos. Nevertheless, history is not exclusively what Gibbon called it, a record of "the crimes, follies, and misfortunes of mankind." Intellectually, the century in which Descartes lived was one of the greatest periods in the history of civilization. It began with Galileo and Kepler; it ended with Newton and Leibniz; and it nurtured an array of remarkable men so gifted and diverse in their talents and achievements that it can only be compared with the golden age of ancient Greece. It is often called the Century of Genius.

René Descartes came from a family of the lesser nobility near Tours, in central France. His mother died soon after his birth, and left him some property, which later made it possible for him to enjoy a leisurely life of travel and study. When the boy was 8, his father sent him to the nearby Jesuit school at La Flèche, an excellent institution for the education of young gentlemen which Henry IV had recently established in one of his favorite palaces. Here Descartes was thoroughly trained in literature and the classical languages, rhetoric, philosophy, theology, science, and mathematics. He was treated with kindly consideration by the Jesuit fathers, and in view of his frail constitution and meditative disposition was allowed to lie in bed in the mornings as late as he pleased, long after the other boys were in their classes. He maintained this habit to the end of his life, and liked to say that many of his best thoughts came to him in those tranquil hours of the late morning. There is even a story that he conceived the basic idea of analytic geometry while lying in bed and watching a fly crawling on the ceiling of his room, by noticing that the path of the fly could be described if only one knew a relation connecting its distances from two adjacent walls.

The young Descartes was a born skeptic, and as he matured he began to suspect that the so-called humanistic learning he was absorbing at La Flèche was almost barren of human significance, with little power to enrich or improve human life. As a polite and circumspect youth, he kept most of his doubts to himself. Nevertheless, he saw more and more clearly that the principles of philosophy and theology taught by the Jesuits were often little more than baseless superstitions cloaked in a few tattered rags of scholastic logic. He was passionately interested in science, but the science he was offered was the worthless physics of Aristotle adjusted to the empty doctrines of Saint Thomas Aquinas. Only mathematics escaped his contempt, "because of the certainty of its proofs and the clarity of its reasoning"; and even here, he was "astonished that foundations so firm and solid should have nothing loftier built upon them." It must be remembered that mathematics at that time consisted of classical geometry and a few primitive fragments of elementary algebra and trigonometry, with almost all of its applications to science still beyond the horizon.

Like many intelligent young people in every century, Descartes left school filled with disgust at the arid emptiness of his studies. His state of mind is best expressed in his own words:

This is why, as soon as my age permitted me to leave the control of my teachers, I completely gave up the study of letters. And resolving to seek no other knowledge than that which I could find in myself, or else in the great book of the world, I employed the rest of my youth in travel, in seeing courts and armies, in

> frequenting men of diverse temperaments and conditions...and above all in trying to learn from what I saw, so that I might derive some profit from my experience.[1]

Naturally, he first went to Paris, where he gambled, wenched, and generally played the dandy. These dissipations quickly palled, and he astonished his raffish friends by enlisting in the Dutch army as an unpaid gentleman volunteer. The army was inactive at that time, and in his enforced leisure he was once again attracted to the study of mathematics. A year later, in 1619, he transferred to the army of the Duke of Bavaria; and while staying in winter quarters in a small town on the Danube, he experienced an "illumination" that for him was comparable with the great mystical revelations in the lives of the saints.

In this frigid village where he was a stranger to everyone, Descartes shut himself up for the winter in a well-heated room and plunged into solitary study and meditation. He thought over the knowledge he had acquired in the various sciences and despairingly noted its confusion and uncertainty. What was needed was a totally fresh start, a new beginning that would sweep away all those systems of thought and belief that had become polluted over the centuries with half-truths, wishful thinking, and false reasoning. Only in mathematics had he found the certainty he desired, and the task that gradually formed in his mind was that of extending this certainty to all other fields of knowledge. But how could such a monumental project be accomplished? On November 10, 1619—a famous day in the history of philosophy—in a state of exhaustion and feverish excitement, he found his method and felt that he had glimpsed "the foundations of a marvelous science." Much more than merely showing him the way in an isolated problem, or even clarifying the principles of a particular science, his illumination revealed to him the essential unity of all the sciences—indeed, of all knowledge. And his method of displaying the various disciplines as branches of his single "marvelous science" would be that of mathematics:

> Those long chains of simple reasoning which geometers use to arrive at their most difficult conclusions made me believe that all things which are the objects of human knowledge are similarly interdependent; and that if we will only abstain from assuming something to be true which is not, and always follow the necessary order in deducing one thing from another, there is nothing so remote that we cannot reach it, nor so hidden that we cannot discover it.[2]

---

[1] *Discourse,* Part I.

[2] *Discourse,* Part II.

On that fateful day he made two decisions that shaped the future course of his life. First, he decided that he must systematically doubt everything he knew or thought he knew about all the sciences, and search for self-evident and certain foundations on which the edifice of knowledge could be rebuilt with confidence. Second, since a great work of art is always the product of a single master artist, he decided that he must carry out the entire project himself.

He left the army and traveled extensively in Germany, Switzerland, Italy, and Holland. He worked on mathematical problems, studied glaciers and avalanches, computed the heights of mountains, and continued to harbor his great secret ambition for a total reform of human knowledge. In one of his private notebooks he wrote the following fragment of self-analysis: "As an actor, ready to appear on the stage, dons a mask to hide his timidity, so I go forward in a mask preparing to mount the stage of the world, which up to now I have known only as a spectator." Returning to Paris in 1625, he spent most of the next three years in the company of men with similar mathematical and scientific interests. In particular, he renewed his acquaintance with the Franciscan friar Marin Mersenne, a former schoolmate at La Flèche. Mersenne knew everyone worth knowing, and was to become and remain his closest friend and most faithful admirer.

Late in 1628 Descartes began to realize that he would never "mount the stage of the world" unless he started climbing, and soon. In order to find the peaceful leisure necessary for thinking and writing, he left the noisy bustle of Paris and went to Holland, where he lived for the next 21 years. He preferred the solitary life and greater intellectual freedom available in that country, and he also hoped to avoid the annoyance of oafish visitors who came to see him in the morning and rousted him out of bed. In order to further ensure his privacy, he changed his address an incredible 24 times during that period and carefully kept it a secret from all but his closest friends. The motto he adopted was more appropriate for a fugitive than a philosopher: *Bene vixit qui bene latuit*—"He has lived well who has hidden well." Though solitary, he was far from isolated, for he set aside one day a week to attend to his voluminous correspondence with learned men all over Europe. His chief correspondent was Mersenne, who acted as his link with Parisian intellectual circles for all manner of scientific news, philosophical questions, and mathematical problems. He read comparatively little because he trusted experiments more than books; and more than either, he trusted his own mind. But he didn't believe in overdoing meditation, either. His custom, he said, was "never to spend more than a few hours a day in thoughts which occupy the imagination, or more than a few hours a year in those which occupy the understanding, and to give all the remaining time to the relaxation of the senses and the repose of the mind." But perhaps this was partly bluff, since gentleman intellectuals in those days weren't supposed to work very hard; if they violated this code, as Descartes certainly did, then decency required a modicum of dissembling.

Descartes's first important book, *Rules for the Direction of the Mind,* was probably written during his first year in Holland. It contains the fullest description of his method for clear and correct thinking, but he left it unfinished, and it was not published until 1701. A few years later he decided to make public three of his shorter scientific treatises, accompanied by an explanatory preface. The result was his *Discourse on Method* (1637), with its appendixes entitled *Dioptric, Meteors,* and *Geometry,* which he intended to be convincing illustrations of the power of his method. These appendixes contained little of permanent interest or value, and have deservedly sunk into oblivion. However, the *Discourse* itself remains a landmark of philosophy which is also a literary classic. It is unique in both form and content—a mixture of philosophy and science, a manifesto and a prospectus, an intellectual autobiography of great charm in which the adventures of reason are narrated as vividly as Homer's account of the wanderings of Odysseus. As a natural outgrowth of his continuing study of science, Descartes developed an interest in the theory of knowledge. What do I know? he asked himself as a youth. And later, How do I know? What does it mean to know? In 1641 he published his ideas on this subject in his *Meditations,* a work that has been described as combining literary excellence and philosophical genius on a level unmatched by any thinker since Plato. We have already mentioned his decision in 1619 to doubt everything he thought he knew about the various sciences. The philosophic doubt of his middle years, as expressed in both the *Discourse* and the *Meditations,* was far more searching. It omitted nothing from its withering blast, not even his conviction of his own existence.

A brief summary will perhaps convey the flavor of his ideas. At least once in his life, Descartes said, a man who seeks truth must summon the courage to doubt everything. Most of our beliefs—about the physical world, religion, society, and ourselves—enmesh us in childhood before our defenses are up; and by the time we reach the age of critical judgment, their bonds have become so comfortable and familiar that we are scarcely aware of their presence. We acquire these beliefs from inadequate senses, credulous parents, unreliable teachers, and self-serving institutions. We must examine all our convictions, he said, in the search for those that resist our most strenuous efforts to doubt them, for only these can provide a solid and certain foundation on which to rebuild the temple of knowledge. This systematic search for bedrock by digging down through the mud and sand of the mind, led him at last to his ultimate verity: *Cogito ergo sum*—"I think, therefore I am"—perhaps the most famous sentence in the history of philosophy. It was not an original thought, for St. Augustine expressed the same idea in almost the same words more than a thousands years earlier: "Who doubts that he lives? For if he doubts, he lives."[3] However, this perception was only incidental for Augustine, while

---

[3] *The Trinity,* Book X, Chapter 10.

Descartes made it the source of his entire system of thought. His epistemology and metaphysics arose from his effort to reconstruct the external world from the primal fact of his own thinking; and for better or worse, Western philosophy has been preoccupied ever since with the problem of whether this "external world" really exists except as an idea.[4]

The constructive part of his system is much less interesting and important than the destructive part just described. For example, his eloquent proof of the existence of God reduces in the end to the flimsiest reasoning one can imagine: "It is not possible that I could have in myself the idea of God, if God did not truly exist."[5] Arguing next that his "clear and distinct ideas" are necessarily true, since God is perfect and therefore cannot stoop to deception, he proceeds to "prove" the existence of the external world, the distinctness of mind and body, and so on. In effect, he throws his old ideas out the front door with resounding fanfare, and then, after a decent interval and with suitable ceremonies, quietly lets them in again at the back door.[6] In spite of their flaws, the books of Descartes were widely read, and their skeptical parts gave a powerful push to the simplest, most obvious, yet most potent of all revolutionary principles: Base your beliefs on evidence, and give them only the degree of support that the evidence justifies. This unpopular principle has smouldered feebly ever since, bursting into flame now and then in a few individuals to touch off the explosions of scientific knowledge that have set apart the last four centuries of European civilization from all other periods of human development. Whatever one thinks of the ideas of Descartes, it is undeniable that a great part of his influence was due to his extraordinary skill as a writer. "When writing about transcendental issues," he said, "be transcendentally clear," and in this he usually followed his own advice. Descartes was one of the great masters of the art of language, and for a thinker who wishes to be remembered, this is often better than having important original ideas.

In his own eyes Descartes was mainly a scientist and mathematician, and only incidentally a philosopher. What was the nature of his scientific activity? In 1633 he completed an ambitious treatise entitled *The World,* in which he undertook to explain "the nature of light, and of the sun and stars which emit it; of the heavens which transmit it; of the planets, the comets, and the earth

---

[4] The modern Chinese philosopher Lin Yutang, educated in a different tradition yet profoundly familiar with Western thought, made the following acid comment on this somewhat absurd preoccupation: "How on earth did Descartes, who could not on *prima facie* evidence accept his existence as real, believe that his thinking was? That was the beginning of the dark ages of European philosophy."

[5] *Meditations,* Part III.

[6] "The human brain is a complex organ with the wonderful power of enabling man to find reasons for continuing to believe whatever it is that he wants to believe."—Voltaire.

which reflect it; of all terrestrial bodies which are colored by it; and of Man its spectator." The condemnation of Galileo by the Inquisition caused him to abandon all thought of publishing this work, and it has survived only in fragments. In addition to physics and astronomy, his scientific interests included meterology, optics, embryology, anatomy, physiology, psychology, geology, and even medicine and nutrition, which he studied in the hope of prolonging his own life. In meteorology he gave fanciful explanations of thunder and lighting, thunder being the sound made when a higher cloud falls on a lower one; nevertheless, his explanation of the rainbow, in terms of refraction and reflection within water droplets in the atmosphere, was quite correct.[7] His attempt to account for the colors of the rainbow was not successful, and this achievement was left for Newton. He was the first to publish the sine law of refraction, but he stated it incorrectly and "proved" his incorrect version by an *ad hoc* argument that suggested to many later scientists (including Huygens and Leibniz) that he did not discover the law himself, but learned of it from Snell, who is now generally given credit for the discovery:[8] In other fields, he dissected a fetus and described its anatomy; removed the back of an ox's eye to examine the image formed by an object placed in front of the eye; welcomed Harvey's discovery of the circulation of the blood, but engaged in an unsuccessful dispute with him over the action of the heart; and dissected the heads of various animals in an effort to locate the sources of memory and imagination. His doctrine that the body is a machine had considerable influence on the later history of physiology and psychology. According to him, animals are nothing but machines and a person is a machine distinguished by the possession of a soul, which probably resides in the pineal gland at the base of the brain. These mechanistic views also permeated his physics and astronomy. He rejected action at a distance and assumed that all physical influences—such as gravity, light, and magnetism—are transmitted mechanically by the pressures of adjacent particles. The entire universe, he said, is filled with these particles, moving ceaselessly in vortices; the earth, for example, moves in its orbit because it is swept around the vortex of the sun like a twig in a whirlpool. These pictorial fancies enjoyed a brief vogue, but were soon destroyed by the rise of Newtonian mathematical physics. In his science Descartes searched for the meaning of everything that exists and the cause of everything that happens, but little or nothing remains of his work. He was crippled as a scientist by philosophic preconceptions, unrestrained speculation, and excessive ambition—diseases to which philosophers are

---

[7] He probably did not know that this phenomenon had been correctly explained more than 300 years earlier. See A. I. Sabra, *Theories of Light from Descartes to Newton*, Cambridge University Press, 1981, p. 62.

[8] *Ibid.*, pp. 99–105.

particularly vulnerable—and as a result his ideas were almost entirely erroneous. As Newton said, probably with Descartes in mind:

> To explain all nature is too difficult a task for any one man or even for any one age. 'Tis much better to do a little with certainty, and leave the rest for others that come after you, than to explain all things.

We come at last to the difficult question of Descartes's mathematics—difficult because one of the commonest items of second-hand gossip among historians of science is that he invented analytic geometry, and yet he did nothing of the kind.[9] Any qualified person who examines Descartes's treatise on geometry will soon convince himself that this work contains nothing about perpendicular axes, or the "Cartesian" coordinates of a point, or equations of lines and circles, or any material at all that bears a recognizable relation to analytic geometry as this subject has been understood for the past 350 years. We find familiar notational conventions appearing here for the first time, such as the use of exponents and the custom of denoting constants and variables by the letters a, b, c and x, y, z, respectively; we find geometry and algebra, and algebra used as a language for discussing geometry; but we do not find analytic geometry, or for that matter any content whatever that justifies Descartes's mathematical reputation. His *Geometry* was little read then and is less read now, and deservedly so, for the entire work is a grotesque betrayal of what he earlier called "the transparency and unsurpassable clarity which are proper to a rightly ordered mathematics." It appears from the many deliberate obscurities and condescending remarks—so foreign to his usual way of writing—that he wrote it more to boast than to explain, and somehow he managed to cow most of his contemporaries and successors into believing against the evidence that he had accomplished something worthwhile.[10]

By 1649 Descartes's books and ideas had spread to all corners of Europe, and Queen Christina of Sweden invited him to come to Stockholm to adorn her court and act as her private tutor in philosophy. Stockholm was a cold and disagreeable city, and at first he was not interested in this opportunity "to live in the land of bears among rocks and ice"; but at last he yielded, and, putting

---

[9] It is said that history repeats itself, and historians repeat each other.

[10] He did not cow Newton, who called Cartesian geometry "the Analysis of the Bunglers in mathematicks." Nevertheless, he was certainly a clever mathematician, as is shown by his trisection of an angle using a straightedge, compass, and fixed parabola; see H. Tietze, *Famous Problems of Mathematics,* Graylock Press, 1965, p. 53. A further discussion of Descartes's mathematics, as well as the subtle devices of his engaging but calculated literary style, can be found in John Fauvel's interesting lecture, "What Did Descartes Do for Mathematics? Bell versus Simmons," delivered at Colorado College in May 1989. This lecture is available, but unfortunately is unpublished.

aside his fears for his habits and independence, he left Holland on a special warship which the Queen had sent to fetch him. This headstrong and strangely masculine young woman of 19 was one of history's most remarkable characters. A passionate huntress, an expert horsewoman capable of staying in the saddle all day without tiring, indifferent to women's dress and rarely combing her hair more than once a week, fluent in five languages, enthusiastic about the study of literature and philosophy, and filled with ambitions for turning Stockholm into "the Athens of the North"—she captured poor Descartes as a spider does a fly. Needing little rest herself and impervious to cold and discomfort, she set the bleak hour of 5 a.m. for her philosophy lessons. With his lifelong routine shattered, the unhappy man was forced to stumble out of his warm bed in the dark and make his way to the palace through the bitterest winter Stockholm had known in years. Worse yet, Christina took advantage of his numbed wits; when he tried to convince her that animals are only machines, she objected that she had never yet seen her watch give birth to baby watches (his reply—if he was able to think of one—is not recorded). Exhausted, weakened, and full of despair over his humiliating predicament, Descartes caught a chill and died of pneumonia only four months after his arrival in Sweden.

Is there anything left of the work of Descartes which still has meaning for the modern world? Very little, if we count only specific doctrines or discoveries in philosophy, science, or mathematics. However, he holds a secure place in the canonical succession of the high priests of thought by virtue of the rational temper of his mind and his vision of the unity of knowledge. He struck the gong, and Western civilization has vibrated ever since with the Cartesian spirit of skepticism and inquiry that he made common currency among educated people.

A.12

MERSENNE
(1588–1648)

There is more in Mersenne than in all the universities together.

Thomas Hobbes

Scientists in the twentieth century are linked together into a worldwide communications network by hundreds of professional organizations whose meetings and journals stimulate a constant interchange of ideas and discoveries. In the early seventeenth century none of these organizations existed. Worse yet, most of the universities were hollow shells of medieval ritual and scholastic rigidity that were incapable of welcoming new knowledge. The intense intellectual ferment of the time found its main outlets in informal discussion groups and private correspondence. The circle of friends of Marin Mersenne was by far the most important of these groups.

Mersenne was a Franciscan friar who lived in a monastery in Paris near the Place Royale, but in middle life his interests shifted away from theology toward philosophy, science, and mathematics. His friends included almost everyone in Western Europe who was active in these fields—Pascal, Roberval, Desargues, Gassendi, and others in Paris; Descartes in Holland; Galileo, Torricelli, and Cavalieri in Italy; Fermat in Toulouse; Hobbes during his frequent long visits in Paris; and many more. The Mersenne circle constituted an "invisible college" in which most of the important intellectual activity of the period took place. Those members who lived in Paris met regularly in his rooms with the approval and support of Cardinal Richelieu, who was famous for his patronage of the sciences. These meetings continued for many years after Mersenne's death, and formed the nucleus of the French Academy of Sciences when it was chartered in 1666. Mersenne acted not only as the informal chairman of the group, but also as its corresponding secretary. He transmitted letters and manuscripts from one member to another according to their interests; and his own enormous correspondence, particularly with Descartes and Fermat, lubricated the flow of ideas so effectively that telling him about an interesting discovery amounted to publishing it throughout the whole of Europe.

In science, Mersenne's own work on sound was of such fundamental importance that he is sometimes called "the father of acoustics." His 1636 treatise entitled *Harmonie Universelle* contains accounts of many ingenious experiments and the conclusions he drew from them. He laid out long hemp cords and brass wires, some more than 100 ft in length, and stretched them taut between two posts by means of weights. He found that their vibratory movements when plucked could easily be followed by the eye, and he timed them by using his own pulse. By varying the lengths and tensions, he discovered the following basic principles, which are now known as Mersenne's laws: The frequency of vibration of a stretched string is (1) inversely proportional to the length if the tension is constant, (2) directly proportional to the square root of the tension if the length is constant, and (3) inversely proportional to the square root of the mass per unit length (linear density) for different strings of the same length and tension. He next shortened a stretched brass wire until its sound became audible and tuned it to the pitch of one of his organ pipes. By applying his laws he found that the frequency of this note was 150 vibrations per second, and also that the frequency of its octave was 300 vibrations per second. This was the first determination of the frequency of a specific musical note, and was a remarkable scientific achievement. He was also the first to determine the speed of sound in air, using a seven-syllable shout ("Benedicam Dominum!") that took a second to utter and immediately echoed back from a measured distance of 519 ft. From this he concluded that sound travels 1038 ft/s, which is quite close to the currently accepted figure of about 1087 ft/s in dry air at 32°F.

Among his mathematical interests were the cycloid and perfect numbers. A cycloid is the archlike curve traced out by a point on the rim of a rolling wheel. Mersenne probably first heard of this beautiful curve from Galileo, and he suggested it to many of his friends as a worthy object for investigation. Over the next two centuries the cycloid turned out to have many remarkable geometric and physical properties, and was studied by Roberval, Torricelli, Pascal, Huygens, John Bernoulli, Leibniz, Newton, Euler, and Abel, among others.

Perfect numbers—those, like $6 = 1 + 2 + 3$, which equal the sum of their proper divisors—have fascinated mathematicians since the time of Pythagoras. Mersenne was aware of Euclid's proof that if $2^n - 1$ is prime, then $2^{n-1}(2^n - 1)$

is perfect (this proof is given in Section B.2). He also knew that $2^n - 1$ cannot be prime if n is not, so he was led to the problem of determining those prime numbers p for which $2^p - 1$ is also prime (the latter are now called *Mersenne primes*). Unfortunately, there are some primes p for which $2^p - 1$ is prime and others for which it is not, so this problem is far from simple. In 1644 Mersenne stated that among the 55 primes $p \le 257$ the only ones for which $2^p - 1$ is also prime are $p = 2, 3, 5, 7, 13, 17, 19, 31, 67, 127, 257$. He did not give any of the evidence that led him to make this statement, but it is now known that he made five errors: 67 and 257 do not belong in the list, and 61, 89, and 107 do belong there. The factorability of $2^{67} - 1$ was discovered in 1903 by the American mathematician F.N. Cole, and was announced by him in a dramatic presentation to a meeting of the American Mathematical Society. When called upon for his lecture, Cole walked to the blackboard, silently calculated $2^{67} - 1$, and, still without saying a word, moved over to a clear space on the board and multiplied out

$$193, 707, 721 \quad \text{and} \quad 761, 838, 257, 287.$$

The results were visibly the same, the audience applauded enthusiastically, and Cole returned to his seat after having delivered the only totally wordless lecture in recorded history. He later told a friend that this factorization had cost him "three years of Sundays."[1] In 1931 it was proved by D. M. Lehmer that $2^{257} - 1$ is factorable, but the proof was theoretical and did not exhibit any specific factors. Additional details about the current status of this subject are given in Section B.2.

---

[1] E. T. Bell, *Mathematics, Queen and Servant of Science*, McGraw–Hill, 1951, p. 228.

A.13

FERMAT
(1601–1665)

... a master of masters.

E.T. Bell

Pierre de Fermat was perhaps the greatest mathematician of the seventeenth century, but his influence was limited by his lack of interest in publishing his discoveries, which are known mainly from letters to friends and marginal notes in his copy of the *Arithmetica* of Diophantus.[1] By profession he was a lawyer and a member of the provincial supreme court in Toulouse, in southwestern France. However, his hobby and private passion was mathematics, and his casual creativity was one of the wonders of the age to the few who knew about it.

His letters suggest that he was a shy and retiring man, courteous and affable, but slightly remote. His outward life was as quiet and orderly as one would expect of a provincial judge with a sense of responsibility toward his work. Fortunately, this work was not too demanding, and left ample leisure for the extraordinary inner life that flourished by lamplight in the silence of his study at night. He was a lover of classical learning, and his own mathematical

---

[1] These marginal notes were reproduced in a new edition of this third-century work on number theory that was published by Fermat's son in 1670.

ideas grew in part out of his intimate familiarity with the works of Archimedes, Apollonius, Diophantus, and Pappus. Though he was a genius of the first magnitude, he seems to have thought of himself as at best a rather clever fellow with a few good ideas, and not at all in the same class with the masters of Greek antiquity.

Father Mersenne in Paris heard about some of Fermat's researches from a mutual friend, and wrote to him in 1636 inviting him to share his discoveries with the Parisian mathematicians. If Fermat was surprised to receive this letter, Mersenne was even more surprised at the reply, and at the cascade of letters that followed over the years, to him and also to other members of his circle. Fermat's letters were packed with ideas and discoveries, and were sometimes accompanied by short expository essays in which he briefly described a few of his methods. These essays were handwritten in Latin and were excitedly passed from one person to another in the Mersenne group. To the mathematicians in Paris, who never met him personally, he sometimes seemed to be a looming, faceless shadow dominating all their efforts, a mysterious magician buried in the country who invariably solved the problems they proposed and in return proposed problems they could not solve—and then genially furnished the solutions on request. He enjoyed challenges himself, and naively took it for granted that his correspondents did too. For instance, Mersenne once wrote to him asking whether the very large number $100,895,598,169$ is prime or not. Such questions often take years to answer, but Fermat replied without hesitation that this number is the product of $112,303$ and $898,423$, and that each of these factors is prime—and to this day no one knows how he did it. The unfortunate Descartes locked horns with him several times, on issues that he considered crucial both to his reputation as a mathematician and to the success of his philosophy. As an outsider Fermat knew nothing about Descartes's monumental egotism and touchy disposition, and with calm courtesy demolished him on each occasion. Wonder, exasperation, and chagrin were apparently common emotions among those who came into contact with Fermat's mind.

He invented analytic geometry in 1629 and described his ideas in a short work entitled *Introduction to Plane and Solid Loci*, which circulated in manuscript form from early 1637 on but was not published during his lifetime.[2] The credit for this achievement has usually been given to Descartes on the basis of his *Geometry*, which was published late in 1637 as an appendix to his famous *Discourse on Method*. However, nothing that we would recognize as analytic geometry can be found in Descartes's essay, except perhaps the idea

---

[2] Translations are given in D. E. Smith, *A Source Book in Mathematics*, McGraw-Hill, 1929, pp. 389–96; and in D. J. Struik, *A Source Book in Mathematics, 1200–1800*, Harvard University Press, 1969, pp. 143–50.

of using algebra as a language for discussing geometric problems. Fermat had the same idea, but did something important with it: He introduced perpendicular axes and found the general equations of straight lines and circles and the simplest equations of parabolas, ellipses, and hyperbolas; and he further showed in a fairly complete and systematic way that every first- or second-degree equation can be reduced to one of these types. None of this is in Descartes's essay; but to give him his due, he did introduce several notational conventions that are still with us—which gives his work a modern appearance—while Fermat used an older and now archaic algebraic symbolism. The result is that superficially Descartes's essay looks as if it might be analytic geometry, but isn't; while Fermat's doesn't look it, but is. Descartes certainly knew some analytic geometry by the late 1630s; but since he had possession of the original manuscript of the *Introduction* several months before the publication of his own *Geometry,* it may be surmised that much of what he knew he learned from Fermat.

The invention of calculus is usually credited to Newton and Leibniz, whose ideas and methods were not published until about 20 years after Fermat's death. However, if differential calculus is considered to be the mathematics of finding maxima and minima of functions and drawing tangents to curves, then Fermat was the true creator of this subject as early at 1629, more than a decade before either Newton or Leibniz was born.[3] With his usual honesty in such matters, Newton stated—in a letter that was discovered only in 1934—that his own early ideas about calculus came directly "from Fermat's way of drawing tangents."[4]

So few curves were known before Fermat's time that no one had felt any need to improve upon the old and comparatively useless idea that a tangent is a line that touches a curve at one and only one point. However, with the aid of his new analytic geometry, Fermat was able not only to find the equations of familiar classical curves, but also to construct a multitude of new curves by simply writing down various equations and considering the corresponding graphs. This great increase in the variety of curves that were available for study aroused his interest in what came to be called "the problem of tangents."

What Newton acknowledged in the remark quoted above is that Fermat was the first to arrive at the modern concept of the tangent line to a given curve at a given point P (see Fig. A.28). In essence, he took a second nearby point Q on the curve, drew the secant line PQ, and considered the tangent at

---

[3] Fermat wrote several accounts of his methods, but as usual he made no effort to publish them. The earliest of these was the very short essay given on pp. 223–24 of Struik's *Source Book*; this was circulating in Paris in 1636, and according to Fermat's own statement was then about 7 years old.

[4] See L. T. More, *Isaac Newton,* Scribner's 1934, p. 185.

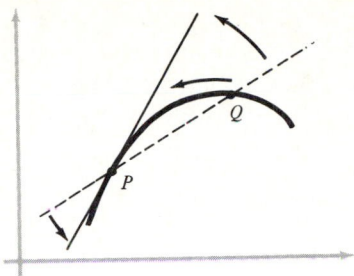

FIGURE A.28   **FIGURE A.29**

P to be the limiting position of the secant as Q slides along the curve toward P. Even more important, this qualitative idea served him as a stepping-stone to quantitative methods for calculating the exact slope of the tangent for specific curves.

Fermat's methods were of such critical significance for the future of mathematics and science that we pause briefly to consider how they arose.

While sketching the graphs of certain polynomial functions $y = f(x)$, he hit upon a very ingenious idea for locating points at which such a function assumes a maximum or minimum (largest or smallest) value. He compared the value $f(x)$ at a point x with the value $f(x + E)$ at a nearby point $x + E$ (see Fig. A.29). For most x's the difference between these values, $f(x + E) - f(x)$, is not small compared with E, but he noticed that at the top or bottom of a curve this difference is much smaller than E and diminishes faster than E does. This idea gave him the approximate equation

$$\frac{f(x + E) - f(x)}{E} \cong 0,$$

which becomes more and more nearly correct as the interval E is taken smaller and smaller. With this in mind, he next put $E = 0$ to obtain the equation

$$\left[\frac{f(x + E) - f(x)}{E}\right]_{E=0} = 0.$$

According to Fermat, this equation is exactly correct at the maximum and minimum points on the curve, and solving it yields the values of x that correspond to these points. The legitimacy of this procedure was a subject of acute controversy for many years. However, students of calculus will recognize that Fermat's method amounts to calculating the derivative

$$f'(x) = \lim_{E \to 0} \frac{f(x + E) - f(x)}{E}$$

and setting this equal to zero, which is just what we do in calculus today, except that we customarily use the symbol Δx in place of his E.

In one of the first tests of his procedure, he gave the following proof of Euclid's theorem that the largest rectangle with a given perimeter is a square. If B is half the perimeter and one side is x, then $B - x$ is the adjacent side, and the area is $f(x) = x(B - x)$. To maximize this area by the process described above, compute

$$f(x + E) - f(x) = (x + E)[B - (x + E)] - x(B - x)$$
$$= EB - 2Ex - E^2,$$

$$\frac{f(x + E) - f(x)}{E} = B - 2x - E,$$

and

$$\left[\frac{f(x + E) - f(x)}{E}\right]_{E=0} = B - 2x.$$

Fermat's equation is therefore $B - 2x = 0$, so $x = \frac{1}{2}B$, $B - x = \frac{1}{2}B$, and the largest rectangle is a square. When he reached this conclusion he remarked with justifiable pride, "We can hardly expect to find a more general method." He also found the shape of the largest cylinder that can be inscribed in a given sphere (ratio of height to diameter of base $= \frac{1}{2}\sqrt{2}$) and solved many similar problems that are familiar in calculus courses today.

Fermat's most memorable application of his method of maxima and minima was his analysis of the refraction of light. The qualitative phenomenon had of course been known for a very long time: that when a ray of light passes from a less dense medium into a denser medium—for instance, from air into water—it is refracted toward the perpendicular (see Fig. A.30). The quantitative description of refraction was apparently discovered experimentally by the Dutch scientist Snell in 1621. He found that when the direction of the incident ray is altered, the ratio of the sines of the two indicated angles remains constant,

$$\frac{\sin \alpha}{\sin \beta} = \text{a constant},$$

but he had no idea why. This sine law was first published by Descartes in 1637

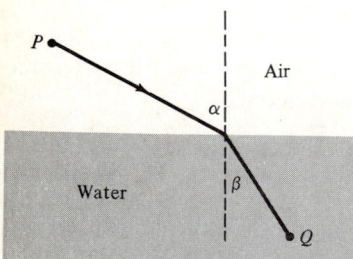

FIGURE A.30

(without any mention of Snell), and he purported to prove it in a form equivalent to

$$\frac{\sin \alpha}{\sin \beta} = \frac{v_w}{v_a},$$

where v_a and v_w are the velocities of light in air and in water. Descartes based his argument on a fanciful model and on the metaphysically inspired opinion that light travels faster in a denser medium. Fermat rejected both the opinion ("shocking to common sense") and the argument ("demonstrations which do not force belief cannot bear this name"). After many years of passive skepticism, he actively confronted the problem in 1657 and proved the correct law himself,

$$\frac{\sin \alpha}{\sin \beta} = \frac{v_a}{v_w}.$$

The foundation of his reasoning was the hypothesis that the actual path along which the ray of light travels from P to Q is that which minimizes the total time of travel—now known as *Fermat's principle of least time*.[5] This principle of least time led to the calculus of variations created by Euler and Lagrange in the next century, and on from this discipline to Hamilton's principle of least action, which has been one of the most important unifying ideas in modern physical science.

Fermat's method of finding tangents developed out of his approach to problems of maxima and minima, and was the occasion of yet another clash with Descartes. When the famous philosopher was informed of Fermat's method by Mersenne, he attacked its generality, challenged Fermat to find the tangent to the curve $x^3 + y^3 = 3axy$, and foolishly predicted that he would fail. Descartes was unable to cope with this problem himself, and was intensely irritated when Fermat solved it easily.[6]

These successes in the early stages of differential calculus were matched by comparable achievements in integral calculus. We mention only one: his calculation of the area under the curve $y = x^n$ from $x = 0$ to $x = b$ for any positive integer n (see Fig. A.31). In modern notation, this amounts to the evaluation of the integral

$$\int_0^b x^n \, dx = \frac{b^{n+1}}{n + 1}.$$

---

[5] A full discussion, including the details of Fermat's proof, can be found in Chapter V of A. I. Sabra, *Theories of Light from Descartes to Newton*, Cambridge University Press, 1981.

[6] The curve $x^3 + y^3 = 3axy$ is now called the *folium of Descartes*.

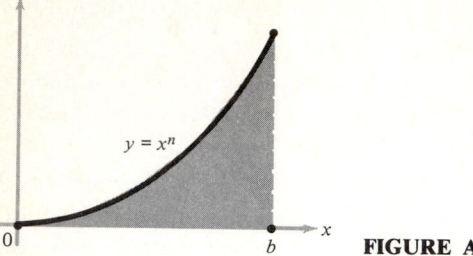

FIGURE A.31

The Italian mathematician Cavalieri had proved this formula by increasingly laborious methods for $n = 1, 2, \ldots, 9$, but bogged down at $n = 10$. Fermat devised a beautiful new approach that worked with equal ease for all n's.[7]

In the light of all these accomplishments, it may reasonably be asked why Newton and Leibniz are commonly regarded as the inventors of calculus, and not Fermat. The answer is that Fermat's activities came a little too early, before the essential features of the subject had fully emerged. He had pregnant ideas and solved many individual calculus problems; but he did not isolate the explicit calculation of derivatives as a formal process, he had no notion of indefinite integrals, he apparently never noticed the Fundamental Theorem of Calculus that binds together the two parts of the subject, and he didn't even begin to develop the rich structure of computational machinery on which the more advanced applications depend. Newton and Leibniz did all these things, and thereby transformed a collection of ingenious devices into a problem-solving tool of great power and efficiency.

The mind of Fermat had as many facets as a well-cut diamond and threw off flashes of light in surprising directions. A minor but significant chapter in his intellectual life began when Blaise Pascal, the precocious dilettante of mathematics and physics, wrote to him in 1654 with some questions about certain gambling games that are played with dice. In the ensuing correspondence over the next several months, they jointly developed the basic concepts of the theory of probability.[8] This was the effective beginning of a subject whose influence is now felt in almost every corner of modern life, ranging from such practical fields as insurance and industrial quality control to the esoteric disciplines of genetics, quantum mechanics, and the kinetic theory of gases. However, neither man carried his ideas very far. Pascal was soon caught up in the paroxysms of piety that blighted the remainder of his short life, and Fermat dropped the subject because he had other, more compelling mathematical interests.

---

[7] The details are given in Section B.5.

[8] See Smith, *Source Book,* pp. 546–65.

The many remarkable achievements sketched here—in analytic geometry, calculus, optics, and the theory of probability—would have sufficed to place Fermat among the outstanding mathematicians of the seventeenth century if he had done nothing else. But to him these activities were all of minor importance compared with the consuming passion of his life, the theory of numbers. It was here that his genius shone most brilliantly, for his insight into the properties of the familiar but mysterious positive integers has perhaps never been equaled. He was the sole and undisputed founder of the modern era in this subject, without any rivals and with few followers until the time of Euler and Lagrange in the next century. Pascal, who called him *le premier homme du monde*—"the foremost man in the world"—wrote to him and said: "Look elsewhere for someone who can follow you in your researches about numbers. For my part, I confess that they are far beyond me, and I am competent only to admire them."

The attractions of number theory are felt by many but are not easy to explain, being mainly aesthetic in nature. On the one hand, the positive whole numbers $1, 2, 3, \ldots$ are perhaps the simplest and most transparent conceptions of the human mind; and on the other, many of their most easily understood properties have roots that strike so deep as to be almost beyond the reach of human ingenuity. A large part of the lure of the subject lies in the fact that its smooth and apparently simple surface conceals depths of the utmost profundity. In order to convey something of the flavor of Fermat's work in this field, we briefly describe several of his most characteristic and influential discoveries. It should be remembered that most of the truths he uncovered are known only because he wrote about them to his friends or jotted them down in the margins of his copy of Diophantus.[9] Unfortunately, many of his proofs went unrecorded and were lost forever when he died.

1. The following is known as *Fermat's theorem*: If p is a prime and n is a positive integer not divisible by p, then p divides $n^{p-1} - 1$. For instance, if $p = 5$ and $n = 4$, then $n^{p-1} - 1 = 4^4 - 1 = 255$, which is divisible by 5; and if $p = 3$ and $n = 8$, then $n^{p-1} - 1 = 8^2 - 1 = 63$, which is divisible by 3. This theorem is of fundamental importance in both number theory and modern algebra.[10] Fermat stated it in a letter in 1640, and the first published proof was given by Euler in 1736.

---

[9] A few were proposed as challenges to certain English mathematicians whom he hoped (in vain) to interest in his ideas. For instance, it is clear that $x = 5$ and $y = 3$ is a positive integer solution of $x^2 + 2 = y^3$, and he asked for a proof that this is the *only* such solution. As E. T. Bell remarked, "It requires more innate intellectual capacity to dispose of this apparently childish thing than it does to grasp the theory of relativity." See J. V. Uspensky and M. A. Heaslet, *Elementary Number Theory*, McGraw-Hill, 1939, p. 398.

[10] See Chapter VI of G. H. Hardy and E. M. Wright, *An Introduction to the Theory of Numbers*, Oxford University Press, 1938.

2. Our second example is his profound and beautiful theorem on polygonal numbers. In his own words, as written in the margin of his copy of Diophantus:

> Every positive integer is triangular or the sum of 2 or 3 triangular numbers; a square or the sum of 2, 3 or 4 squares; a pentagonal number or the sum of 2, 3, 4 or 5 pentagonal numbers; and so on to infinity, whether it is a question of hexagonal, heptagonal, or any polygonal numbers.

To understand this statement, recall that triangular numbers are the numbers $1, 3, 6, 10, \ldots$ that can be obtained by building up triangular arrays of dots,

that the squares $1, 4, 9, 16, \ldots$ arise from square arrays of dots,

and similarly with pentagonal numbers and the rest. After the statement quoted, Fermat continued as follows: "I cannot give the proof here, for it depends on many abstruse mysteries of numbers; but I intend to devote an entire book to this subject, and to present in this part of number theory astonishing advances beyond previously known boundaries." It will not come as a surprise to anyone that this book was never written, though no one doubts that he could have done so. Euler struggled off and on for nearly 40 years to find a proof of the part of Fermat's theorem relating to squares, but he didn't quite reach his goal.[11] Lagrange at last succeeded in 1772, with an argument based heavily on Euler's ideas. Gauss established the part about triangular numbers in 1796, and Cauchy proved the complete theorem in 1815.

3. A far more famous marginal note—as familiar to mathematicians as the activities of Napoleon are to historians—occurs next to a passage in Diophantus dealing with positive integer solutions of the equation $x^2 + y^2 = z^2$. It is easy to see that $3^2 + 4^2 = 5^2$ and $5^2 + 12^2 = 13^2$, so the triples, 3, 4, 5 and 5, 12, 13 are obvious solutions. There are infinitely many such triples. They have been completely known since the time of Euclid, and were discussed by Diophantus.[12] Fermat's note in its entirety reads as follows:

---

[11] He did manage to prove Fermat's two squares theorem (every prime of the form $4n + 1$ is expressible as the sum of two squares) after a mere 7 years of effort.

[12] The general solution of this problem is given in Section A.9. See also H. Rademacher and O. Toeplitz, *The Enjoyment of Mathematics*, Princeton University Press, 1957, Chapter 14; or Chapter XIII of H. Tietze, *Famous Problems of Mathematics*, Graylock Press, 1965.

In contrast to this, it is impossible to separate a cube into two cubes, a fourth power into two fourth powers, or, generally, any power above the second into two powers of the same degree. I have discovered a truly wonderful proof which this margin is too narrow to contain.

This simple statement is now known as *Fermat's last theorem*: In modern notation, the equation $x^n + y^n = z^n$ has no positive integer solutions whatever for any exponent $n > 2$. Generations of mathematicians have cursed the narrowness of that margin, for in spite of intense efforts by some of the most penetrating minds in the world for more than 300 years, no proof has ever been found by anyone else.[13] In another place, Fermat himself left a sketch of a proof for the case $n = 4$. Euler published a proof for $n = 4$ (1747) and also for the more difficult case $n = 3$ (1770).[14] Gauss, Legendre, Dirichlet, and others settled the cases $n = 5$ and $n = 7$, and at the present time the theorem is known to be true for all exponents $n \leq 125,000$.[15] No one doubts its truth for all n, but its interest and unique reputation lie in the resistance it offers to complete and rigorous proof. Instant immortality awaits anyone who can find such a proof, but those considering this as a research project for their next free weekend should remember what David Hilbert, perhaps the greatest mathematician of the twentieth century, said when asked why he did not try: "Before beginning I would have to put in three years of intensive study, and I haven't that much time to waste on a probable failure." Some experts believe that Fermat deceived himself in thinking he had a proof. However, he was a man of complete integrity and a number theorist of unsurpassed ability. It should also be remembered that he has never been caught in a mistake; with this single exception—which has not been refuted—others have succeeded in proving every theorem of which he definitely stated that he had a proof. Fermat's last theorem remains to this day the most celebrated of the enigmatic legacies which he left to his baffled posterity.[16]

---

[13] Not even the Devil. See the charming short story "The Devil and Simon Flagg," by Arthur Porges, in Clifton Fadiman's anthology, *Fantasia Mathematica*, Simon and Schuster, 1958.

[14] Struik, *Source Book*, pp. 26–40.

[15] The current status of the subject is discussed in H. M. Edwards, "Fermat's Last Theorem," *Scientific American*, October 1978. See also H. S. Vandiver's article *Fermat's Last Theorem* in the *Encylopaedia Britannica* (any recent edition before 1974) or the article "Fermat Still Has the Last Laugh," *Discover* magazine, Jan. 1989, pp. 48–50. Perhaps the most enjoyable source for up-to-date information is Underwood Dudley's notes to his revised edition of E. T. Bell's last book (1961), *The Last Problem*, Mathematical Association of America, 1990.

[16] We feel obliged to mention that there does exist a book about Fermat and his mathematical work; however, this book was dissected and destroyed in a famous review by the eminent French mathematician André Weil (*Bulletin of the American Mathematical Society*, vol. 79 (1973), pp. 1138–1149).

A.14

CAVALIERI
(1598–1647)

I think sincerely that few men (perhaps, indeed, no one) from the time of Archimedes on have attained a greater knowledge of geometry...He has discovered a new method for the study of mathematical truths; by it he proves in a shorter manner many of the theorems of Archimedes and other mathematicians.

Galileo

At the time Galileo wrote these words Cavalieri was preparing his great work, the *Geometria Indivisibilium* ("The Geometry of Indivisibles"), which was published in 1635 in Bologna, where Cavalieri taught mathematics from 1629 until his death. This book contained a systematic development of the ideas of Kepler about finding areas and volumes. It had great influence in Italy, France and England, and was a major step toward the differential and integral calculus of Newton and Leibniz that arose a few decades later.

Bonaventura Cavalieri was a disciple of Galileo in physics and mathematics, and a member of the order of the *Gesuati* (or Jesuats, not the Jesuits as is often incorrectly stated in reference books). This order was established in 1367 to care for and bury victims of the Black Death, the plague that killed more than one-fourth the population of Europe. In the early 17th century the order attempted to revive its declining material and spiritual vigor by producing and selling wines and liquors, apparently in a manner contrary to Canon Law. This, and the order's declining membership, led to its suppression by Pope Clement IX in 1668, which partly explains why so little biographical material relating to Cavalieri has survived.

His method in geometry is to consider an area as made up of an indefinite number of parallel line segments, and a volume as composed of an indefinite number of parallel plane sections; these segments and sections he calls the *indivisibles* of the given area or volume. As he expressed it himself, "A line is made up of points as a string is of beads; a plane area is made up of lines as a cloth is of threads; and a solid is made up of plane sections as a book is made up of pages." An infinite number of these constituent elements is allowed, but Cavalieri does not discuss this in any detail.

As an illustration of his ideas we shall find the area of an ellipse with semiaxes a and b, where $a > b$. First we sketch the ellipse and circle whose equations are

$$\frac{x^2}{a^2} + \frac{y^2}{b^2} = 1 \qquad \text{and} \qquad x^2 + y^2 = a^2$$

on the same coordinate axes, as shown in Fig. A.32. By solving these equations for y, we obtain

$$y = \pm \frac{b}{a}\sqrt{a^2 - x^2} \qquad \text{and} \qquad y = \pm\sqrt{a^2 - x^2}.$$

These formulas show that each ordinate of the ellipse is b/a times the corresponding ordinate of the circle, and since the same thing is true of the vertical chords, we conclude that

$$\text{Area of ellipse} = \frac{b}{a}(\text{area of circle}) = \frac{b}{a}(\pi a^2) = \pi ab.$$

This is essentially the same procedure Kepler used, except that he thought of a large number of thin vertical rectangles instead of an infinite number of

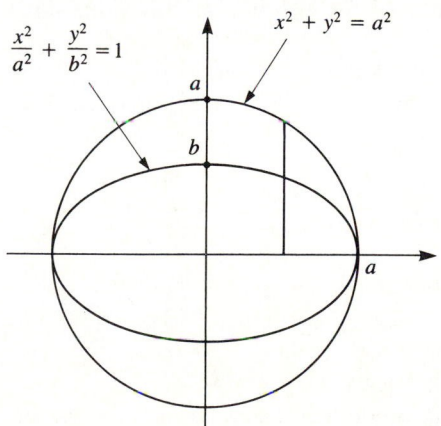

FIGURE A.32

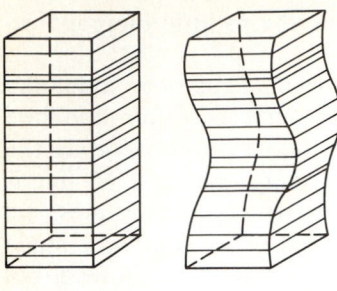

FIGURE A.33

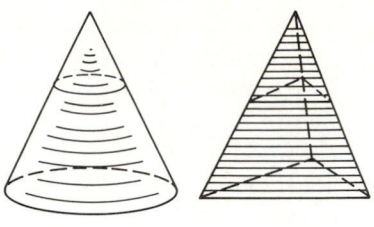

FIGURE A.34

vertical chords. Archimedes knew this area formula and almost certainly discovered it, but he simply gave a rigorous proof, without providing intuitive explanations of the kind discussed here.

To understand Cavalieri's method as it applies to solids, we consider a rectangular solid (Fig. A.33, left) consisting of a stack of thin cards, all with the same dimensions. The shape of this stack can easily be altered without changing its volume, by gently pushing at it horizontally (Fig. A.33, right). The volume before is clearly the same as the volume after, since each card in the stack is unchanged except in its position relative to nearby cards. Next, consider two solids with different shapes but the same height (Fig. A.34), made up of equal numbers of thin cards. If we assume that each card in one stack has the same face area as the corresponding card in the other stack, regardless of the different shapes of these cards, then it seems reasonable to conclude that the two solids have the same volume. The idea developed here is known as *Cavalieri's Principle* (or *Theorem*): If two solids have equal heights and if sections made by planes parallel to the bases and at equal distances from them always have equal areas, then the volumes of the two solids are equal (Fig. A.35). By a slight extension, we can say that if the areas of the sections always have a given ratio, then the volumes of the solids also have the same ratio.

The classic example of the application of this principle is in finding the formula for the volume of a sphere of radius r. Consider a cylinder of base

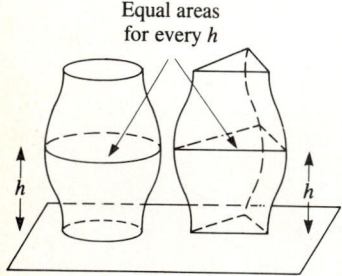

Equal areas
for every h

FIGURE A.35

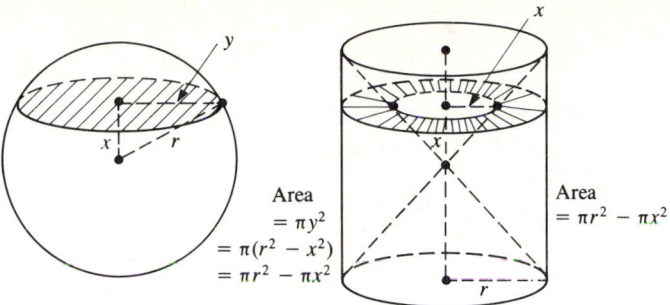

Area
$= \pi y^2$
$= \pi(r^2 - x^2)$
$= \pi r^2 - \pi x^2$

Area
$= \pi r^2 - \pi x^2$

FIGURE A.36

radius r and height $2r$; the comparison solid (Fig. A.36) is what remains of this cylinder after the removal of the two cones shown in the figure, that is, it is the cylinder with two conical hollows on the ends. If we calculate the areas of corresponding cross-sections of these solids, as indicated in the figure, we find that they are equal. By Cavalieri's Principle, the solids have equal volumes. The volume V of the sphere, being equal to the volume of the cylinder minus the volumes of the two cones, is therefore given by the formula

$$V = \pi r^2(2r) - 2(\tfrac{1}{3}\pi r^2 \cdot r)$$
$$= 2\pi r^3 - \tfrac{2}{3}\pi r^3$$
$$= \tfrac{4}{3}\pi r^3.$$

Archimedes discovered this formula in the 3rd century B.C. by a very beautiful method quite unknown to Cavalieri, that involved weighing cross-sections of solids against each other by means of his own principle of the lever, as explained in Section B.6. However, he considered his method to have intuitive value but little logical force; it encouraged belief, but did not compel it. He therefore gave an entirely different and fully rigorous proof in his treatise *On the Sphere and Cylinder,* which was well known to Cavalieri and his contemporaries. Archimedes would surely have admired Cavalieri's method, but his sense of mathematical rigor was so highly developed that he would certainly have assigned it no more logical weight than his own.

To understand the potential difficulties with Cavalieri's method, let us consider the following simple geometric fact, which of course can be proved in other ways (see Fig. A.37). To show that the diagonal BD in the parallelogram $ABCD$ divides the parallelogram into two triangles with equal areas, Cavalieri would argue that when $DE = BH,$ then $EF = GH.$ Since triangles ABD and BCD are made up of an equal number of equal lines, like EF and $GH,$ the triangles must have equal areas.

But Cavalieri himself was well aware that this method of summing lines into areas and areas into volumes can be dangerous. In a letter to his younger friend Torricelli he used exactly the argument just given, but this time to

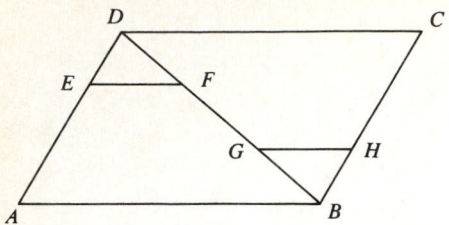

FIGURE A.37

produce an absurd conclusion: Start with a nonisosceles triangle *ABC* with altitude *CD,* as shown in Fig. A.38. Draw an arbitrary line *EF* parallel to *AB,* and then draw *EG* and *FH* parallel to *CD.* Then *EG = FH,* and by the theory of indivisibles the sum of the *EG*'s, which is the area of triangle *ADC,* equals the sum of the *FH*'s, which is the area of triangle *DBC.* In this way Cavalieri reached a conclusion that he knew very well was wrong regardless of the reasoning taking him there: the areas of the two triangles are simultaneously equal and unequal. Nowadays we settle this paradox very easily by the properties of integrals, but Cavalieri had no such weapons with which to respond.

Logical difficulties like this were recognized and discussed by philosophers—sometimes with intelligence; recognized and set aside by mathematicians—who preferred to get on with exploring the uses of their powerful but dangerous tools; and finally settled satisfactorily by mathematicians—not philosophers—two centuries later, in the early 19th century. Cavalieri himself knew the risks but preferred to go forward and take his chances.

We mention two more of his memorable achievements. First, he demonstrated by prolix and cumbersome geometric methods that, in our notation,

$$\int_0^b x^n \, dx = \frac{b^{n+1}}{n+1}$$

for the positive integers $n = 1, 2, \ldots, 9$. At $n = 9$ he gave up in exhaustion, but of course he conjectured that this formula is true for every positive integer n without restriction. A few years later Fermat established the formula for all

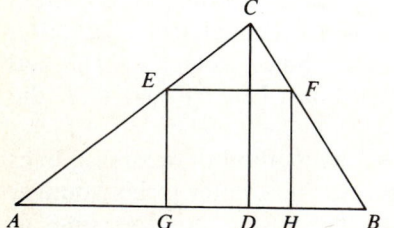

FIGURE A.38

positive integers and fractions n by a simple but ingenious argument whose details are given in Section B.5.

Second, it was Cavalieri the disciple and not Galileo the master who first published the correct parabolic law of projectile motion, which most scientists assume is due to Galileo. In fact, Galileo was not much of a mathematician, and managed to confuse himself about the geometry. He was naturally very upset to have his pupil take the last simple step, because finding this curve was a major motive behind the years he had spent studying motion under gravity. Cavalieri freely conceded Galileo's moral priority and even contritely offered to suppress his own little book in which the correct curve was first published.

Galileo was often testy and difficult to get along with, and usually he had good reasons; but when he described wine—which he loved—as "light held together by moisture," he revealed a personality that no civilized man or woman could long resist.

PROBLEMS

(The following problems are intended for those readers who wish to explore for themselves a few of the remarkable applications of Cavalieri's Principle.)

1. Use Cavalieri's Principle to find the volume of a *spherical segment* of one base and thickness h if the radius of the sphere is r (Fig. A.39). Answer: $\pi h^2(r - \frac{1}{3}h)$.

2. In Problem 1, find the volume of the *spherical sector* (the solid shown in the figure, resembling a filled ice cream cone). By comparing areas and volumes show that the area of the curved surface on top of the sector is $2\pi rh$. Hint: Use the fact that the area of the surface of the sphere is $4\pi r^2$. Answer: volume of sector $= \frac{2}{3}\pi r^2h$.

3. A *spherical ring* is the solid that remains after removing from a solid sphere of radius r a cylindrical boring whose axis passes through the center of the sphere. If h is the height of the ring, use the result of Problem 2 to show that the volume of the ring is $\frac{1}{6}\pi h^3$. Also obtain this result by applying Cavalieri's Principle and using a sphere of diameter h as the comparison solid. (Notice how remarkable it is that the volume of the ring depends only on h, and not on the radius r of the sphere.)

4. Use Cavalieri's Principle to show that $2\pi^2 a^2 b$ is the volume of the *torus* formed by revolving a circle of radius a about a line in the plane of the circle at a distance $b \geq a$ from the center of the circle. Hint: Rest the torus on a plane perpendicular to the

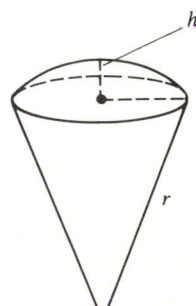

FIGURE A.39

axis of the torus and use as a comparison solid a cylinder of radius a and height $2\pi b$ that rests lengthwise on the same plane.

5. A *cylindrical wedge* (or *hoof*) is the solid cut from a cylinder by a tilted plane passing through a diameter of the base. Apply Cavalieri's Principle to find the volume of such a wedge if its height is $2r$, where r is the radius of the base. (Hint: Use as a comparison solid a rectangular box having edges r, r, $2r$ with two square pyramids removed, where the pyramids have the square ends of the box as bases and common vertex at the center of the box. Stand the box on one of its square ends and place the wedge so that the bounding diameter of its base is vertical.) Adapt your thinking to find the volume of the wedge if its height is h instead of $2r$. Answers: $\frac{4}{3}r^3$, $\frac{2}{3}r^2 h$.

6. Show that the following tetrahedron ABCD can be used as a comparison solid for obtaining the volume of a sphere of radius r by means of Cavalieri's Principle: the segments AB and CD are each of length $2r\sqrt{\pi}$ and lie in parallel planes a distance $2r$ apart; the line joining their midpoints is perpendicular to both planes; and the segments are perpendicular to each other. Now use the known volume of the sphere to find the volume of the tetrahedron. Answer: $\frac{4}{3}\pi r^3$.

A.15

TORRICELLI
(1608–1647)

To us his incredible genius seems almost miraculous.

Marin Mersenne

In 1623 the eminent Italian physicist Galileo issued his call-to-arms of modern science: "The Great Book of Nature is written in mathematical symbols." Ten years later, in 1633, he was condemned for heresy by the Inquisition, ordered to recant his vigorous support of Copernican astronomy—which was believed to contradict the Holy Scriptures of Christianity—and sentenced to house arrest and seclusion on his little estate at Arcetri, near Florence, where he spent the remainder of his life. In 1638–39 the English poet John Milton visited him there and later wrote: "There it was that I found and visited the famous Galileo, grown old, a prisoner to the Inquisition for thinking in Astronomy otherwise than the Franciscan and Dominican licensers of thought." On the day of Galileo's condemnation, Florentine civilization—which had led the Western world since the time of Dante in the thirteenth century—almost vanished from history.

Almost, but not quite. The last brilliant chapter of Italian Renaissance science was the brief seven-year career of Evangelista Torricelli from 1641 to 1647. He was assistant and secretary to the blind and aged Galileo during the last months before his death in 1642, and succeeded him as mathematician to the Grand Duke of Tuscany. He was a close friend of Cavalieri and corresponded with Mersenne and Roberval in Paris. In mathematics, he

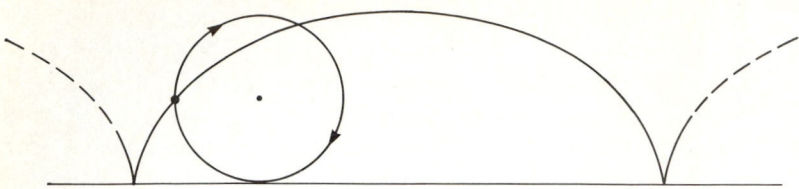

FIGURE A.40
The cycloid.

calculated many areas, volumes and tangents by an improved form of Cavalieri's method of indivisibles, and in particular discovered the area of the cycloid (Fig. A.40); he found the exact finite length of the infinitely coiled inner portion of the equiangular spiral (Fig. A.42); he glimpsed the inverse relationship between area and tangent problems, which is the central fact of differential and integral calculus; and much else. In physics, he advanced the first correct ideas about atmospheric pressure and the nature of vacuums, and invented the barometer as an application of his theories; and in doing this, he destroyed the Aristotelian notion that "nature abhors a vacuum."[1] Also, he studied projectiles and fluid motion, and discovered *Torricelli's law* on the speed of discharge of liquids through orifices in the bottoms of containers, which initiated the science of hydrodynamics.[2] Altogether, a promising young man.

Galileo was a very noticing person: his net of awareness was flung wide and had a fine mesh. He was interested in whatever he saw, often for the sake of the mathematical ideas that might be brought to light. He was not a mathematician himself, but he idolized Archimedes and encouraged mathematical thinking among his friends and disciples. At some time around the year 1600 he noticed that the archlike path traced out by a point on the rim of a rolling wheel (Fig. A.40) is a curve of exceptional beauty and interest. This curve, later known as the *cycloid,* obsessed the mathematicians of the 17th century and later. Among those who played a part in developing its properties were Galileo himself, Torricelli, Mersenne, Roberval, Descartes, Fermat, Pascal, Christopher Wren, Huygens, John Bernoulli, Leibniz, and Newton— almost a complete list of the stars and superstars of the Scientific Revolution.

Galileo's own study of the cycloid was primitive but direct. He attempted to find the area under one arch by balancing a cycloidal template against circular templates the same size as the generating circle. This experiment led

---

[1] See James B. Conant, *Science and Common Sense,* Yale University Press, 1951, pp. 63–71.

[2] See W. F. Magie, *A Source Book In Physics,* McGraw-Hill, 1935, pp. 111–13 and 70–73. For Torricelli's law and some applications, see George F. Simmons, *Differential Equations,* 2nd ed., McGraw-Hill, 1991, p. 44.

him to conclude—incorrectly—that the area of the cycloid is nearly, but not exactly, three times the area of the generating circle. The first published proof that the area is *exactly* three times that of the circle was given by Torricelli in 1644. No doubt it was Galileo who urged him to study the problem.

Mersenne was virtually Galileo's scientific representative in Paris, and through him an interest in the cycloid spread northward out of Italy, to Descartes and others.[3] Descartes found the tangent to the cycloid by the following ingenious argument. If a polygon $ABCD$ rolls (awkwardly) on a straight line $A'B'C'D'$ as shown in Fig. A.41, then the point A will trace out in succession several arcs of circles with centers B' C', D'. The tangent to any such arc is evidently perpendicular to the line joining the point of tangency to the corresponding center. Therefore, if the rolling circle that generates a cycloid is thought of as a polygon with an infinite number of sides, then the tangent to the cycloid at any point is the line perpendicular to the line joining the point of tangency to the bottom of the rolling circle, as shown on the right in the figure.

The cycloid continued to yield its secrets. The next major step was taken in 1658 by Christopher Wren in England.[4] Wren discovered that the length of the cycloid is exactly four times the diameter of the generating circle. He did

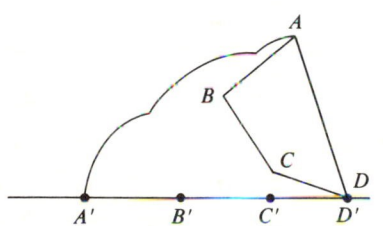
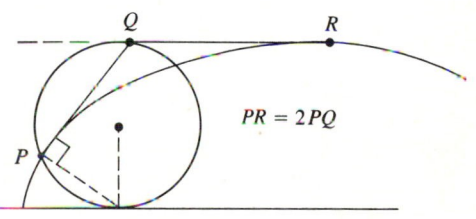

$$PR = 2PQ$$

FIGURE A.41
The ideas of Descartes and Wren.

---

[3] One of the "others" was Gilles Persone de Roberval (1602–1675), a brilliant but quarrelsome man of waspish disposition who was one of Mersenne's circle of friends. Roberval seems to have discovered the area of the cycloid before Torricelli, but lost credit for his discovery because he refused to publish it. He held the chair of mathematics at the Collège Royale in Paris. This chair automatically became vacant every three years, and was filled by open competition based on examinations whose questions were set by the incumbent. Roberval won this contest in 1634 by developing a method of indivisibles similar to Cavalieri's, and by keeping his method a secret held off all challengers and retained his professorship until the end of his life. Unfortunately for him, this strategy for keeping his job meant that he lost credit for most of his discoveries and spent much of his life in ill-tempered priority disputes.

[4] At this time Wren had not yet started his career as the greatest of English architects (St. Paul's Cathedral, etc.), but was only a talented astronomer and mathematician—in fact, Savilian Professor of Astronomy at Oxford.

this (see Fig. A.41) by showing that the arc *PR* from *P* to the point *R* at the top of the cycloid equals twice the tangential segment *PQ*.[5]

The next important properties of the cycloid to be discovered were in the realm of physics, and are described below in our accounts of Huygens, Newton and the Bernoullis. Full mathematical details are provided in the appropriate sections of Part B. In the 18th century these ideas led to the emergence of an entirely new branch of mathematics called the Calculus of Variations, which in the 19th and 20th centuries made crucial contributions to analytical mechanics, quantum mechanics, and the general theory of relativity. All in all, the study of the cycloid—in which Torricelli took the first solid step— became one of the golden threads in the tapestry of mathematics and has retained its luster for 400 years.

We now move on to a few of Torricelli's other mathematical achievements. Wren found the length of the cycloid in 1658, but in 1645 Torricelli performed the very first calculation of the length of a curve other than a circle. By applying refinements of the infinitesimal methods he learned from Cavalieri, he showed that the length of the equiangular spiral in Fig. A.42, as it winds in from *P* an infinite number of times around its asymptotic point, is exactly equal to the length of the tangent *PQ*. The equiangular spiral is characterized by the property that at each point on it the angle ψ from the radius to the tangent has the same value. In modern notation the polar equation of the spiral is $r = ae^{b\theta}$, and Torricelli's theorem is equivalent to the calculation of

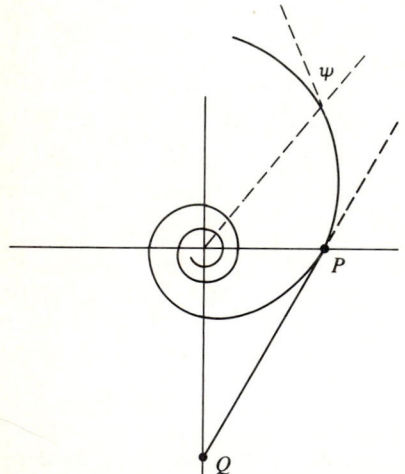

FIGURE A.42
The equiangular spiral.

---

[5] Wren's proof was published in 1659 by his friend John Wallis (see Wallis's *Opera*, vol. 1, pp. 550–69).

a certain fairly simple integral. Until this discovery became known, few people believed that the length of a curve could exactly equal the length of a line. In fact, in Descartes' *Geometry* (1637) he asserts that the relation between curves and straight lines is not known and can never be known:

> Geometry should not include lines that are like strings, in that they are sometimes straight and sometimes curved, since the ratios between straight and curved lines are not known, and I believe cannot be discovered by human minds.

Galileo responded, "Who is so blind as not to see that, if there are two equal straight lines, one of which is then bent into a curve, that curve will be equal to the straight line?" However, Torricelli's exact determination of the length of the spiral was more crushing than even Galileo's scornful reply.

Torricelli took great delight in problems involving the use of infinitesimal methods. Another of his discoveries, which pleased him very much and caused great astonishment at the time, was that a solid can have finite volume even though it has infinite extent. More specifically, in 1643 he proved that the hyperbolic solid of revolution (see Fig. A.43) generated by revolving $y = 1/x$ about the x-axis from $x = 1$ to $x = \infty$ has finite volume.[6] Again, in the notation of calculus this is a straightforward problem in setting up and evaluating a certain simple integral. In 1672 the English political philosopher Thomas Hobbes attacked this result by saying, "To understand this for sense, it is not required that a man should be a geometrician or a logician, but that he should be mad."[7] Hobbes had the curious belief that mathematical theorems can be

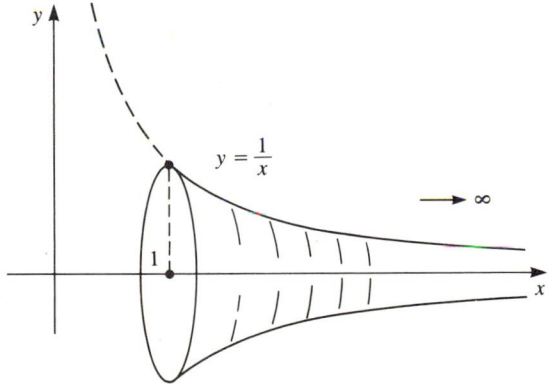

FIGURE A.43
The hyperbolic solid.

[6] A translation of Torricelli's original paper is given on pp. 227–31 of D. J. Struik, *A Source Book In Mathematics, 1200–1800*, Harvard University Press, 1969.

[7] *The English Works of Thomas Hobbes*, vol. 7, pp. 443–48.

attacked by ridicule and invective as if they were obnoxious planks in an opponent's political platform.[8]

In mathematics, the 17th century above everything else was the century of the rise of calculus. The invention of calculus is usually credited to the independent work of two men, Isaac Newton and Gottfried Wilhelm Leibniz, in the 1660s and 1670s. However, great advances in science and mathematics are rarely the work of single individuals. Newton later said, "If I have seen farther than others, it is because I have stood on the shoulders of giants." Torricelli was one of these giants. He had reached the threshold of the new world of mathematics—like several others—and could have crossed it more easily than any other man of his time. His early death at the age of 39, just as he was reaching the full flower of his maturity, was a tragedy for European thought but a disaster for Italy. After Torricelli, the center of gravity of mathematics and science left Italy and moved northward, to France, Holland, England, and Germany.

---

[8] This belief caused him to make a fool of himself on other occasions as well. See the footnote in Section B.12.

A.16

PASCAL
(1623–1662)

Pascal stamped his works with the passionate conviction of a man in love with the absolute.

Jean Orcibal

A modern office worker, setting off in the morning, may glance at a wrist watch, inspect the barometer, buy a newspaper at the corner store and receive change from the cash register, and board a bus for the trip downtown to the business district. What has all this to do with a French mathematician who was mixed up in musty theological disputes when Louis XIV was still a teenager? Pascal invented that wrist watch, originated that barometer, invented that calculating machine, and was the first to think of a bus system and organize a public transportation company.[1]

Blaise Pascal was one of the most gifted and tragic figures in the whole history of Western thought. A child prodigy, he was even more prodigious as a man. And yet his life was twisted and stunted by mystic visions and religious neuroses, and of all the great things he had it in him to do, none progressed beyond memorable beginnings.

Pascal was born in Clermont-Ferrand, in the Auvergne region of central

---

[1] See the excellent biography by Ernest Mortimer, *Blaise Pascal*, Harper, 1959, p. 12.

119

France. His mother died when he was only 3 years old, so that he and his two sisters were brought up by their father Étienne, a man of strong character and broad learning. In 1631 the family moved to Paris for the sake of the children's development. Blaise never attended ordinary schools, but instead was taught exclusively by his father.

Étienne Pascal became a member of Mersenne's weekly discussion group, which was sponsored by Cardinal Richelieu and later developed into the French Academy. Its purpose was to encourage interest in scientific matters, especially mathematics and physics. It included the philosopher Descartes (when he happened to be in town), the mathematicians Roberval and Desargues, the Englishman Hobbes, who was present in the winter of 1636–1637, and several others. From the age of 12 or 13, Pascal often participated in these gatherings, where he listened avidly and sometimes entered into disputes himself. There is an old saying: "Genius is like a fire; a single burning log will smoulder or go out, while a heap of logs piled loosely together will flame fiercely." And so it was with the young Pascal and his father's eminent friends.

When he was 16 he published his famous *Essai sur les coniques* ("Essay on Conic Sections"). Descartes was jealous of its success, and refused to believe that it was produced by a mere boy. This brief work contains what is still the most important theorem of projective geometry, known as *Pascal's theorem*: If a hexagon is inscribed in a conic section (Fig. A.44), then the three points of intersection of its opposite sides always lie on a straight line.

At 17, while watching his father's arithmetical drudgery over tax assessments, he conceived the possibility of a calculating machine, by 18 or 19 he had completed the first working model, and in the next few years he manufactured over 50 machines which he offered for sale. He hoped that his project would make him rich, but it never did. The manufacturing costs were too high, and the increasing use of logarithms reduced the demand.

When he was 23 he heard a sketchy report of Torricelli's experiment involving a 3-ft glass tube closed at one end and filled with mercury. If the open end is closed with a thumb and immersed in a bowl of mercury, and then

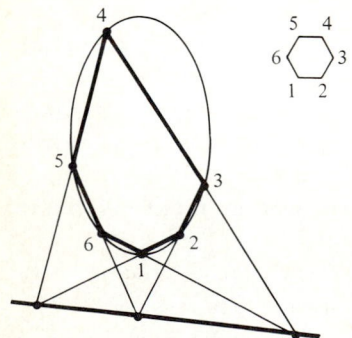

FIGURE A.44
Pascals theorem.

the thumb is removed, the mercury in the tube sinks to a level about 30 inches above the level of the mercury in the bowl. Pascal repeated this experiment with many variations and great care, and gave a complete and correct explanation of the results. His explanation agreed with Torricelli's, which he had not heard: namely, that the empty space in the top of the tube is a vacuum, and the mercury is held up in the tube by the weight of the ocean of air that presses down on the surface of the mercury in the bowl. Also, as a byproduct of his work, Pascal invented the syringe. The conclusions established by these investigations stirred up a storm of controversy with the scholastic philosophers, who tried vainly to maintain the Aristotelian doctrine that "nature abhors a vacuum."

A year or two later Pascal conceived the famous Puy-de-Dôme experiment, in which a Torricelli tube was carried up a mountain in a single day, so that the fall of the mercury level in the tube with increasing altitude could actually be observed. Further, among the notes found after his death was a series of observations of the variations of the Torricelli tube according to changes in the weather. He writes, "This knowledge can be very useful to farmers, travellers, etc., to learn the present state of the weather, and that which is to follow immediately, but not to know that which is to come in three weeks." Thus was the barometer born.

During this period he also studied hydrostatics, and discovered what is now called *Pascal's Principle* for the transmission of pressure through an enclosed fluid. This led him to the idea of the hydraulic press, which he described very clearly even though technical difficulties prevented him from making a successful working model.

There is a French saying that "Many people know the whole history of human thought without ever having had one." It is clearly true that most educated people in any period think other people's thoughts and little else. However, Pascal was trained by his father from infancy in the art of original thought, which is a rare and precious thing. He wanted to know the reason for everything, and reasons consisting of mere words threw his mind into a turmoil of frustration. In his scientific work he was strongly committed to the experimental method, with its emphasis on empirical facts combined with logical thinking, and with its total disregard for the appeals to authority that constituted most of the act of reasoning for the scholastic philosophers. However, he was also convinced that evidence and thought are not enough in the domain of religion: In this part of human experience, faith is necessary in order to arrive at the truth.

Religion played an important part in Pascal's life from the time of his "first conversion" in 1646, when he felt a strong compulsion to turn away from the world toward God. This impulse soon weakened, and he became absorbed again by his scientific interests and his fashionable friends. It was near the end of this period, in 1654, that in correspondence with Fermat he assisted in formulating some of the ideas that led to the mathematical theory of probability. At this time he also wrote his *Traité du triangle arithmétique* ("Treatise on the Arithmetical Triangle"), in which he studied a triangular arrangement of the binomial coefficients and discovered and proved many of their properties.[2] It is in this work that he gives what seems to be the first satisfactory statement of the principle of proof by mathematical induction.[3]

In the years 1653 and 1654 Pascal worked harder and harder on science and mathematics, as a desperate diversion from the increasing emptiness within. Late in 1654 he had the decisive experience of his life, an overwhelming mystic vision that caused him to turn away from his worldly, free-thinking friends and immerse himself permanently in religious contemplation. He returned to mathematics only once again, in 1658. While suffering from a severe toothache, he began to think about some problems concerning the cycloid, the archlike curve traced out by a point on the rim of a rolling wheel. His tooth suddenly stopped aching, and he took this as a sign of divine approval. Over the next few months he worked feverishly on this topic and solved a number of problems in a series of small treatises. The main problems were to find the area and length of one arch, but unfortunately for him these had already been solved by Torricelli and Christopher Wren. Nevertheless, these works had a consequence of very great importance, for about 15 years later they suggested to Leibniz an idea that was crucial for his own invention of differential and integral calculus. Leibniz later wrote that "a great light" burst upon him when he read a particular passage, and he wondered how Pascal could have missed the idea.[4]

It was toward the end of his brief life that he wrote the works that earned him a place among the very greatest figures of French literature. His *Provincial Letters* is a series of polemical pamphlets against the Jesuits. They took the form of "Letters Written to a Provincial by One of His Friends," and were signed with a fictitious name. In these letters we encounter for the first time the variety, brevity, and tautness of style that distinguish the best modern French prose. As Voltaire said, "The first work of genius in prose that we find is the collected edition of the *Provincial Letters.* All kinds of eloquence are displayed here. It is in this work that our language takes its final form." At the end of Letter XVI, Pascal makes his memorable apology for the length of the letter he has just written, his reason being that "I don't have time enough to make it shorter."

The *Provincial Letters* has only a limited historical interest for us today, but Pascal's *Pensées* ("Thoughts") will presumably endure as long as the

---

[2] See pp. 21–26 of D. J. Struik (ed.), *A Source Book in Mathematics, 1200–1800,* Harvard University Press, 1969.

[3] See also G. Polya, *Mathematical Discovery,* Wiley, 1962, vol. 1, pp. 70–75, especially p. 74.

[4] See Struik, *Source Book,* pp. 239–41.

French language lasts. His purpose was to write a monumental and irresistible defense of the Christian religion against the unbelievers. However, during the last years of his life he was often weak, semidelirious, and racked with the pain of his illnesses, which after his death were determined to be a brain tumor and cancer of the stomach. He was incapable of connected work, and took to scribbling down on any bit of paper that came to hand the thoughts that flashed into his mind. Thus we have about a thousand scraps of paper containing ideas and fragments of ideas for his intended great work—phrases, single sentences, sometimes several whole paragraphs together.[5] The specific subjects are very diverse, but in general the theme is the grandeur and misery of man. Here are a few of Pascal's *pensées*:

The heart has its reasons that the reason does not know. (277)

The eternal silence of these infinite spaces terrifies me. (206)

I have discovered that all the misfortune of men comes from one single thing, not knowing how to remain quiet in a room. (139)

Men never do evil so completely and cheerfully as when they do it from religious conviction. (894)

What a chimera is man! What a novelty! What a monster, what a chaos, what a contradiction, what a prodigy! Judge of all things, feeble worm of the earth, depository of truth, a sink of uncertainty and error, the glory and shame of the universe. (434)

Man is only a reed, the feeblest thing in nature, but he is a thinking reed. There is no need for the whole universe to arm itself to crush him; one vapor, one drop of water, will suffice to kill him. But when the universe crushes him, man is still nobler than that which kills him, because he knows what kills him and he understands the advantage the universe has over him. The universe knows nothing of this. (347)

As his biographer says, "In the general estimation of his own countrymen Pascal occupies not only a high place but the top place. 'He is to France,' writes Professor Chevalier, 'what Plato is to Greece, Dante to Italy, Cervantes and St. Thersa to Spain, Shakespeare to England.'"[6]

---

[5] About 300 years ago these scraps of paper were glued helter-skelter (upside-down, sideways, at all angles) into a large album that is now protected as one of the most precious cultural treasures of France in the Bibliothèque Nationale in Paris. The present writer has had the privilege of spending a couple of hours examining this album, looking for familiar passages and occasionally finding them.

[6] Ernest Mortimer, *Pascal*, p. 183.

The achievements of Pascal's short life were remarkable enough. However, if he had not died at the early age of 39, if he had not been ill of body and mind during those last years, if he had not deliberately rejected mathematics and science for a handful of dust, there is little doubt that he could have discovered calculus—and probably much else besides—years before Newton and Leibniz. He was surely the greatest "might-have-been" in the history of mathematics.

A.17

HUYGENS
(1629–1695)

The extraordinary astronomer–mathematician–physicist Christiaan Huygens was undoubtedly Holland's greatest scientist, and he deserves to be much more widely known among people who are interested in the great thinkers of our past.

Even apart from its scientific immortals like Huygens and his friend Leeuwenhoek, the inventor of the microscope, Holland in the mid-seventeenth century was a rich garden of civilization. Its worldwide trading empire provided peace and comfort for its people, its serene light inspired great artists like Rembrandt and Vermeer, and its religious liberty provided safe haven for philosophers and free-thinkers like Spinoza and Descartes. Also, as an outward sign of intense intellectual ferment, the Dutch cities were teeming with publishers and books. In the whole world at that time there were not more than ten or a dozen cities where books were printed on any substantial scale. England had only two centers of the publishing trade, London and Oxford; France also had only two, Paris and Lyons; but in Holland there were five—Amsterdam, Rotterdam, Leiden, The Hague, and Utrecht—all printing books in Greek, Latin, English, French, German, Italian, and Hebrew as well as in Dutch. In Amsterdam alone there were more than four hundred printers or booksellers. A touchstone like this is an almost infallible guide to the quality of a society.[1]

As we see, Holland was probably the most civilized nation in Europe at that time, and Constantijn Huygens, the father of Christiaan, was surely its most civilized citizen. He was a statesman and diplomat who spent most of his life in the service of the Princes of Orange; a scholar who knew seven languages; a poet, playwright, musician, composer, and amateur scientist; friend of Francis Bacon, Descartes, and Mersenne; friend and translator of the English poet John Donne; knighted by James I of England; friend and patron of Rembrandt, whom he persuaded to move from Leiden to Amsterdam; and head of one of the great families of his country. Descartes described his reaction after first meeting Constantijn: "I could not believe that a single mind could occupy itself so well with so many things." Eminent thinkers and travelers from other nations were often guests at the Huygens home in The Hague. As a young man growing up in such an environment, it was almost inevitable that Christiaan Huygens would become skilled in languages, art, music, science, and mathematics. "The world is my country," he said, "and science is my religion."[2]

Huygens made his first notable discovery in 1646, at the age of 17. Galileo had maintained that a flexible chain hangs in the shape of a parabola, but Huygens proved that this is not correct. His father informed Mersenne, who responded with gratifying enthusiasm. In 1691 Huygens returned to this problem and determined the true shape of this interesting curve.[3]

In 1655 he developed an improved method of grinding and polishing lenses for telescopes, and a flood of new knowledge quickly followed. He discovered that Saturn is surrounded by rings that nowhere touch the planet.[4] He discovered Titan, the moon of Saturn that is now known to be the largest moon in the solar system. He was the first to notice a surface feature on Mars, and by observing the movement of this feature as the planet rotates, he was the first to determine that the Martian day is approximately 24 hours long, very nearly the same as our own. In 1656–1657 he invented the pendulum clock, which sprang from the need for a more precise way of measuring time in astronomical observations.

In 1657 he published a little booklet that was the first formal treatise on the theory of probability, *De rationiis in ludo aleae* ("On Reasoning in Games

---

[1] See p. 88 of P. Hazard, *The European Mind, 1680–1715,* Yale University Press, 1953.

[2] See p. 251 of G. N. Clark, *The Seventeenth Century,* Oxford University Press, 1931.

[3] Leibniz and John Bernoulli also found the equation of this curve in the same year, independently of Huygens and each other, and Leibniz named it the *catenary.* See Section B.11.

[4] Galileo had earlier seen these rings through his own more primitive telescope, but he had no idea what they were. To him they appeared to be two strange protuberances attached to the planet like ears.

of Dice"). He had visited Paris and heard about the 1654 correspondence between Fermat and Pascal on this subject; and since neither of these men seemed inclined to write up their ideas, he sought his own answers. Among other things, he introduced the important concept of "mathematical expectation."

In 1666 Huygens moved to Paris at the urging of Colbert, the great minister of Louis XIV who was chiefly responsible for much of the economic and political power of France over the next several centuries. He became one of the first salaried members of Colbert's newly created French Academy of Sciences, and made his home in Paris for the next 15 years.

In 1663 he was elected a fellow of the Royal Society of London, and in 1669 he presented that organization with the first clear and correct statement of the laws of impact for elastic bodies. His laws refuted Descartes's erroneous laws of impact as set forth in his *Principia Philosophiae* (1644).[5] One of Huygens's laws states that in the mutual impact of two bodies, the sum of the products of the masses and the squares of their velocities is the same before and after impact. This appears to be the earliest version of the principle of conservation of energy.[6]

In 1673 Huygens published his greatest work, his treatise *Horologium oscillatorium* ("The Pendulum Clock"). Here he dug deeply into the theory of the pendulum clock that he had invented 16 years earlier, and uncovered many valuable nuggets of mathematics and physics. He had long been acutely aware of the so-called circular error inherent in such clocks, namely, the fact that the period of oscillation is not determined strictly by the length of the pendulum alone, but also depends on the magnitude of the swing. To express this differently, if a frictionless ball is placed on the side of a smooth hemispherical bowl and released, the time it takes to reach the lowest point will be almost, but not quite, independent of the height from which it starts. It happened that various properties of the cycloid were widely discussed in Western Europe in the late 1650s, and it occurred to Huygens to wonder what would happen if the hemispherical bowl were replaced by one whose vertical cross section is an inverted arch of a cycloid. He was overjoyed to discover that in this case the ball will reach the lowest point in exactly the same time no matter where it is

---

[5] For example, one of Descartes's laws states that if a small ball collides with a large ball at rest, then the small ball will rebound while the large ball remains immobile, which is obviously false. Some devastating criticisms of Descartes's attitudes toward science and scientific experiment are given at the end of Chapter 6 of H. Butterfield, *The Origins of Modern Science, 1300–1800,* G. Bell & Sons, 1957.

[6] Some of his reasoning is described on pp. 16–19 of C. Lanczos, *Albert Einstein and the Cosmic World Order,* Interscience, 1965. On p. 19 Lanczos remarks, "He [Huygens] employs here in his proof for the first time that 'principle of relativity' which in Einstein's hands attained such fundamental importance."

released on the side of the bowl. This is the *tautochrone* ("same time") *property* of the cycloid, and is the main theorem in the second part of his treatise.[7] In the third part he introduced the concepts of the evolute and involute of a plane curve and determined the evolutes of a parabola and a cycloid.[8] And in the last part he applied his mathematical discoveries to formulating the theory of a cycloidal pendulum clock, in which the pendulum bob is compelled to move along a cycloidal path instead of a circular path, and in which the period of oscillation is therefore exactly the same regardless of the magnitude of the swing, thereby eliminating the circular error. Huygens actually built several of these cycloidal clocks, but because of construction difficulties they turned out to be impractical as a way of obtaining greater accuracy. At the end of his treatise he gave a number of theorems about circular motion, proving, among other things, that for a body moving around a circular path with constant speed, the centripetal force is directly proportional to the square of the speed and inversely proportional to the radius of the path. Newton greatly respected Huygens, and used many of these discoveries in his own work a few years later.

Early in 1673 Huygens had some memorable conversations with Leibniz that had very far-reaching consequences. Leibniz was then 26 years old, a young diplomat on a mission to Paris for his employer in Germany, and largely ignorant of contemporary mathematics.

Huygens came to like the studious and intelligent young German more and more, gave him a copy of the *Horologium* as a present and talked to him about this latest work of his, the fruit of ten years of study, of the deep theoretical research to which he had been led in connection with the problem of pendular motion, and how eventually everything went back to Archimedes' methods for centers of gravity. Leibniz listened intently; at the close he felt he had to say something, but what he brought up was clumsy to a degree; surely a straight line drawn through the centroid of a plane (convex) area will always bisect the area, will it not? This was nearly too much: if it had been one of his mathematical rivals like Gregory or Newton then Huygens would probably never have condoned such a remark, but what this innocent young German had to say one could not really take amiss; good-humoredly, Huygens corrected his error and advised him to seek out further details from the relevant works of Pascal, etc. . . . Leibniz very readily and willingly took refuge in science. He procured the books named by Huygens and a few more from the Royal Library, made excerpt after excerpt and went really deeply into mathematics. As he learned, his personality rapidly matured,

---

[7] We prove this at the end of Section B.21, where the property is stated in terms of a bead sliding down a frictionless wire.

[8] These results are established in Section B.23. See also pp. 263–69 of D. J. Struik, *A Source Book in Mathematics, 1200–1800,* Harvard University Press, 1969.

digesting what he read and systematically penetrating its essence; he was concerned to acquire not facility in calculation or a mere catalog of results, but basic insights and methods, and what he took in inspired continually in turn a surge of creative activity within himself . . . It was, to begin with, a diversion for a mind deprived of its customary field of action, but soon became a unified passion for knowledge.[9]

Even though Huygens discovered many fine things in mathematics, his most important discovery was certainly the mind of Leibniz.

In the late 1670s Huygens began to feel an atmosphere of increasing intolerance for Protestants, and in 1681 he decided to leave Paris and move back to his home in The Hague. The next years were spent partly on microscopy in loose association with his friend Leeuwenhoek, and partly working on his wave theory of light.

His originality as a protozoologist was entirely unknown until a few years ago. As a modern expert says:

> Christiaan Huygens never himself published any serious contributions to protozoology; and the records of his own observations, which were made in an attempt to repeat Leeuwenhoek's experiments, remained in manuscript and unknown until only a few years ago. Consequently, his private work had no influence whatsoever upon the progress of protozoology. Had it been published in his lifetime, it would have assured him a place in the very forefront of the founders of the science.[10]

Among other things, he explained how microorganisms develop in water previously sterilized by boiling. He suggested that these creatures are small enough to float through the air and reproduce when they fall into the water, a speculation that was proved correct by Louis Pasteur two centuries later.

In 1690 he published his *Traité de la lumière* ("Treatise on Light"), in which he propounded his wave theory and used this as a basis for deducing geometrically the laws of reflection and refraction, and explained the phenomenon of double refraction in Iceland Crystal.[11]

Huygens's last work, and his most popular, was his posthumously published *Cosmotheoros*, in which he summed up man's knowledge of the universe at that time and frankly and freely speculated about the nature of

---

[9] See pp. 47–48 of J. E. Hofmann, *Leibniz in Paris, 1672–1676*, Cambridge University Press, 1974. An account of these conversations in Leibniz's own words is given on p. 215 of J. M. Child, *The Early Mathematical Manuscripts of Leibniz*, Open Court, 1920.

[10] See pp. 163–164 of C. Dobell, *Antony van Leeuwenhock and his "Little Animals,"* Dover, 1960.

[11] This treatise has been translated into English by S.P. Thompson, University of Chicago Press, 1945.

possible inhabitants of other planets.[12] He declined to allow this book to be published during his lifetime because he had no wish to be attacked for his unorthodox religious ideas. As he said to his sister-in-law, "If people knew my opinions and sentiments on religion, they would tear me apart." Toward the end of this book we find what is perhaps his most brilliant contribution to astronomy, the first reasonable estimate of the distance to a fixed star. He compared the remembered brightness of the star Sirius from the previous night with the observed brightness of a tiny piece of sun as seen through a very small hole; and by calculating the fraction of the sun's diameter visible through the hole, he concluded that Sirius is 27,664 times as far away as the sun. This result was slightly in error, because Sirius turned out to have greater intrinsic brightness than the sun, but Huygens had the right idea, and his estimate was the best that was available for more than a century.

The eminent modern philosopher Alfred North Whitehead has written:

A brief, and sufficiently accurate, description of the intellectual life of the European races during the succeeding two centuries and a quarter up to our own times [that is, from about 1700 to the early 20th century] is that they have been living upon the accumulated capital of ideas provided for them by the genius of the seventeenth century. The men of this epoch inherited a ferment of ideas attendant upon the historical revolt of the sixteenth century, and they bequeathed formed systems of thought touching every aspect of human life. It is the one century which consistently, and throughout the whole range of human activities, provided intellectual genius adequate for the greatness of its occasions.[13]

Christiaan Huygens was one of the brightest stars in that galaxy of brilliant men whose light still shines undiminished over the world of our own century.

---

[12] A charming English translation was published in 1698 under the title *The Celestial Worlds Discover'd*. This was reprinted in 1968 by Frank Cass & Co.

[13] See pp. 57–58 of Whitehead's *Science and the Modern World*, Macmillan, 1946.

A.18

NEWTON
(1642–1727)

Nature to him was an open book, whose letters he could read without effort.
Albert Einstein

Most people are acquainted in some degree with the name and reputation of Isaac Newton, for his universal fame as the discoverer of the law of gravitation has continued undiminished over the two and a half centuries since his death. It is less well known, however, that in the immense sweep of his vast achievements he virtually created modern physical science, and in consequence has had a deeper influence on the direction of civilized life than the rise and fall of nations. Those in a position to judge have been unanimous in considering him one of the very few supreme intellects that the human race has produced.

Newton was born to a farm family in the village of Woolsthorpe in northern England. Little is known of his early years, and his undergraduate life at Cambridge seems to have been outwardly undistinguished. In 1665 an outbreak of the plague caused the universities to close, and Newton returned to his home in the country, where he remained until 1667. There, in 2 years of rustic solitude—from age 22 to 24—his creative genius burst forth in a flood of discoveries unmatched in the history of human thought: the binomial series for negative and fractional exponents; differential and integral calculus; universal gravitation as the key to the mechanism of the solar system; and the resolution of sunlight into the visual spectrum by means of a prism, with its implications for understanding the colors of the rainbow and the nature of light in general. In his old age he reminisced as follows about this miraculous period of his youth: "In those days I was in the prime of my age for invention and minded Mathematicks and Philosophy [i.e., science] more than at any time since."[1]

Newton was always an inward and secretive man, and for the most part kept his monumental discoveries to himself. He had no itch to publish, and most of his great works had to be dragged out of him by the cajolery and persistence of his friends. Nevertheless, his unique ability was so evident to his teacher, Isaac Barrow, that in 1669 Barrow resigned his professorship in favor of his pupil (an unheard-of event in academic life), and Newton settled down at Cambridge for the next 27 years. His mathematical discoveries were never really published in connected form; they became known in a limited way almost by accident, through conversations and replies to questions put to him in correspondence. He seems to have regarded his mathematics mainly as a fruitful tool for the study of scientific problems, and of comparatively little interest in itself. Meanwhile, Leibniz in Germany had also invented calculus independently; and by his active correspondence with the Bernoullis and the later work of Euler, leadership in the new analysis passed to the Continent, where it remained for 200 years.[2]

Not much is known about Newton's life at Cambridge in the early years of his professorship, but it is certain that optics and the construction of telescopes were among his main interests. He experimented with many techniques for grinding lenses (using tools which he made himself), and about 1670 built the first reflecting telescope, the earliest ancestor of the great instruments in use today at Mount Palomar and throughout the world. The pertinence and simplicity of his prismatic analysis of sunlight have always marked this early work as one of the timeless classics of experimental science. But this was only the beginning, for he went further and further in penetrating the mysteries of light, and all his efforts in this direction continued to display experimental genius of the highest order. He published some of his discoveries, but they were greeted with such contentious stupidity by the leading scientists of the day that he retired back into his shell with a strengthened resolve to work thereafter for his own satisfaction alone. Twenty years later he unburdened himself to Leibniz in the following words: "As for the phenomena

---

[1] The full text of this autobiographical statement (probably written sometime about 1714) is given in the Appendix to this section. The present writer owns a photograph of the original document.

[2] It is interesting to read Newton's correspondence with Leibniz (via Oldenburg) in 1676 and 1677 (see *The Correspondence of Isaac Newton*, Cambridge University Press, 1959—1977, 7 volumes). In Items 165, 172, 188, and 209, Newton discusses his binomial series but conceals in anagrams his ideas about calculus and differential equations, while Leibniz freely reveals his own version of calculus. Item 190 is also of considerable interest, for in it Newton records what is probably the earliest statement and proof of the Fundamental Theorem of Calculus.

of colours . . . I conceive myself to have discovered the surest explanation, but I refrain from publishing books for fear that disputes and controversies may be raised against me by ignoramuses."[3]

In the late 1670s Newton lapsed into one of his periodic fits of distaste for science, and directed his energies into other channels. As yet he had published nothing about dynamics or gravity, and the many discoveries he had already made in these areas lay unheeded in his desk. At last, however, under the skillful prodding of the astronomer Edmund Halley (of Halley's Comet), he turned his mind once again to these problems and began to write his greatest work, the *Principia*.[4]

It all seems to have started in 1684 with three men in deep conversation in a London inn—Halley, and his friends Christopher Wren and Robert Hooke. By thinking about Kepler's third law of planetary motion, Halley had come to the conclusion that the attractive gravitational force holding the planets in their orbits was probably inversely proportional to the square of the distance from the sun.[5] However, he was unable to do anything more with the idea than formulate it as a conjecture. As he later wrote (in 1686):

> I met with Sir Christopher Wren and Mr. Hooke, and falling in discourse about it, Mr. Hooke affirmed that upon that principle all the Laws of the celestiall motions were to be demonstrated, and that he himself had done it. I declared the ill success of my attempts; and Sir Christopher, to encourage the Inquiry, said that he would give Mr. Hooke or me two months' time to bring him a convincing demonstration therof, and besides the honour, he of us that did it, should have from him a present of a book of 40 shillings. Mr. Hooke then said that he had it, but that he would conceale it for some time, that others triing and failing, might know how to value it, when he should make it publick; however I remember Sir Christopher was little satisfied that he could do it, and tho Mr. Hooke then promised to show it him, I do not yet find that in that particular he has been as good as his word.[6]

---

[3] *Correspondence*, Item 427.

[4] The full title is *Philosophiae Naturalis Principia Mathematica* (*Mathematical Principles of Natural Philosophy*).

[5] At that time this was quite easy to prove under the simplifying assumption—which contradicts Kepler's other two laws—that each planet moves with constant speed v in a circular orbit of radius r. [Proof: In 1673 Huygens had shown, in effect, that the acceleration a of such a planet is given by $a = v^2/r$. If T is the periodic time, then

$$a = \frac{(2\pi r/T)^2}{r} = \frac{4\pi^2}{r^2} \cdot \frac{r^3}{T^2}.$$

By Kepler's third law, T^2 is proportional to r^3, so r^3/T^2 is constant, and a is therefore inversely proportional to r^2. If we now suppose that the attractive force F is proportional to the acceleration, then it follows that F is also inversely proportional to r^2.]

[6] *Correspondence*, Item 289.

It seems clear that Halley and Wren considered Hooke's assertions to be merely empty boasts. A few months later Halley found an opportunity to visit Newton in Cambridge, and put the question to him: "What would be the curve described by the planets on the supposition that gravity diminishes as the square of the distance?" Newton answered immediately, "An ellipse." Struck with joy and amazement, Halley asked him how he knew that. "Why," said Newton, "I have calculated it." Not guessed, or surmised, or conjectured, but *calculated.* Halley wanted to see the calculations at once, but Newton was unable to find the papers. It is interesting to speculate on Halley's emotions when he realized that the age-old problem of how the solar system works had at last been solved—but that the solver hadn't bothered to tell anybody and had even lost his notes. Newton promised to write out the theorems and proofs again and send them to Halley, which he did. In the course of fulfilling his promise he rekindled his own interest in the subject, and went on, and greatly broadened the scope of his researches.[7]

In his scientific efforts Newton somewhat resembled a live volcano, with long periods of quiescence punctuated from time to time by massive eruptions of almost superhuman activity. The *Principia* was written in 18 incredible months of total concentration, and when it was published in 1687 it was immediately recognized as one of the supreme achievements of the human mind. It is still universally considered to be the greatest contribution to science ever made by one man. In it he laid down the basic principles of theoretical mechanics and fluid dynamics; gave the first mathematical treatment of wave motion; deduced Kepler's laws from the inverse square law of gravitation, and explained the orbits of comets; calculated the masses of the earth, the sun, and the planets with satellites; accounted for the flattened shape of the earth, and used this to explain the precession of the equinoxes; and founded the theory of tides. These are only a few of the splendors of this prodigious work.[8] The *Principia* has always been a difficult book to read, for the style has an inhuman quality of icy remoteness, which perhaps is appropriate to the grandeur of the theme. Also, the densely packed mathematics consists almost entirely of

---

[7] For additional details and the sources of our information about these events, see pp. 47–54 of I. Bernard Cohen's *Introduction to Newton's 'Principia,'* Harvard University Press, 1971.

[8] A valuable outline of the contents of the *Principia* is given in Chapter VI of W. W. Rouse Ball, *An Essay on Newton's Principia* (first published in 1893; reprinted in 1972 by Johnson Reprint Corp.) The emphasis throughout is on calculation and mathematical proof. This expressed Newton's deepest conviction about the nature of the scientific enterprise, that if it begins with observations (natural history), only the derivation of exact quantitative relations hidden in the observations deserves the name of science (natural philosophy). One of his favorite sayings was, "Natural History might indeed furnish materials for Natural Philosophy; but, however, Natural History is not Natural Philosophy."

classical geometry, which was little cultivated then and is less so now.[9] In his dynamics and celestial mechanics, Newton achieved the victory for which Copernicus, Kepler, and Galileo had prepared the way. This victory was so complete that the work of the greatest scientists in these fields over the next two centuries amounted to little more than footnotes to his colossal synthesis. It is also worth remembering in this context that the science of spectroscopy, which more than any other has been responsible for extending astronomical knowledge beyond the solar system to the universe at large, had its origin in Newton's spectral analysis of sunlight.

After the mighty surge of genius that went into the creation of the *Principia,* Newton again turned away from science. However, in a famous letter to Bentley in 1692, he offered the first solid speculations on how the universe of stars might have developed out of a primordial featureless cloud of cosmic dust:

> It seems to me, that if the matter of our Sun and Planets and all the matter in the Universe was evenly scattered throughout all the heavens, and every particle had an innate gravity towards all the rest . . . some of it would convene into one mass and some into another, so as to make an infinite number of great masses scattered at great distances from one to another throughout all that infinite space. And thus might the Sun and Fixt stars be formed, supposing the matter were of a lucid nature.[10]

This was the beginning of scientific cosmology, and later led, through the ideas of Thomas Wright, Kant, Herschel, and their successors, to the elaborate and convincing theory of the nature and origin of the universe provided by late twentieth-century astronomy.

In 1693 Newton suffered a severe mental illness accompanied by delusions, deep melancholy, and fears of persecution. He complained that he could not sleep, and said that he lacked his "former consistency of mind." He lashed out with wild accusations in shocking letters to his friends Samuel Pepys and John Locke. Pepys was informed that their friendship was over and that Newton would see him no more; Locke was charged with trying to entangle him with women and with being a "Hobbist" (a follower of Hobbes, i.e., an

---

[9] The nineteenth century British philosopher Whewell has a vivid remark about this: "Nobody since Newton has been able to use geometrical methods to the same extent for the like purposes; and as we read the *Principia* we feel as when we are in an ancient armoury where the weapons are of gigantic size; and as we look at them we marvel what manner of man he was who could use as a weapon what we can scarcely lift as a burden."

[10] *Correspondence,* Item 398.

atheist and materialist).[11] Both men feared for Newton's sanity. They responded with careful concern and wise humanity, and the crisis passed.

In 1696 Newton left Cambridge for London to become Warden (and soon Master) of the Mint, and during the remainder of his long life he entered a little into society and even began to enjoy his unique position at the pinnacle of scientific fame. These changes in his interests and surroundings did not reflect any decrease in his unrivaled intellectual powers. For example, late one afternoon, at the end of a hard day at the Mint, he learned of a now-famous problem that the Swiss scientist John Bernoulli had posed as a challenge "to the most acute mathematicians of the entire world." The problem can be stated as follows: Suppose two nails are driven at random into a wall, and let the upper nail be connected to the lower by a wire in the shape of a smooth curve. What is the shape of the wire down which a bead will slide (without friction) under the influence of gravity so as to pass from the upper nail to the lower in the least possible time? This is Bernoulli's *brachistochrone* ("shortest time") *problem*. Newton recognized it at once as a challenge to himself from the Continental mathematicians; and in spite of being out of the habit of scientific thought, he summoned his resources and solved it that evening before going to bed. His solution was published anonymously, and when Bernoulli saw it, he wryly remarked, "I recognize the lion by his print."

Of much greater significance for science was the publication of his *Opticks* in 1704. In this book he drew together and extended his early work on light and color. As an appendix he added his famous Queries, or speculations on areas of science that lay beyond his grasp in the future. In part the Queries relate to his lifelong preoccupation with chemistry (or alchemy, as it was then called). He formed many tentative but exceedingly careful conclusions— always founded on experiment—about the probable nature of matter; and though the testing of his speculations about atoms (and even nuclei) had to await the refined experimental work of the late nineteenth and early twentieth centuries, he has been proven absolutely correct in the main outlines of his ideas.[12] So, in this field of science too, in the prodigious reach and accuracy of his scientific imagination, he passed far beyond not only his contemporaries but also many generations of his successors. In addition, we quote two astonishing remarks from Queries 1 and 30, respectively: "Do not Bodies act upon Light at a distance, and by their action bend its Rays?" and "Are not gross Bodies and Light convertible into one another?" It seems as clear as words can be that Newton is here conjecturing the gravitational bending of light and the equivalence of mass and energy, which are prime consequences of

---

[11] *Correspondence*, Items 420, 421, and 426.

[12] See S. I. Vavilov, "Newton and the Atomic Theory," in *Newton Tercentenary Celebrations*, Cambridge University Press, 1947.

the theory of relativity. The former phenomenon was first observed during the total solar eclipse of May 1919, and the latter is now known to underlie the energy generated by the sun and the stars. On other occasions as well he seems to have known, in some mysterious intuitive way, far more than he was ever willing or able to justify, as in this cryptic sentence in a letter to a friend: "It's plain to me by the fountain I draw it from, though I will not undertake to prove it to others."[13] Whatever the nature of this "fountain" may have been, it undoubtedly depended on his extraordinary powers of concentration. When asked how he made his discoveries, he said, "I keep the subject constantly before me and wait till the first dawnings open little by little into the full light." This sounds simple enough, but everyone with experience in science or mathematics knows how very difficult it is to hold a problem continuously in mind for more than a few seconds or a few minutes. One's attention flags; the problem repeatedly slips away and repeatedly has to be dragged back by an effort of will. From the accounts of witnesses, Newton seems to have been capable of almost effortless sustained concentration on his problems for hours and days and weeks, with even the need for occasional food and sleep scarcely interrupting the steady squeezing grip of his mind.

In 1695 Newton received a letter from his Oxford mathematical friend John Wallis, containing news that cast a cloud over the rest of his life. Writing about Newton's early mathematical discoveries, Wallis warned him that in Holland "your Notions" are known as "Leibniz's *Calculus Differentialis*," and he urged Newton to take steps to protect his reputation.[14] At that time the relations between Newton and Leibniz were still cordial and mutually respectful. However, Wallis's letters soon curdled the atmosphere, and initiated the most prolonged, bitter, and damaging of all scientific quarrels: the famous (or infamous) Newton–Leibniz priority controversy over the invention of calculus.

It is now well established that each man developed his own form of calculus independently of the other, that Newton was first by 8 or 10 years but did not publish his ideas, and that Leibniz's papers of 1684 and 1686 were the earliest publications on the subject. However, what are now perceived as simple facts were not nearly so clear at the time. There were ominous minor rumblings for years after Wallis's letters, as the storm gathered:

> What began as mild innuendoes rapidly escalated into blunt charges of plagiarism on both sides. Egged on by followers anxious to win a reputation under his auspices, Newton allowed himself to be drawn into the centre of the fray; and, once his temper was aroused by accusations of dishonesty, his anger was beyond

---

[13] *Correspondence*, Item 193.

[14] *Correspondence*, Items 498 and 503.

constraint. Leibniz's conduct of the controversy was not pleasant, and yet it paled beside that of Newton. Although he never appeared in public, Newton wrote most of the pieces that appeared in his defense, publishing them under the names of his young men, who never demurred. As president of the Royal Society, he appointed an "impartial" committee to investigate the issue, secretly wrote the report officially published by the society [in 1712], and reviewed it anonymously in the *Philosophical Transactions*. Even Leibniz's death could not allay Newton's wrath, and he continued to pursue the enemy beyond the grave. The battle with Leibniz, the irrepressible need to efface the charge of dishonesty, dominated the final 25 years of Newton's life. Almost any paper on any subject from those years is apt to be interrupted by a furious paragraph against the German philosopher, as he honed the instruments of his fury ever more keenly.[15]

All this was bad enough, but the disastrous effect of the controversy on British science and mathematics was much more serious. It became a matter of patriotic loyalty for the British to use Newton's geometrical methods and clumsy calculus notations, and to look down their noses at the upstart work being done on the Continent. However, Leibniz's analytical methods proved to be far more fruitful and effective, and it was his followers who were the moving spirits in the richest period of development in mathematical history. What has been called "the Great Sulk" continued; for the British, the work of the Bernoullis, Euler, Lagrange, Laplace, Gauss, and Riemann remained a closed book; and British mathematics sank into a coma of impotence and irrelevancy that lasted through most of the eighteenth and nineteenth centuries.

Newton has often been thought of and described as the ultimate rationalist, the embodiment of the Age of Reason. His conventional image is that of a worthy but dull absent-minded professor in a foolish powdered wig. But nothing could be further from the truth. This is not the place to discuss or attempt to analyze his psychotic flaming rages; or his monstrous vengeful hatreds that were unquenched by the death of his enemies and continued at full strength to the end of his own life; or the 58 sins he listed in the private confession he wrote in 1662; or his secretiveness and shrinking insecurity; or his peculiar relations with women, especially with his mother, who he thought had abandoned him at the age of 3. And what are we to make of the bushels of unpublished manuscripts (millions of words and thousands of hours of thought!) that reflect his secret lifelong studies of ancient chronology, early Christian doctrine, and the prophecies of Daniel and St. John? Newton's desire to know had little in common with the smug rationalism of the eighteenth century; on the contrary, it was a form of desperate self-preservation against the dark forces that he felt pressing in around him. As an

---

[15] Richard S. Westfall, in the *Encyclopaedia Britannica*.

original thinker in science and mathematics he was a stupendous genius whose impact on the world can be seen by everyone; but as a man he was so strange in every way that normal people can scarcely begin to understand him.[16] It is perhaps most accurate to think of him in medieval terms—as a consecrated, solitary, intuitive mystic for whom science and mathematics were means of reading the riddle of the universe.

APPENDIX:
NEWTON'S 1714(?) MEMORANDUM OF THE TWO PLAGUE YEARS OF 1665 AND 1666

In the beginning of the year 1665 I found the Method of approximating series and the Rule for reducing any dignity [i.e., power] of any Binomial into such a series. The same year in May I found the method of Tangents of Gregory and Slusius, and in November had the direct method of fluxions and the next year in January had the Theory of colours and in May following I had entrance into ye inverse method of fluxions. And the same year I began to think of gravity extending to ye orb of the Moon, and having found out how to estimate the force with w^{ch} (a) globe revolving within a sphere presses the surface of the sphere, from Keplers Rule of the periodical times of the Planets being in a sesquialternate proportion of their distances from the centers of their Orbs, deduced that the forces wch keep the Planets in their Orbs must (be) reciprocally as the squares of their distances from the centers about wch they revolve: and thereby compared the force requisite to keep the Moon in her Orb with the force of gravity at the surface of the earth, and found them to answer pretty nearly. All this was in the two plague years of 1665 and 1666, for in those days I was in the prime of my age for invention and minded Mathematicks and Philosophy [i.e., science] more than at any time since. What Mr Hugens has published since about centrifugal forces I suppose he had before me. At length in the winter between the years 1676 and 1677 I found the Proposition that by a centrifugal force reciprocally as the square of the distance a Planet must revolve in an Ellipsis about the center of the force placed in the lower umbilicus of the Ellipsis and with the radius drawn to that

---

[16] The best effort is Frank E. Manuel's excellent book, *A Portrait of Isaac Newton,* Harvard University Press, 1968. Manuel's book is just what it claims to be—a portrait—directed at the psychology of its subject and not particularly concerned with his scientific career. Since the present account was first written, Richard S. Westfall has published a superb, definitive biography of over 900 pages entitled *Never at Rest,* Cambridge University Press, 1980.

center describe areas proportional to the times. And in the winter between the years 1683 and 1684 this Proposition wth the Demonstration was entered in the Register book of the R. Society. And this is the first instance upon record of any Proposition in the higher Geometry found out by the Method in dispute. In the year 1689 Mr Leibnitz endeavoring to rival me published a Demonstration of the same Proposition upon another supposition but his Demonstration proved erroneous, for want of skill in the Method.[17]

[17] Cambridge University Library, MS Add. 3968 No. 41. A fascinating volume of articles analyzing Newton's activities during these years has been edited by Robert Palter, under the title *The Annus Mirabilis of Sir Isaac Newton, 1666–1966,* The M.I.T. Press, 1970. See also pp. 290–92 of I. Bernard Cohen, *Introduction to Newton's 'Principia,'* Harvard University Press, 1971.

A.19

LEIBNIZ
(1646–1716)

It would be difficult to name a man more remarkable for the greatness and universality of his intellectual powers than Leibniz.

John Stuart Mill

The ideas of calculus were "in the air" in the 1650s and 1660s. The ingenious area calculations and tangent constructions of Cavalieri, Torricelli, Fermat, Pascal, Barrow, and others were so suggestive that the final discovery of calculus as an autonomous discipline was almost inevitable within a very few years. The last steps of putting it all together were taken by two men of great genius working independently of each other: by Isaac Newton in what he called "the two plague years of 1665 and 1666," and also by Gottfried Wilhelm Leibniz during his sojourn in Paris from 1672 to 1676.

Leibniz is probably better known to most people as a philosopher than as a mathematician. The history of philosophy has long recognized him as one of its greatest system builders, and also as the producer of most of the ammunition with which Kant later attacked Hume. But this too was only a small fraction of his total thought. He made memorable creative contributions across the entire spectrum of intellectual life, from mathematics and logic through the various sciences to history, law, diplomacy, politics, philology, metaphysics, and theology. No other thinker except Aristotle has rivaled him in the range and variety of his abilities and achievements. Leibniz lived in a period when it was still possible—as his own astounding career demonstrated—

for a very highly intelligent and hard-working scholar to absorb all the knowledge of his time. To Oswald Spengler he was "without doubt the greatest intellect in Western philosophy"; and Admiral Mahan—perhaps the most influential historian of modern times—called him "one of the world's great men." What manner of man was he, and how did he live and what did he think?

Leibniz was born in 1646 at Leipzig, where his father was professor of moral philosophy at the university. He was sent to a good school, but after his father's death in 1652 he seems to have acted for the most part as his own teacher, leading a self-propelled intellectual life even as a small child. The German books that were available to him were quickly read through. He began teaching himself Latin at the age of 8, and soon mastered it sufficiently to read it with ease and compose acceptable Latin verse; he started the study of Greek a few years later. He had acquired a love of history from his father, and he spent most of his childhood eagerly devouring the large library of choice books that his father had collected, including Herodotus, Xenophon, Homer, Plato, Aristotle, Cicero, Quintilian, Seneca, Pliny, Polybius, and many others. By his early teens Leibniz had thus become familiar with a broad range of classical literature, and was well embarked on the omnivorous reading that was to be his custom throughout his life.[1]

At this stage of his mental development, classical studies no longer satisfied him. He turned his attention to logic, zealously reading the scholastic philosophers and attempting already to reform the doctrines of Aristotle. In a letter written in 1696 he recalled this period of his life as follows:

> As soon as I began to learn logic, I was fascinated by the classification and order which I perceived in its principles. I soon observed, as much as a boy of 13 could, that there must be something great to the subject. My strongest pleasure lay in the categories, which seemed to me to call the roll of all the things in the world.[2]

He had an insatiable appetite for discovering the meaning and purpose of everything around him. The thoughts of youth stoke secret fires, and few such fires could have burned as intensely as his.[3]

---

[1] His willingness to read almost anything led Fontenelle to remark of him that he bestowed the honor of reading them on a great mass of bad books. However, as Leibniz himself said, "When a new book reaches me, I search for what I can learn, not for what I can criticize in it." For more on his reading habits, see p. 237 of L. E. Loemker (ed. and trans.), *Leibniz: Philosophical Papers and Letters,* University of Chicago Press, 1956.

[2] Loemker, *op. cit.,* p. 756.

[3] It is worth noting that Leibniz's IQ has been estimated by experts as 180 at the very least and probably much higher, "close to the maximum for the human race." See pp. 155 and 702–705 of Lewis M. Terman (ed.), *Genetic Studies of Genius*: Vol. II, *The Early Mental Traits of Three Hundred Geniuses,* Stanford University Press, 1926.

At the age of 15, Leibniz entered the University of Leipzig as a law student. During the first two years he studied mainly philosophy and mathematics as far as Euclid. At that time the University was firmly congealed in the sterile Aristotelian tradition and did nothing to encourage science. It was by his own efforts that he became acquainted with those thinkers who had already launched the modern age in science and philosophy: Francis Bacon, Kepler, Galileo, and Descartes. As with most men of really great intellect, Leibniz's formal education was only a minor eddy in the torrent of thought and study and learning that was the essence of his life.

The next 3 years were devoted to legal studies, and in 1666 he applied for the degree of doctor of law with the aim of seeking appointment to a judicial position. This application was refused, ostensibly on the grounds of his youth but probably because of the small-minded jealousy of the faculty, and he left Leipzig in disgust. At Altdorf, the university town of the free imperial city of Nuremberg, his brilliant dissertation *De casibus perplexis in jure* ("On Perplexing Cases in the Law") procured him the doctor's degree at once and the immediate offer of a professorship in the University. He declined this offer, having, as he said, "very different things in view." He called the universities "monkish," and charged that they possessed learning but little common sense and were preoccupied with empty trivialities. His purpose was to enter public rather than academic life. It is remarkable how few of the major philosophers have been professors in universities. Leibniz spent the next year in Nuremberg, which was then a center of the secret mystical order of the Rosicrucians, and he made himself so familiar with the ideas and writings of the alchemists—much as Newton was doing at Cambridge—that he was elected secretary of the local Rosicrucian society.

At the age of 20, Leibniz had not only earned his doctorate in law, but had also published several highly original essays on logic and jurisprudence. His *Dissertatio de arte combinatoria* ("Dissertation on the Art of Combinations") initiated his lifelong project of reducing all knowledge and reasoning to what he called a "universal characteristic." By this he meant a precise system of notation—a symbolic mathematical language analogous to algebra—in which the symbols themselves and their rules of combination would automatically analyze all concepts into their ultimate constituents in such a way as to provide the means for obtaining knowledge of the essential nature of all things.[4] Needless to say, this grandiose project was not realized in his lifetime or thereafter, but its spirit continued to propel and guide his thinking, and later led to the invention of his differential and integral calculus and his first tentative steps toward the creation of symbolic logic. There was also his *Nova*

---

[4] For Leibniz's own explanation of what he had in mind, see Loemker, *op. cit.*, pp. 339–46; or pp. 12–25 of Philip P. Wiener (ed.), *Leibniz Selections*, Scribner's, 1951.

methodus docendae discendaeque jurisprudentiae ("A New Method for Teaching and Learning Jurisprudence"), which he wrote during the rest stops of his journey from Leipzig to Altdorf. This essay is remarkable for containing the first clear recognition of the importance of the historical approach to law.[5] It also had the practical effect of securing him a position (in 1667) as legal advisor to the Prince Elector of Mainz.[6]

Leibniz remained at Mainz for 5 years, at first as an assistant in recodifying the laws and later as a trusted councilor and diplomat serving the political goals of the Elector. The most important of these goals was survival, for at that time the swollen arrogance of Louis XIV was like a boil on the face of Europe, and his armies were threatening the Low Countries and the small German states along the Rhine. Leibniz conceived a plan to divert Louis from Germany by persuading him to conquer Egypt and build a colonial empire in North Africa, thereby satisfying his imperial ambitions at little cost to his European neighbors. A detailed memorandum was sent to the French government; and in March of 1672, at the invitation of the French foreign minister, the young diplomat traveled to Paris to present his proposals to the King. Unfortunately, however, Louis had an irrational hatred for the Dutch, and declared war on them a few weeks later. Leibniz never met the King, and his plan for a French conquest of Egypt disappeared from practical politics until the time of Napoleon, who revived it in 1798.[7] The important thing for Leibniz himself was the visit to the great city of Paris, where he spent most of the next four years. This experience was crucial for his intellectual development, for he mastered the French language, became personally acquainted with the leaders in science and philosophy, and immersed himself in the mainstream of European thought.

The world Leibniz entered in 1672—in France, England, and Holland— was bubbling with the ferment of new ideas and teeming with men of genius. It was a garden of intellectual civilization, in comparison with which his former

---

[5] See the chapter on Leibniz in H. Cairns, *Legal Philosophy from Plato to Hegel,* Johns Hopkins University Press, 1949.

[6] At that time Mainz was more than just a city on the Rhine; it was one of the most powerful member states of the Holy Roman Empire, that strange agglomeration that Voltaire characterized as "neither holy, nor Roman, nor an empire." The ruler of Mainz was one of the seven regional princes empowered to elect the Emperor.

[7] It was a great misfortune for France that Louis ignored Leibniz's proposals; for if they had been adopted and pursued vigorously, it would probably have been France instead of England that captured India and the mastery of the seas, and the subsequent history of Europe would have been very different. As it was, the King's folly "ruined the prosperity of France, and was felt in its consequences from generation to generation afterward." See pp. 106–107 and 141–43 of A. T. Mahan, *The Influence of Sea Power upon History: 1660–1783,* Little, Brown, 12th ed., 1944: first published in 1890.

world of Leipzig, Nuremberg, and Mainz was little better than dull barbarism. Philosophy and theology were in a state of upheaval, and the champions of both the new and the old ways of regarding man and God—Hobbes, Spinoza, Locke, Arnauld, Malebranche, Bossuet—were well known in all learned circles. Leibniz met Arnauld and Malebranche in Paris, visited Spinoza in Holland, and corresponded with the others. The sciences were enjoying a period of unprecedented growth, and Leibniz kept himself informed of all the latest discoveries and made a number of contributions of his own. The Dutch physicist Huygens—creator of the wave theory of light and inventor of the pendulum clock—became Leibniz's friend and mathematical mentor during his years in Paris. Another of his friends was the Danish astronomer Roemer, who in 1675 first calculated the speed of light from the observations of the moons of Jupiter that he made at the Paris observatory. And in 1676 Leibniz entered into a mathematical correspondence with Newton that had fateful consequences for the future development of European science. Other eminent men who were active during the second half of the seventeenth century were Boyle, Hooke, von Guericke, and Halley in chemistry, physics, and astronomy; Leuuwenhoek, Malpighi, and Swammerdam in biology; and Wallis, Wren, Roberval, Tschirnhaus, and the Bernoullis in mathematics—and Leibniz knew and corresponded with almost all of them. Western Europe was drunk with the wine of reason, and Leibniz enthusiastically joined the party when he moved to Paris at the age of 26.

In January of 1673 Leibniz crossed the Channel to England on a diplomatic mission for the Elector of Mainz. In London he quickly became acquainted with Henry Oldenburg, the German-born first secretary of the Royal Society, and also with others of its members, including Boyle and Hooke.[8] He had invented a calculating machine for performing more complicated operations than the earlier machine of Pascal—multiplying and dividing as well as adding and subtracting. He exhibited a rough model of his invention to the Royal Society, and was elected a member of the Society shortly after his return to Paris in March.[9]

---

[8] Oldenburg knew almost everyone worth knowing in Western Europe at the time, from Milton and Cromwell to Newton, Leibniz, Spinoza, Leeuwenhoek, and many others. He was a major crossroads in the intellectual life of the period, and deserves a full-scale scholarly biography. See the 9 volumes of *The Correspondence of Henry Oldenburg,* ed. and trans. by A. Rupert Hall and Marie Boas Hall, University of Wisconsin Press, 1965–1973. The introductions to these volumes provide a good running biography of Oldenburg.

[9] For descriptions of several stages in the development of this machine, see pp. 23, 79, and 126 of J. E. Hofmann, *Leibniz in Paris: 1762–1676,* Cambridge University Press, 1974. A picture, together with Leibniz's own explanation, can be found in pp. 173–81 of D. E. Smith, *A Source Book in Mathematics,* McGraw-Hill, 1929. See also pp. 7–9 of H. H. Goldstine, *The Computer from Pascal to von Neumann,* Princeton University Press, 1972.

When Leibniz first arrived in Paris in 1672, he had little knowledge of mathematics beyond the simplest parts of Euclid and some fragmentary ideas from Cavalieri. He quickly became aware that in that age to be ignorant of mathematics was to be negligible in the eyes of most educated men, and he began his mathematical studies with the aim of establishing his credibility as a serious thinker.[10] However, once started, he was irresistibly drawn to the subject. When he returned to Paris from London, he spent more and more of his time on higher geometry, under the general guidance of Huygens, and began the series of investigations that led over the next few years to his invention of the differential and integral calculus. In 1673 he made one of his most remarkable discoveries, the infinite series expansion

$$\frac{\pi}{4} = 1 - \frac{1}{3} + \frac{1}{5} - \frac{1}{7} + \frac{1}{9} - \cdots.$$

This beautiful formula reveals a striking relation between the mysterious number π and the familiar sequence of all the odd numbers.[11]

During this entire period Leibniz read, wrote, and thought continually, pursuing ideas with a strength and intensity known to ordinary people only in their pursuit of wealth and power. His motto at the time was, "With every lost hour a part of life perishes." In 1673 the Elector of Mainz died, and Leibniz tentatively entered the service of the learned John Frederick, Duke of Brunswick-Lüneburg, with whom he had already been corresponding for several years, as curator of the ducal library at Hanover. However, he was held by the magnetic atttraction of Paris and continued to hope that he could find some way of staying on there indefinitely.[12] Nothing turned up, and late in 1676, at the Duke's insistence, he left Paris and traveled slowly to Hanover by way of London and Holland. At The Hague he had several long conversations with Spinoza, who permitted him to read and copy passages from his unpublished *Ethics*.[13] He also visited Antony van Leeuwenhoek, discoverer of the tiny forms of life that can be seen only through a microscope. The miniature universes that Leeuwenhoek was able to find in every drop of pond

---

[10] Loemker, *op. cit.*, pp. 400–401.

[11] A modern expert on infinite series (K. Knopp) has said: "It is as if, by this expansion, the veil which hung over that strange number had been drawn aside." Leibniz found his formula—of which he was justifiably proud all his life—by a very ingenious calculation of the area of one quarter of a circle of radius 1. We show how he did it in Section B.13. See also Items 123, 126, 130, and 134 in *The Correspondence of Isaac Newton*, Cambridge University Press, 1959–1977, 7 volumes.

[12] Hofmann, *op. cit.*, pp. 46–47 and 160–63.

[13] Little is known of these tantalizing conversations between the two greatest metaphysical thinkers of the time; see pp. 37–39 of F. Pollock, *Spinoza: His Life and Philosophy*, 2nd ed., Duckworth, 1899. A reconstruction—vivid and dramatic, but mostly imaginary—is attempted in pp. 281–92 of R. Kayser, *Spinoza: Portrait of a Spirtual Hero*, Greenwood Press, 1968.

water made a deep impression on Leibniz, and years later contributed to the metaphysical system in which he portrayed the whole world as consisting of tiny, invisible centers of awareness called monads.[14]

Leibniz reached Hanover by the end of November, and through the remaining 40 years of his life served three successive dukes as librarian, family historian, and informal minister in charge of scientific and cultural affairs. Though he lived in the atmosphere of the petty local politics of a small German principality, his activities and outlook were always constructive and cosmopolitan. He supervised the mint, and suggested various improvements in the coinage and the economic theory behind it. He reorganized the Harz silver mines, the basis of the currency, and acted as an engineer in designing windmill-powered pumps intended to protect the mines against the seeping water that threatened them. He was both an engineer and a landscape architect in planning the fountains for the great formal garden of the summer palace at Herrenhausen. He wrote many pamphlets and position papers to support various rights and claims of his patrons. He also wrote a masque that was performed at court by the nobility; and his memorial poem on the occasion of John Frederick's death in 1679 contained a description of the recently discovered element phosphorus that was considered one of the finest passages in modern Latin poetry.

In the midst of all this miscellaneous activity, Leibniz's main continuing responsibilities were as librarian and historian. His ideas about the purposes, organization, and administration of scholarly libraries were so far-sighted that he has been called "the greatest librarian of his age."[15] His historical work began with an assignment to compile a genealogy of the Brunswick family for use as a weapon in the dynastic political struggles of the day. The necessary research meant that Leibniz had to travel, which has been one of the main benefits of the profession of history since the time of Herodotus. He spent three years (1687–1690) examining the archives and private libraries of Southern Germany and Italy, and at last was able to prove the ancestral connection between the ducal houses of Brunswick and Este. This achievement was influential in winning for Hanover (Brunswick-Lüneburg) the status of an electorate of the Empire (in 1692). Leibniz's collection of historical documents enabled him not only to undertake an extensive history of the House of Brunswick (*Annales Brunsvicenses*), but also to publish two important volumes of source material for a code of international law (*Codex juris gentium diplomaticus*, 1693 and 1700). Over the years his history expanded into an exhaustive study of the German Empire in the Middle Ages, and was later used by Gibbon.

---

[14] Loemker, *op. cit.*, p. 1056 ("The Monadology," 66–69).

[15] By Sir Frank C. Francis, Director and Principal Librarian of the British Museum, 1959–1968.

But all this was only the froth on the surface of Leibniz's life, the visible career of the courtier and public official. Beneath was a churning sea of private intellectual activity, on so vast and varied a scale as to be almost beyond belief. We briefly summarize his main interests and achievements (apart from mathematics) under four headings: (1) logic, (2) theology, (3) metaphysics, (4) science.

1. He was the first to perceive that the laws of thought are essentially algebraic in nature, and by this insight and his subsequent efforts he founded symbolic logic. He imagined a distant future when philosophical discussions would be carried on by means of logical symbolism and would reach conclusions as certain as those of mathematics. He expressed his vision as follows:

> If controversies were to arise, there would be no more need of disputation between two philosophers than between two accountants. For it would suffice to take their pencils in their hands, to sit down to their slates, and to say to each other (with a friend as witness if they liked): Let us calculate.[16]

Philosophy has not yet reached this stage, and perhaps it never will, but much of what Leibniz foresaw can be recognized in the computerized decision-making processes of modern business, government, and military strategy. During the 1670s and 1680s he made considerable progress on his project to deal with logic by algebraic methods.[17] In modern terminology, he stated the main formal properties of logical addition, multiplication, and negation; he considered the empty set and set inclusion; and he pointed out the similarity between certain properties of set inclusion and the relation of implication for propositions. Though unfortunately most of this work was not published until two centuries later, it was the historical source of the symbolic logic (Boolean algebra) developed by George Boole in the nineteenth century and carried forward by Whitehead and Russell in the early part of the twentieth century. There is evidently little exaggeration in the judgment, "Leibniz deserves to be ranked among the greatest of logicians."[18]

2. Leibniz had been a close student of Protestant theology ever since his early teens, and his interest in such matters was greatly stimulated by his contact with Catholic clergymen and ritual at the court of Mainz. He observed that

---

[16] See p. 170 of Bertrand Russell, *A Critical Exposition of the Philosophy of Leibniz,* 2nd ed., George Allen and Unwin, 1937; first published in 1900.

[17] See pp. 123–32 of D. J. Struik, *A Source Book in Mathematics, 1200–1800,* Harvard University Press, 1969.

[18] See pp. 320–45 of W. and M. Kneale, *The Development of Logic,* Oxford University Press, 1962.

Catholic and Protestant doctrines differ only in minor ways, and he began to dream of reuniting the divided creeds of Christianity into a monolithic Christendom. In 1686 he wrote his *Systema theologicum* as a statement of basic belief on which he hoped all Christians could agree, but it was only in the late twentieth century that such ecumenical ideas began to find a favorable climate. His main abstract theological purpose was to establish logical proofs for the existence of God, and he put the four standard arguments—one of which he invented himself—into their final form.[19] However, this project also had little influence, since most people find it difficult to feel affection or reverence for a deity whose main role is to fill a gap in a metaphysical jigsaw puzzle. The idea of proving religious creeds as if they were theorems of geometry has never been widely popular, for in our moments of clarity we recognize that such creeds have always been matters of culture, custom, and preference, not logical truth or falsity. He also tried to give a rational explanation for the presence of evil in the world, and this led him to write a long, dull book (*Essais de Théodicée*, 1710) that brought him European fame for his doctrine that this is "the best of all possible worlds."[20] This unfortunate production stimulated Voltaire to satirize his ideas through the character of Doctor Pangloss in his most famous work, *Candide* (1758).[21]

3. Leibniz lived at a time when the passion for metaphysics was deep and strong, when it was still believed possible to understand the world purely by thought. He struggled for years to penetrate the mysteries of nature and God by the use of reason alone—by constructing a great metaphysical system that explains all things by the *a priori* method of deducing necessary consequences from a few self-evident principles. His main starting point was his Principle of Sufficient Reason: "Nothing happens without a reason, there is no effect without a cause." From this slender beginning he purported to demonstrate by irresistible logic the chief tenets of his metaphysical system.[22] However, by equally irresistible logic Spinoza had earlier established the main features of his own totally different metaphysical system. And so with most of the other eminent philosophers of the seventeenth and eighteenth centuries—each

---

[19] See pp. 585–89 of Bertrand Russell, *A History of Western Philosophy*, Simon and Schuster, 1945.

[20] Because God, being all-wise, must know all possible worlds, being all-powerful, must be able to create whatever kind of world he chooses, and being all-good, must choose the best.

[21] Leibniz was also the object of Voltaire's laughter in the wittiest of his satirical romances, *Micromégas* (1752), in which a visitor from Sirius comes to the earth and amuses himself by listening to the arguments of philosophers.

[22] This system is mostly concerned with the nature, activities, and interrelations of the invisible percipient points called monads, whose existence Leibniz deduced and which he believed to be the ultimate constituents of reality and the causes of all phenomena.

thought he was looking through a window at the great outside world of reality, but instead looked into a mirror and saw only his own face. As Spinoza said in another context, "What St. Paul tells us about God tells us more about St. Paul than it does about God." The growth of empiricism and the rise of science over the past three centuries have made it almost impossible to take seriously the extravagant pretensions of the *a priori* philosopher, who sits in his study and spins a web of words, fanciful imaginings, and empty speculations out of the material of his own consciousness. Faith in reason alone is alien to us, and we believe that only careful observation and experiment can reveal anything of substance about the actual universe. We no longer study philosophy for the old reason, the hope of learning the truth about the nature of things, but rather, for the fascination of learning what people have thought, and if possible why they have thought it.[23]

4. Leibniz was torn between the contradictory claims of science and philosophy as ways of knowing reality, but he leaned in the direction of science. In 1691 he wrote as follows to his friend Huygens: "I prefer a Leeuwenhoek who tells me what he sees to a Cartesian who tells me what he thinks. It is, however, necessary to add reasoning to observation." He often urged the pursuit of real knowledge—chemistry, physics, geology, botany, zoology, anatomy, history, and geography—in contrast to the learned nonsense of the academics. He was fascinated by every aspect of the developing sciences and technology, and his own contributions to these fields alone would have filled several distinguished lifetimes. In 1693 he published his ideas about the beginnings of the earth, in a paper in the *Acta Eruditorum*.[24] He explained his geological theories more fully in his remarkable treatise *Protogaea,* which unfortunately was not published until 1749, long after his death. The earth, he believed, was originally an incandescent globe; it slowly cooled, contracted, and formed a crust; and as it cooled, the surrounding vapor condensed into oceans, which gradually became salty by dissolving salts in the crust.[25] He was the first to distinguish igneous from sedimentary rocks. He also gave a good

---

[23] For readers who wish to sample Leibniz's metaphysics at the source, we suggest three brief essays given in Loemker, *op. cit.,* pp. 346–50, 411–17 and 1033–43. A useful general commentary is provided by L. Couturat in his article "On Leibniz's Metaphysics," pp. 19–45 in H. G. Frankfurt (ed.), *Leibniz: A Collection of Critical Essays,* Doubleday, 1972. In addition, we recommend the article on Metaphysics by Gilbert Ryle in the *Encyclopaedia Britannica*; here the reader is treated to the enthralling spectacle of an eminent Professor of Metaphysical Philosophy at Oxford University suavely demonstrating that metaphysical philosophy does not exist.

[24] This periodical was the most influential European journal of the time in science and mathematics. It was founded by Leibniz in 1682, and he was its editor-in-chief for many years.

[25] See p. 352 of A. Wolf, *A History of Science, Technology and Philosophy in the 16th and 17th Centuries,* George Allen and Unwin, 1935.

explanation for fossils, suggested that the different kinds of fossils found in different layers of the crust might be clues to the earth's history, gave the earliest reasonably satisfactory definition of the concept of species, and— foreshadowing the evolutionists of the eighteenth and nineteenth centuries— thought it likely that species have undergone many drastic changes through the long history of the earth. All this, in an age in which even the most intelligent and well-educated people considered *Genesis* to be the final authority in such matters. No new idea or discovery escaped his attention, and he had a hand in many of the notable achievements of the day: Papin's steam engine, for which he suggested a self-regulating device; the discovery of European porcelain by his friend Tschirnhaus at Meissen; the use of microscopes in biological research; the use of vital statistics in coping with problems of public health, on which he wrote several articles; and the principle of the aneroid barometer, which he was the first to propose. It was at his urging that a continuous series of observations of barometric pressures and weather conditions was carried out at Kiel from 1679 to 1714, with the purpose of testing the value of the barometer in forecasting the weather. The science of linguistics originated in his efforts to construct a comparative system of linguistic genealogy for the main languages of Europe and Asia.[26] Also, it was he who destroyed the prevailing belief that Hebrew was the primordial language of the human race.[27] His *New Essays on Human Understanding* (written in French in 1704 but not published until 1765) was one of the most influential books in the history of psychology, for in it he introduced for the first time the concept of subconscious or unconscious mental processes, and thereby permanently altered the development of psychology in ways that are well known to everyone in the twentieth century.[28] In physics, his contributions to the emerging concept of kinetic energy were so significant that "he stood beside Newton as one of the creators of modern dynamics."[29] In fact, the very word "dynamics" is due to Leibniz, in its French form "dynamique." He also solved the problem of the catenary curve, gave the first analysis of the tension in the interior fibers of a loaded beam, and studied many other physical problems with the aid of his own differential and integral calculus.[30] Leibniz was

---

[26] This work was published in 1710 in the first volume of *Miscellania Berolinensia*, the official journal of the Berlin Academy of Sciences. This volume contained 58 papers on science and mathematics, of which 12 were by Leibniz himself. He had founded the Berlin Academy in 1700 with the support of the Queen of Prussia, whom he had tutored when she was a child.

[27] See pp. 9–10 of H. Pedersen, *Linguistic Science in the 19th Century*, Harvard University Press, 1931.

[28] See Wolf, *op. cit.*, pp. 579–81.

[29] See pp. 283–322 of Richard S. Westfall, *Force in Newton's Physics*, American Elsevier, 1971.

[30] See many index references in C. Truesdell, *Essays in the History of Mechanics*, Springer Verlag, 1968.

obviously a scientific thinker of rare talent and vision, and there was so much of real value to do and learn at that time that it seems a pity he didn't spend more of his time and energy on these projects and less on the fantasies of theology and metaphysics.

Aristotle said that by their nature all men desire to know. In the case of Leibniz this desire was intensified into an overwhelming passion. The great modern philosopher A. N. Whitehead once remarked, "There is a book to be written, and its title should be *The Mind of Leibniz*." But who could write such a book and do justice to the subject? The prodigious variety of his interests was an essential part of his genius, but he paid a price, for he scattered himself in so many directions that he left mostly fragments behind him. In a letter written in 1695 he expressed his occasional despair in these words:

> How extremely distracted I am cannot be described. I dig up various things from the archives, examine ancient documents, and collect unpublished manuscripts. From these I strive to throw light on the history of Brunswick. I receive and send letters in great numbers. I have, indeed, so much that is new in mathematics, so many thoughts in philosophy, so many other literary observations which I do not wish to have perish, that I am often bewildered as to where to begin.[31]

With Leibniz the mind and the hand went together so closely that thinking and writing were almost a single act, and he coped with the flood of his thoughts by writing them down. Since he rarely discarded written material, he accumulated over the years an immense chaotic mass of papers as big as a haystack, which was hastily packed away in crates soon after his death and stored in the Royal Library at Hanover. Only a small fraction of this mountain of material was published in his lifetime. Partial excavations have been made by various scholars, but most of it remains unpublished to this day.[32] There are rough notes and early drafts and almost-finished projects; innumerable essays and memoranda; book-length manuscripts; dialogues with himself; and letters— more than 15,000 letters written by Leibniz in correspondence with 1063 different people.

All his life Leibniz collected correspondence on topics in science and philosophy as other people collect stamps or works of art. The letters exchanged with a particular person often constitute an absorbing intellectual drama. Thus we have, among others, the famous correspondence of 1715–1716 between Leibniz and Dr. Samuel Clarke, Newton's disciple and spokesman,

---

[31] Loemker, *op. cit.,* p. 21.

[32] In 1900 the Berlin Academy began planning a complete critical edition of Leibniz's works in 40 volumes. Only about a dozen of these volumes have so far been published, and it seems unlikely that this edition will be finished until sometime in the twenty-first century.

which quickly heated up into a pitched battle over the validity of Newton's ideas about absolute space and absolute time.[33] Leibniz believed that these ideas were devoid of meaning, and that space and time are purely relative concepts; and to Newton's intense annoyance, he claimed to have proved his contention.[34] Whatever we may think of Leibniz's arguments today, the fact remains that modern science has discovered serious defects in the Newtonian concepts of the framework of the physical universe, and these defects were remedied only by the discoveries of Einstein in the early twentieth century.

On looking back at Leibniz through the haze of 300 years of history, we seem to see a multitude of men—philosopher, mathematician, scientist, logician, diplomat, lawyer, historian, etc.—instead of a single individual. But what were his qualities as a living, breathing human being?

Like many of his great contemporaries he never married, and we know little of his personal life. At the age of 50, according to Fontenelle, he proposed to a certain lady, but "the lady asked for time to consider the matter, so Leibniz had a chance to think again, and he withdrew his offer." He had an astonishing capacity for rapid and sustained work, often spending days on end at his desk, except for occasional hurried meals or brief naps; even when traveling in the jolting, uncomfortable carriages of the day, he used his time to work on mathematical problems; and to the end of his life he preserved the indomitable energy without which all his ambitions and plans and projects would have come to nothing. He is described as a man of moderate habits in everything but work, quick of temper but easily appeased, very self-assured, and tolerant of differences of opinion though confident of the correctness of his own opinions. He enjoyed social life of all kinds, and was firmly convinced that there was something interesting to be learned from everyone he met. According to his secretary, he spoke well of everybody and made the best of everything. On the less amiable side, he is said to have been fond of money to the point of avarice, and he had a reputation as a pennypincher.

When Queen Anne died in August 1714, Leibniz's master, the Elector George Louis of Hanover, succeeded her as George I, the first German King of England. Leibniz was in Vienna at the time and returned to Hanover as quickly as he could, but George and his entourage had already departed. Leibniz hoped to join them in London as court historian and councilor of state, but the new King refused to consider it and ordered him to remain at Hanover

---

[33] In the *Principia* Newton wrote: "Absolute space, in its own nature, without regard to anything external, remains always similar and immovable . . . Absolute, true, and mathematical time, of itself, and from its own nature, flows equably and without regard to anything external."

[34] See H. G. Alexander (ed.), *The Leibniz-Clarke Correspondence*, Manchester University Press, 1956, especially paragraphs 5 and 6 of Leibniz's third letter. For a further commentary of great interest and value, see Koyré's article "Leibniz and Newton," in H. G. Frankfurt, *op. cit.*, pp. 239–79.

and finish his history of the House of Brunswick. It appears that George disliked Leibniz and had no interest in any of his ideas or projects except those that might add luster to his own family background.[35] Leibniz's last years were made difficult by neglect and the agonies of gout and kidney stones, but he struggled on with his work. His death in November 1716 was ignored by the court, both in London and Hanover. Only his secretary followed him to his grave, and an eyewitness of his funeral wrote that "he was buried more like a robber than what he really was, the ornament of his country." But the books are now balanced, for he is recognized today as one of the few universal geniuses in human history, and the first in the line of those eminent Germans who were also towering figures of world culture: Leibniz, Bach, Goethe, Beethoven, Gauss, Einstein.

It remains to add a few words of a more detailed nature about Leibniz's greatest creation, his differential and integral calculus, for this incomparable tool of thought is the means by which his genius continues to make itself felt on a day-by-day basis in every civilized country of the modern world.

Leibniz published many sketchy papers on his calculus, beginning in 1684, and we shall say more about these below. However, the development of his ideas and the sequence of his discoveries can be followed in full detail through the hundreds of pages of his private notes made from 1673 on.[36]

It all seems to have started in a memorable conversation Leibniz had with Huygens in the spring of 1673, which he referred to repeatedly in later years.[37] As a result of this encounter, and on Huygens's advice, he began an intensive study of some of the mathematical writings of Pascal and others. In particular, it was on reading Pascal's brief paper *Traité des sinus du quart de cercle* ("Treatise on the Sines of a Quadrant of a Circle") that Leibniz later reported that "a great light" burst upon him. He suddenly realized that the tangent to (or slope of) a given curve can be found by forming the ratio of the *differences* in the ordinates and abscissas of two neighboring points on the curve as these differences become infinitely small (see Fig. A.45). He also saw that the area under the curve is the *sum* of the infinitely thin rectangles making up this area. Most important of all, he observed that the two processes of differencing and summing—in our terminology, differentiating and integrating—are inverses of each other, and are linked together by means of the infinitesimal or differential triangle (dx, dy, ds) shown in the figure.

---

[35] This dislike was perhaps natural, for George I was ridiculous and repulsive both as a man and as a monarch, and inferior men in positions of power often dislike superior men who happen to be their subordinates. Jonathan Swift called him "the ruling Yahoo," and a later British prime minister (Winston Churchill, in his role as historian) described him as "a humdrum German martinet with dull brains and coarse tastes."

[36] These notes were found in the Hanover Library in the middle of the nineteenth century. See J. M. Child (ed.), *The Early Mathematical Manuscripts of Leibniz*, Open Court, 1920.

[37] Hofmann, *op. cit.*, pp. 47–48.

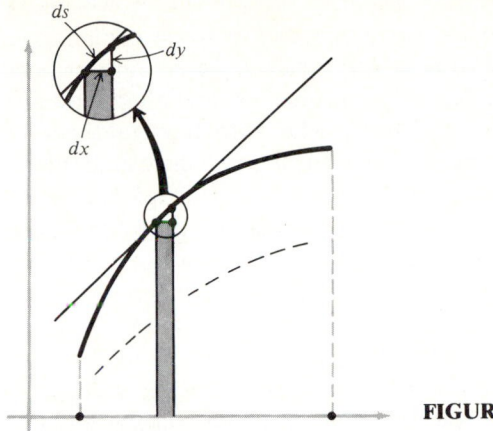

FIGURE A.45

It was in a famous manuscript dated October 29, 1675, that he first introduced the modern integral sign, a long letter S suggesting the first letter of the Latin word *summa* (sum). He was doing integrations by forming sums of Cavalieri's indivisibles, and he abbreviated "omnes lineae"—all lines—to "omn l." He then remarked, "It will be useful to write ∫ for omn, thus ∫ l for omn l, that is, the sum of those l's." On the same day he introduced the differential symbol d, and soon he was writing dx, dy, and dy/dx as we do today, as well as integrals like $\int y\,dy$ and $\int y\,dx$. All this time he was formulating and solving new problems and learning to use the machinery he was developing. Throughout his notes for these years he carried on a fascinating conversation with himself, filled with warnings, reminders, and efforts to cheer himself up, as well as occasional expressions of triumph.

Leibniz's first published account of his differential calculus was in a seven-page paper in the *Acta Eruditorum* of 1684.[38] The meaning of the differentials dx and dy is far from clear, and in fact he never did clarify this issue to anyone's satisfaction. He states the formulas $d(xy) = x\,dy + y\,dx$, $d(x/y) = (y\,dx - x\,dy)/y^2$, and $d(x^n) = nx^{n-1}\,dx$, but he makes no attempt to explain or justify them. As we know from his notes and letters, he thought of dx and dy as infinitely small increments, or "infinitesimals," and he derived these formulas by discarding infinitesimals of higher order, but none of this is in the paper.[39] However, he does give the condition $dy = 0$ for maxima and minima and $ddy = 0$ for points of inflection, and he makes several geometric

---

[38] A translation is given in Struik, *op. cit.*, pp. 271–80.

[39] Thus, for example, $d(xy) = (x + dx)(y + dy) - xy = x\,dy + y\,dx + dx\,dy$, and since dx and dy are infinitely small, the product $dx\,dy$ is infinitely infinitely small and—according to Leibniz—can be dropped, giving the correct formula $d(xy) = x\,dy + y\,dx$.

applications. This paper was followed in 1686 by a second,[40] in which he casually introduces his integral symbol $\int$ with what amounts to no explanation at all and claims that $\int$ and d "are each other's converse." These early papers appear to be hastily and carelessly written, and are so unclear that they are barely intelligible. Even the Bernoulli brothers, who somehow understood Leibniz's intentions and realized that something profound was being born, spoke of these papers as presenting "an enigma rather than an explanation."

This early work of Leibniz, though obscure and fragmentary, was a fertile seed of great potential. It aroused no interest in Germany or England, but James and John Bernoulli of Basle found it exciting and richly suggestive. They eagerly absorbed Leibniz's ideas and methods and contributed many of their own; and before the end of the century these three men, in constant correspondence and stimulating one another like athletes in a race, had discovered much of the content of our modern college calculus courses. In fact, between 1695 and 1700 every monthly issue of the *Acta Eruditorum* contained at least one paper—and often several—by Leibniz or the Bernoulli brothers in which they treated, using notation almost identical with that used today, a great variety of problems in differential and integral calculus, differential equations, infinite series, and even the calculus of variations. In this headlong rush to exploit the wealth of applications of the new analysis, there was little interest in pausing to indulge in leisurely examinations of the basic ideas. This uncritical spirit prevailed throughout the eighteenth century, and it was not until the early decades of the nineteenth century that serious attention was given to the logical foundations of the subject.

In addition to the actual content of his work, Leibniz was also one of the greatest inventors of mathematical symbols. Few people have understood so well that a really good notation smooths the way and is almost capable of doing our thinking for us. He wrote about this to his friend Tschirnhaus as follows:

> In symbols one observes an advantage in discovery which is greatest when they express the exact nature of a thing briefly and, as it were, picture it; then indeed the labor of thought is wonderfully diminished.[41]

---

[40] Struik, *op. cit.,* pp. 281–82.

[41] See F. Cajori, *A History of Mathematical Notations,* Open Court, 1929, vol. II, p. 184. On pp. 180–96 and 201–205 Cajori gives a comprehensive discussion of Leibniz's use of mathematical symbols. See also the long quotation from A. N. Whitehead on pp. 332–33 ("By relieving the brain of all unnecessary work, a good notation sets it free to concentrate on more advanced problems, and in effect increases the mental power of the race . . . Civilization advances by extending the number of important operations which we can perform without thinking about them.").

His flexible and suggestive calculus notations dx, dy, dy/dx, and $\int y\,dx$ are perfect illustrations of this remark and are still in standard use, as are the English versions of his descriptive phrases "calculus differentialis" and "calculus integralis."[42] It was mainly through his influence that the symbol $=$ is universally used for equality, and he advocated the dot ($\cdot$) instead of the cross ($\times$) for multiplication.[43] His colon for division ($x{:}y$ for x/y) and his symbols for geometric similarity and congruence ($\sim$ and $\simeq$) are still widely used. He introduced the terms "constant," "variable," "parameter," and "transcendental" (in the sense of "nonalgebraic"), as well as "abscissa" and "ordinate," which together he called "coordinates." Also, it was he who first used the word "function" with essentially its modern meaning.

Leibniz is sometimes criticized for not producing any great work that can be pointed to and admired, like Newton's *Principia*. But he did produce such a work, even though it was not a book. The line of descent for all the greatest mathematicians of modern times begins with him—not with Newton—and extends in unbroken succession down to the twentieth century. He was the intellectual father of the Bernoullis; John Bernoulli was Euler's teacher; Euler adopted Lagrange as his scientific protégé; then came Gauss, Riemann, and the rest—all direct intellectual descendants of Leibniz. He had predecessors, of course, as every great thinker does. But apart from this, he was the true founder of modern European mathematics.

---

[42] Leibniz first suggested "calculus summatorius," but in 1696 he and John Bernoulli agreed on "calculus integralis."

[43] The equality sign was first introduced by the Englishman Robert Recorde in 1557, as follows: "To avoide the tediouse repetition of these woordes: is equalle to: I will sette as I doe often in woorke use, a paire of paralleles, or Gemowe [twin] lines of one lengthe, thus :=, bicause noe .2. thynges, can be moare equalle."

A.20

John Bernoulli

THE BERNOULLI BROTHERS

> With justice we admire Huygens because he first discovered that a heavy particle falls down along a cycloid in the same time no matter from what point on the cycloid it begins its motion. But you will be petrified with astonishment when I say that precisely this cycloid, the *tautochrone* of Huygens, is our required *brachistochrone*.
>
> *John Bernoulli*

Most people are aware that Johann Sebastian Bach was one of the greatest composers of all time. However, it is less well known that his prolific family was so consistently talented in this direction that several dozen Bachs were eminent musicians from the sixteenth to the nineteenth centuries. In fact, there were parts of Germany where the very word *bach* meant a musician. What the Bach clan was to music, the Bernoullis were to mathematics and science. In three generations this remarkable Swiss family produced eight mathematicians—two of them outstanding—who in turn had a swarm of descendants who distinguished themselves in many fields.[1] These two were the brothers James (1654–1705) and John (1667–1748), who played indispensable roles in the development of modern European mathematics.

---

[1] See Francis Galton, *Hereditary Genius*, Macmillan, 1892, pp. 195–96.

At the insistence of their merchant father, James studied theology and John studied medicine. However, they found their true vocation when Leibniz's early papers of 1684 and 1686 were published in the *Acta Eruditorum*. They taught themselves the new calculus, entered into extensive correspondence with Leibniz, and became his most important students and disciples. James was professor of mathematics at Basel from 1687 until his death. John became a professor at Groningen in Holland in 1695, and on James's death succeeded his brother in the chair at Basel, where he flourished for another 43 years.

James was interested in infinite series, and among other things established the divergence of the sum of the reciprocals of the positive integers,

$$1 + \frac{1}{2} + \frac{1}{3} + \frac{1}{4} + \cdots ,$$

and the convergence of the sum of the reciprocals of the squares,

$$1 + \frac{1}{4} + \frac{1}{9} + \frac{1}{16} + \cdots .$$

He often speculated about the sum of the latter series, but the question wasn't settled until 1736, when Euler discovered that this sum is $\pi^2/6$. James invented polar coordinates, studied many special curves (including the catenary, the tractrix, the leminscate, and the exponential spiral), and introduced the Bernoulli numbers that appear in the power series expansion of the function $\tan x$. It was he (in 1690) who first used the word "integral." In his book *Ars Conjectandi* (published posthumously in 1713) he formulated the basic principle in the theory of probability known as *Bernoulli's theorem* or the *law of large numbers*: If the probability of a certain event is *p,* and if *n* independent trials are made with *k* successes, then $k/n \rightarrow p$ as $n \rightarrow \infty$. At first sight this statement may seem to be a triviality, but beneath its surface lies a tangled thicket of philosophical (and mathematical) problems that have been a source of controversy down to the present day.

Like his older brother, John was fascinated by the almost magical power of Leibniz's calculus. He quickly mastered it and applied it to many problems in geometry, differential equations, and mechanics. Many of the ideas of Leibniz and the Bernoulli brothers were given wide circulation in 1696 in the first calculus textbook, *Analyse des infiniment petits,* by G. F. A. de l'Hospital (1661–1701). This man, a French nobleman and good amateur mathematician, openly acknowledged his debt to his teachers: "I have made free use of their discoveries, so that I frankly return to them whatever they please to claim as their own." L'Hospital's book is best known for its rule about indeterminate forms of the type 0/0. After l'Hospital's death John Bernoulli stated that much of the content of the book, and in particular this rule, was his own property. This minor mystery was cleared up in 1955 by the publication of the correspondence between the two men. L'Hospital was interested in mathematics but lacked confidence in his ability to learn calculus by himself. John was willing to tutor him, and in return for a yearly allowance agreed to sell him some of his own discoveries. The arrangement was discussed in a letter from l'Hospital of March 17, 1694, and the rule for 0/0 is contained in a letter from Bernoulli of July 22, 1694.[2]

The Bernoulli brothers sometimes worked on the same problems, which was unfortunate in view of their suspicious natures and surly dispositions. On occasion the friction between them flared up into a bitter and abusive public feud, as it did over the brachistochrone problem. In 1696 John proposed the problem (which is stated in our note on Newton) as a challenge to the mathematicians of Europe. It aroused great interest, and was solved by Newton and Leibniz as well as by the two Bernoullis. John's solution was more elegant, while James's—though rather clumsy and laborious—was more general. This situation started an acrimonious quarrel that dragged on for several years and was often conducted in rough language more suited to a street brawl than a scientific discussion.

After the deaths of his brother and Leibniz, John Bernoulli became the acknowledged leader of the Continental mathematicians in their battle against the English. He continued to produce good mathematical ideas and biting invective for many years, and was a major force in the ultimate triumph of Leibniz's calculus over the fluxions of Newton. Perhaps his greatest contribution was his student, the prodigious Euler, whose incredible flood of discoveries dominated mathematics through most of the eighteenth century.

---

[2] See D. J. Struik's article in *The Mathematics Teacher*, vol. 56 (1963), pp. 257–60.

A.21

EULER
(1707–1783)

Read Euler: he is our master in everything.

Pierre Simon de Laplace

Leonhard Euler was Switzerland's foremost scientist and one of the three greatest mathematicians of modern times (the other two being Gauss and Riemann).

He was perhaps the most prolific author of all time in any field. From 1727 to 1783 his writings poured out in a seemingly endless flood, constantly adding knowledge to every known branch of pure and applied mathematics, and also to many that were not known until he created them. He averaged about 800 printed pages a year throughout his long life, and yet he almost always had something worthwhile to say and never seems long-winded. The publication of his complete works was started in 1911, and the end is not in sight. This edition was planned to include 887 titles in 72 volumes, but since that time extensive new deposits of previously unknown manuscripts have been unearthed, and it is now estimated that more than 100 large volumes will be required for completion of the project. Euler evidently wrote mathematics with the ease and fluency of a skilled speaker discoursing on subjects with which he is intimately familiar. His writings are models of relaxed clarity. He never condensed, and he reveled in the rich abundance of his ideas and the vast scope of his interests. The French physicist Arago, in speaking of Euler's incomparable mathematical facility, remarked that "He calculated without apparent effort, as men breathe, or as eagles sustain themselves in the wind." He suffered total blindness during the last 17 years of his life, but with the aid of his powerful memory and fertile imagination, and with helpers to write his books and scientific papers from dictation, he actually increased his already prodigious output of work.

Euler was a native of Basel and a student of John Bernoulli at the University, but he soon outstripped his teacher. His working life was spent as a member of the Academies of Science at Berlin and St. Petersburg, and most of his papers were published in the journals of these organizations. His business was mathematical research, and he knew his business. He was also a man of broad culture, well versed in the classical languages and literatures (he knew the *Aeneid* by heart), many modern languages, physiology, medicine, botany, geography, and the entire body of physical science as it was known in his time. However, he had little talent for metaphysics or disputation, and came out second best in many good-natured verbal encounters with Voltaire at the court of Frederick the Great. His personal life was as placid and uneventful as is possible for a man with 13 children.

Though he was not himself a teacher, Euler has had a deeper influence on the teaching of mathematics than any other person. This came about chiefly through his three great treatises: *Introductio in Analysin Infinitorum* (1748); *Institutiones Calculi Differentialis* (1755); and *Institutiones Calculi Integralis* (1768–1794). There is considerable truth in the old saying that all elementary and advanced calculus textbooks since 1748 are essentially copies of Euler or copies of copies of Euler.[1] These works summed up and codified the discoveries of his predecessors, and are full of Euler's own ideas. He extended and perfected plane and solid analytic geometry, introduced the analytic approach to trigonometry, and was responsible for the modern treatment of the functions $\ln x (= \log_e x)$ and e^x. He created a consistent theory of logarithms of negative and imaginary numbers, and discovered that $\ln x$ has an infinite number of values. It was through his work that the symbols e, π, and i $(=\sqrt{-1})$ became common currency for all mathematicians, and it was he who linked them together in the astonishing relation $e^{\pi i} = -1$. This is merely a special case (put $\theta = \pi$) of his famous formula $e^{i\theta} = \cos\theta + i\sin\theta$, which connects the exponential and trigonometric functions and is absolutely indispensable in higher analysis.[2] Among his other contributions to standard

---

[1] See C. B. Boyer, "The Foremost Textbook of Modern Times," *American Mathematical Monthly*, vol. 58, 1951, pp. 223–226.

[2] An even more astonishing consequence of his formula is the fact that an imaginary power of an imaginary number can be real, in particular $i^i = e^{-\pi/2}$; for if we put $\theta = \pi/2$, we obtain $e^{\pi i/2} = i$, so

$$i^i = (e^{\pi i/2})^i = e^{\pi i^2/2} = e^{-\pi/2}.$$

Euler further showed that i^i has infinitely many values, of which this calculation produces only one.

mathematical notation were $\sin x$, $\cos x$, the use of $f(x)$ for an unspecified function, and the use of Σ for summation.[3] Good notations are important, but the ideas behind them are what really count, and in this respect, Euler's fertility was almost beyond belief. He preferred concrete special problems to the general theories in vogue today, and his unique insight into the connections among apparently unrelated formulas blazed many trails into new areas of mathematics which he left for his successors to cultivate.

He was the first and greatest master of infinite series, infinite products, and continued fractions, and his works are crammed with striking discoveries in these fields. James Bernoulli (John's older brother) found the sums of several infinite series, but he was not able to find the sum of the reciprocals of the squares, $1 + \frac{1}{4} + \frac{1}{9} + \frac{1}{16} + \cdots$. He wrote, "If someone should succeed in finding this sum, and will tell me about it, I shall be much obliged to him." In 1736, long after James's death, Euler made the wonderful discovery that

$$1 + \frac{1}{4} + \frac{1}{9} + \frac{1}{16} + \cdots = \frac{\pi^2}{6}.$$

He also found the sums of the reciprocals of the fourth and sixth powers,

$$1 + \frac{1}{2^4} + \frac{1}{3^4} + \cdots = 1 + \frac{1}{16} + \frac{1}{81} + \cdots = \frac{\pi^4}{90}$$

and

$$1 + \frac{1}{2^6} + \frac{1}{3^6} + \cdots = 1 + \frac{1}{64} + \frac{1}{729} + \cdots = \frac{\pi^6}{945}.$$

When John heard about these feats, he wrote, "If only my brother were alive now."[4] Few would believe that these formulas are related—as they are—to Wallis's infinite product (1656),

$$\frac{\pi}{2} = \frac{2}{1} \cdot \frac{2}{3} \cdot \frac{4}{3} \cdot \frac{4}{5} \cdot \frac{6}{5} \cdot \frac{6}{7} \cdots.$$

Euler was the first to explain this in a satisfactory way, in terms of his infinite product expansion of the sine,

$$\frac{\sin x}{x} = \left(1 - \frac{x^2}{\pi^2}\right)\left(1 - \frac{x^2}{4\pi^2}\right)\left(1 - \frac{x^2}{9\pi^2}\right) \cdots.$$

---

[3] See F. Cajori, *A History of Mathematical Notations*, Open Court, 1929.

[4] The world is still waiting—more than 200 years later—for someone to discover the sum of the reciprocals of the cubes.

Wallis's product is also related to Brouncker's remarkable continued fraction,

$$\frac{\pi}{4} = \cfrac{1}{1 + \cfrac{1^2}{2 + \cfrac{3^2}{2 + \cfrac{5^2}{2 + \cfrac{7^2}{2 + \cdots}}}}},$$

which became understandable only in the context of Euler's extensive researches in this field.[5]

His work in all departments of analysis strongly influenced the further development of this subject through the next two centuries. He contributed many important ideas to differential equations, including substantial parts of the theory of second-order linear equations and the method of solution by power series. He gave the first systematic discussion of the calculus of variations, which he founded on his basic differential equation for a minimizing curve. He introduced the number now known as *Euler's constant,*

$$\gamma = \lim_{n \to \infty} \left(1 + \frac{1}{2} + \frac{1}{3} + \cdots + \frac{1}{n} - \ln n \right) = 0.5772\ldots,$$

which is the most important special number in mathematics after π and e. He discovered the integral defining the gamma function,

$$\Gamma(x) = \int_0^\infty t^{x-1} e^{-t}\, dt,$$

which is often the first of the so-called higher transcendental functions that students meet beyond the level of calculus, and he developed many of its applications and special properties.[6] He also worked with Fourier series, encountered the Bessel functions in his study of the vibrations of a stretched circular membrane, and applied Laplace transforms to solve differential equations—all before Fourier, Bessel, and Laplace were born. Even though Euler died about 200 years ago, he lives everywhere in analysis.

E. T. Bell, the well-known historian of mathematics, observed that "One of the most remarkable features of Euler's universal genius was its equal strength in both of the main currents of mathematics, the continuous and the discrete." In the realm of the discrete, he was one of the originators of number theory and made many far-reaching contributions to this subject throughout

---

[5] The ideas described in this paragraph are explained more fully in Sections B.14 and B.20.

[6] A few of the simpler properties of the gamma function are discussed on pp. 351–53 of George F. Simmons, *Differential Equations,* 2nd ed., 1991.

his life. In addition, the origins of topology—one of the dominant forces in modern mathematics—lie in his solution of the Königsberg bridge problem and his formula $V - E + F = 2$ connecting the numbers of vertices, edges, and faces of a simple polyhedron. In the following paragraphs, we briefly describe his activities in these fields.

In number theory, Euler drew much of his inspiration from the challenging marginal notes left by Fermat in his copy of the works of Diophantus, and some of his achievements are mentioned in our account of Fermat. He also initiated the theory of partitions, a little-known branch of number theory that turned out much later to have applications in statistical mechanics and the kinetic theory of gases. A typical problem of this subject is to determine the number $p(n)$ of ways in which a given positive integer n can be expressed as a sum of positive integers, and if possible to discover some properties of this function. For example, 4 can be partitioned into $4 = 3 + 1 = 2 + 2 = 2 + 1 + 1 = 1 + 1 + 1 + 1$, so $p(4) = 5$, and similarly $p(5) = 7$ and $p(6) = 11$. It is clear that $p(n)$ increases very rapidly with n, so rapidly, in fact, that[7]

$$p(200) = 3,972,999,029,388.$$

Euler began his investigations by noticing (only geniuses notice such things) that $p(n)$ is the coefficient of x^n when the function $[(1-x)(1-x^2)(1-x^3) \cdots]^{-1}$ is expanded in a power series:

$$\frac{1}{(1-x)(1-x^2)(1-x^3) \cdots} = 1 + p(1)x + p(2)x^2 + p(3)x^3 + \cdots.$$

By building on this foundation, he derived many other remarkable identities related to a variety of problems about partitions.[8]

The Königsberg bridge problem originated as a pastime of Sunday strollers in the town of Königsberg (now Kaliningrad) in what was formerly East Prussia. There were seven bridges across the river that flows through the

---

[7] This evaluation required a month's work by a skilled computer in 1918. His motive was to check an approximate formula for $p(n)$, namely

$$p(n) \cong \frac{1}{4n\sqrt{3}} e^{\pi\sqrt{2n/3}}$$

(the error was extremely small).

[8] See Chapter XIX of G. H. Hardy and E. M. Wright, *An Introduction to the Theory of Numbers*, Oxford, 1938; or Chapters 12–14 of G. E. Andrews, *Number Theory*, W. B. Saunders, 1971. These treatments are "elementary" in the technical sense that they do not use the high-powered machinery of advanced analysis, but nevertheless they are far from simple. For students who wish to experience some of Euler's most interesting work in number theory at first hand, and in a context not requiring much previous knowledge, we recommend Chapter VI of G. Polya's fine book, *Induction and Analogy in Mathematics*, Princeton University Press, 1954.

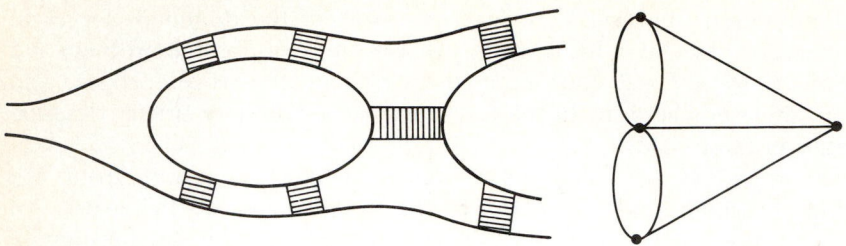

FIGURE A.46
The Konigsberg bridges.

town (see Fig. A.46). The residents used to enjoy walking from one bank to the islands and then to the other bank and back again, and the conviction was widely held that it is impossible to do this by crossing all seven bridges without crossing any bridge more than once. Euler analyzed the problem by examining the schematic diagram given on the right in the figure, in which the land areas are represented by points and the bridges by lines connecting these points. The points are called vertices, and a vertex is said to be odd or even according to whether the number of lines leading to it is odd or even. In modern terminology, the entire configuration is called a *graph,* and a path through the graph that traverses every line but no line more than once is called an *Euler path.* An Euler path need not end at the vertex where it began, but if it does, it is called an *Euler circuit.* By the use of combinatorial reasoning, Euler arrived at the following theorems about any such graph: (1) there are an even number of odd vertices; (2) if there are no odd vertices, there is an Euler circuit starting at any point; (3) if there are two odd vertices, there is no Euler circuit, but there is an Euler path starting at one odd vertex and ending at the other; (4) if there are more than two odd vertices, there are no Euler paths.[9] The graph of the Königsberg bridges has four odd vertices, and therefore, by the last theorem, has no Euler paths.[10] The branch of mathematics that has developed from these ideas is known as *graph theory*; it has applications to chemical bonding, economics, psychosociology, the properties of networks of roads and railroads, and other subjects.

In Section A.4 on Euclid we remarked that a polyhedron is a solid whose surface consists of a number of polygonal faces, and Fig. A.10 of that section

---

[9] Euler's original paper of 1736 is interesting to read and easy to understand; it can be found on pp. 573–80 of J. R. Newman (ed.), *The World of Mathematics,* Simon and Schuster, 1956.

[10] It is easy to see—without appealing to any theorems—that this graph contains no Euler circuit, for if there were such a circuit, it would have to enter each vertex as many times as it leaves it, and therefore every vertex would have to be even. Similar reasoning shows also that if there were an Euler path that is not a circuit, there would have to be two odd vertices.

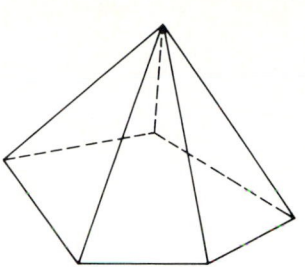 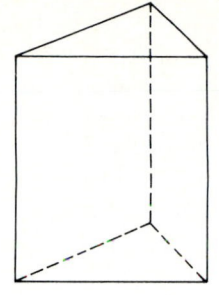

FIGURE A.47

displays the five regular polyhedra. The Greeks studied these figures assiduously, but it remained for Euler to discover the simplest of their common properties: If V, E, and F denote the numbers of vertices, edges, and faces of any one of them, then in every case we have

$$V - E + F = 2.$$

This fact is known as *Euler's formula for polyhedra*, and it is easy to verify from the data summarized in the following table.

	V	E	F
Tetrahedron	4	6	4
Cube	8	12	6
Octahedron	6	12	8
Dodecahedron	20	30	12
Icosahedron	12	30	20

This formula is also valid for any irregular polyhedron as long as it is *simple*—which means that it has no "holes" in it, so that its surface can be deformed continuously into the surface of a sphere. Figure A.47 shows two simple irregular polyhedra for which $V - E + F = 6 - 10 + 6 = 2$ and $V - E + F = 6 - 9 + 5 = 2$. However, Euler's formula must be extended to

$$V - E + F = 2 - 2p$$

in the case of a polyhedron with p holes (a simple polyhedron is one for which $p = 0$). Figure A.48 illustrates the cases $p = 1$ and $p = 2$; here we have $V - E + F = 16 - 32 + 16 = 0$ when $p = 1$, and $V - E + F = 24 - 44 + 18 = -2$ when $p = 2$. The significance of these ideas can best be understood by

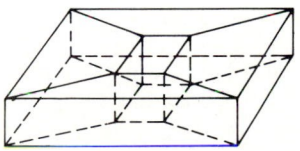 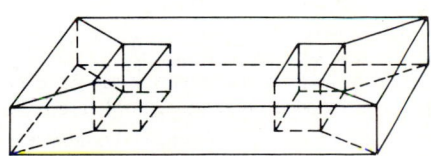

FIGURE A.48

imagining a polyhedron to be a hollow figure with a surface made of thin rubber, and inflating it until it becomes smooth. We no longer have flat faces and straight edges, but instead a map on the surface consisting of curved regions, their boundaries, and points where boundaries meet. The number $V - E + F$ has the same value for all maps on our surface, and is called the *Euler characteristic* of this surface. The number p is called the *genus* of the surface. These two numbers, and the relation between them given by the equation $V - E + F = 2 - 2p$, are evidently unchanged when the surface is continuously deformed by stretching or bending. Intrinsic geometric properties of this kind—which have little connection with the type of geometry concerned with lengths, angles, and areas—are called *topological*. The serious study of such topological properties has greatly increased during the past century, and has furnished valuable insights to many branches of mathematics and science.[11]

The distinction between pure and applied mathematics did not exist in Euler's day, and for him the entire physical universe was a convenient object whose diverse phenomena offered scope for his methods of analysis. The foundations of classical mechanics had been laid down by Newton, but Euler was the principal architect. In his treatise of 1736 he was the first to explicitly introduce the concept of a mass-point or particle, and he was also the first to study the acceleration of a particle moving along any curve and to use the notion of a vector in connection with velocity and acceleration. His continued successes in mathematical physics were so numerous, and his influence was so pervasive, that most of his discoveries are not credited to him at all and are taken for granted by physicists as part of the natural order of things. However, we do have Euler's equations of motion for the rotation of a rigid body, Euler's hydrodynamic equation for the flow of an ideal incompressible fluid, Euler's law for the bending of elastic beams, and Euler's critical load in the theory of the buckling of columns. On several occasions the thread of his scientific thought led him to ideas his contemporaries were not ready to assimilate. For example, he foresaw the phenomenon of radiation pressure, which is crucial for the modern theory of the stability of stars, more than a century before Maxwell rediscovered it in his own work on electromagnetism.

Euler was the Shakespeare of mathematics—universal, richly detailed, and inexhaustible.[12]

---

[11] Proofs of Euler's formula and its extension are given on pp. 236–40 and 256–59 of R. Courant and H. Robbins, *What Is Mathematics?*, Oxford, 1941. See also G. Polya, *op. cit.*, pp. 35–43.

[12] For further information, see C. Truesdell, "Leonhard Euler, Supreme Geometer (1707–1783)," in *Studies in Eighteenth-Century Culture*, Case Western Reserve University Press, 1972. Also, the November 1983 issue of *Mathematics Magazine* is wholly devoted to Euler and his work.

A.22

LAGRANGE
(1736–1813)

The "generalized coordinates" of our mechanics of today were conceived and installed by Lagrange, and this was an achievement of unmatchable magnitude.

Salomon Bochner

Joseph Louis Lagrange detested geometry, but made outstanding discoveries in the calculus of variations and analytical mechanics. He also contributed to number theory and algebra, and fed the stream of thought that later nourished Gauss and Abel. His mathematical career can be viewed as a natural extension of the work of his older and greater contemporary, Euler, which in many respects he carried forward and refined.

Lagrange was born in Turin of mixed French–Italian ancestry. As a boy, his tastes were more classical than scientific, but his interest in mathematics was kindled while he was still in school by reading a paper by Edmund Halley on the uses of algebra in optics. He then began a course of independent study, and progressed so rapidly that at the age of 19 he was appointed professor at the Royal Artillery School in Turin.[1]

Lagrange's contributions to the calculus of variations were among his earliest and most important works. In 1755 he communicated to Euler his

---

[1] See George Sarton's valuable essay, "Lagrange's Personality," *Proceedings of the American Philosophical Society,* vol. 88 (1944), pp. 457–96.

method of multipliers for solving isoperimetric problems. These problems had baffled Euler for years, since they lay beyond the reach of his own semi-geometrical techniques. Euler was immediately able to answer many questions he had long contemplated, but he replied to Lagrange with admirable kindness and generosity, and withheld his own work from publication "so as not to deprive you of any part of the glory which is your due." Lagrange continued working for a number of years on his analytic version of the calculus of variations, and both he and Euler applied it to many new types of problems, especially in mechanics.

In 1766, when Euler left Berlin for St. Petersburg, he suggested to Frederick the Great that Lagrange be invited to take his place. Lagrange accepted and lived in Berlin for 20 years until Frederick's death in 1786. During this period he worked extensively in algebra and number theory and wrote his masterpiece, the treatise *Mécanique Analytique* (1788), in which he unified general mechanics and made of it, as Hamilton later said, "a kind of scientific poem." Among the enduring legacies of this work are Lagrange's equations of motion, generalized coordinates, and the concept of potential energy.

Men of science found the atmosphere of the Prussian court rather uncongenial after the death of Frederick, so Lagrange accepted an invitation from Louis XVI to move to Paris, where he was given apartments in the Louvre. Lagrange was extremely modest and undogmatic for a man of his great gifts; and though he was a friend of aristocrats—and indeed an aristocrat himself—he was respected and held in affection by all parties throughout the turmoil of the French Revolution. His most important work during these years was his leading part in establishing the metric system of weights and measures. In mathematics, he tried to provide a satisfactory foundation for the basic processes of analysis, but these efforts were largely abortive. Toward the end of his life, Lagrange felt that mathematics had reached a dead end, and that chemistry, physics, biology, and other sciences would attract the ablest minds of the future. His pessimism might have been relieved if he had been able to foresee the coming of Gauss and his successors, who made the nineteenth century the richest in the long history of mathematics.

A.23

LAPLACE
(1749–1827)

Laplace is the great example of the wisdom of directing all of one's efforts to a single central objective worthy of the best that a man has in him.

E. T. Bell

Pierre Simon de Laplace was a French mathematician and theoretical astronomer who was so famous in his own time that he was known as the Newton of France. His main interests throughout his life were celestial mechanics, the theory of probability, and personal advancement.

At the age of 24 he was already deeply engaged in the detailed application of Newton's law of gravitation to the solar system as a whole, in which the planets and their satellites are not governed by the sun alone, but interact with one another in a bewildering variety of ways. Even Newton had been of the opinion that divine intervention would occasionally be needed to prevent this complex mechanism from degenerating into chaos. Laplace decided to seek reassurance elsewhere, and succeeded in proving that the ideal solar system of mathematics is a stable dynamical system that will endure unchanged for all time. This achievement was only one of the long series of triumphs recorded in his monumental treatise *Mécanique Céleste* (published in five volumes from 1799 to 1825), which summed up the work on gravitation of several generations of illustrious mathematicians. Unfortunately for his later reputation, he omitted all reference to the discoveries of his predecessors and contemporaries, and left it to be inferred that the ideas were entirely his own. Many anecdotes are associated with this work. One of the best known

171

describes the occasion on which Napoleon tried to get a rise out of Laplace by protesting that he had written a huge book on the system of the world without once mentioning God as the author of the universe. Laplace is supposed to have replied, "Sire, I had no need of that hypothesis." The principal legacy of the *Mécanique Céleste* to later generations lay in Laplace's wholesale development of potential theory, with its far-reaching implications for a dozen different branches of physical science ranging from gravitation and fluid mechanics to electromagnetism and atomic physics. Even though he lifted the idea of the potential from Lagrange without acknowledgment, he exploited it so extensively that ever since his time the fundamental equation of potential theory has been known as Laplace's equation.

His other masterpiece was the treatise *Théorie Analytique des Probabilités* (1812), in which he incorporated his own discoveries in probability from the preceding 40 years. Again he failed to acknowledge the many ideas of others he mixed in with his own; but even discounting this, his book is generally agreed to be the greatest contribution to this part of mathematics by any one person. In the introduction he says, "At bottom, the theory of probability is only common sense reduced to calculation." This may be so, but the following 700 pages of intricate analysis—in which he freely used Laplace transforms, generating functions, and many other highly nontrivial tools—has been said by some to surpass in complexity even the *Mécanique Céleste*.

After the French Revolution, Laplace's political talents and greed for position came to full flower. His compatriots speak ironically of his "suppleness" and "versatility" as a politician. What this really means is that each time there was a change of regime (and there were many), Laplace smoothly adapted himself by changing his principles—back and forth between fervent republicanism and fawning royalism—and each time he emerged with a better job and grander titles. He has been aptly compared with the apocryphal Vicar of Bray in English literature, who was twice a Catholic and twice a Protestant. The Vicar is said to have replied as follows to the charge of being a turncoat: "Not so, neither, for if I changed my religion, I am sure I kept true to my principle, which is to live and die the Vicar of Bray."

To balance his faults, Laplace was always generous in giving assistance and encouragement to younger scientists. From time to time he helped forward in their careers such men as the chemist Gay-Lussac, the traveler and naturalist Humboldt, the physicist Poisson, and—appropriately—the young Cauchy, who was destined to become one of the chief architects of 19th century mathematics.

A.24

FOURIER
(1768–1830)

The profound study of nature is the most fruitful source of mathematical discoveries.

Joseph Fourier

Joseph Fourier, an excellent mathematical physicist, was a friend of Napoleon (so far as such people have friends) and accompanied his master to Egypt in 1798. On his return he became prefect of the district of Isère in southeastern France, and in this capacity built the first real road from Grenoble to Turin. He also befriended the boy Champollion, who later deciphered the Rosetta Stone as the first long step toward understanding the hieroglyphic writing of the ancient Egyptians.

During these years he worked on the theory of the conduction of heat, and in 1822 published his famous *Théorie Analytique de la Chaleur,* in which he made extensive use of the series that now bear his name. These series were of profound significance in connection with the evolution of the concept of a function. The general attitude at that time was to call $f(x)$ a function if it could be represented by a single expression like a polynomial, a finite combination of elementary functions, a power series $\sum_{n=0}^{\infty} a_n x^n$, or a trigonometric series of the form

$$\tfrac{1}{2}a_0 + \sum_{n=1}^{\infty} (a_n \cos nx + b_n \sin nx).$$

If the graph of $f(x)$ were "arbitrary"—for example, a polygonal line with a

173

number of corners and even a few gaps—then $f(x)$ would not have been accepted as a genuine function. Fourier claimed that "arbitrary" graphs can be represented by trigonometric series and should therefore be treated as legitimate functions, and it came as a shock to many that he turned out to be right. It was a long time before these issues were completely clarified, and it was no accident that the definition of a function that is now almost universally used was first formulated by Dirichlet in 1837 in a research paper on the theory of Fourier series. Also, the classical definition of the definite integral due to Riemann was first given in his fundamental paper of 1854 on the subject of Fourier series. Indeed, many of the most important mathematical discoveries of the nineteenth century are directly linked to the theory of Fourier series, and the applications of this subject to mathematical physics have been scarcely less profound.

Fourier himself is one of the fortunate few: his name has become rooted in all civilized languages as an adjective that is well known to physical scientists and mathematicians in every part of the world.[1]

---

[1] For more details on Fourier's life and work, see John Herivel's *Joseph Fourier, the Man and the Physicist,* Oxford University Press, 1975.

A.25

GAUSS
(1777–1855)

The name of Gauss is linked to almost everything that the mathematics of our century [the nineteenth] has brought forth in the way of original scientific ideas.

L. Kronecker

Carl Friedrich Gauss was the greatest of all mathematicians and perhaps the most richly gifted genius of whom there is any record. This gigantic figure, towering at the beginning of the nineteenth century, separates the modern era in mathematics from all that went before. His visionary insight and originality, the extraordinary range and depth of his achievements, his repeated demonstrations of almost superhuman power and tenacity—all these qualities combined in a single individual present an enigma as baffling to us as it was to his contemporaries.

Gauss was born in the city of Brunswick in northern Germany. His exceptional skill with numbers was clear at a very early age, and in later life he joked that he knew how to count before he could talk. It is said that Goethe wrote and directed little plays for a puppet theater when he was 6 and that Mozart composed his first childish minuets when he was 5, but Gauss corrected an error in his father's payroll accounts at the age of 3.[1] His father was a gardener and bricklayer without either the means or the inclination to help

---

[1] See W. Sartorius von Waltershausen, *Gauss zum Gedächniss*. These personal recollections appeared in 1856, and a translation by Helen W. Gauss (the mathematician's great-granddaughter) was privately printed in Colorado Springs in 1966.

develop the talents of his son. Fortunately, however, Gauss's remarkable abilities in mental computation attracted the interest of several influential men in the community, and eventually brought him to the attention of the Duke of Brunswick. The Duke was impressed with the boy and undertook to support his further education, first at the Caroline College in Brunswick (1792–1795) and later at the University of Göttingen (1795–1798).

At the Caroline College, Gauss completed his mastery of the classical languages and explored the works of Newton, Euler, and Lagrange. Early in this period—perhaps at the age of 14 or 15—he discovered the prime number theorem, which was finally proved in 1896 after great efforts by many mathematicians (see our account of Chebyshev in Section A.31). He also invented the method of least squares for minimizing the errors inherent in observational data, and conceived the Gaussian (or normal) law of distribution in the theory of probability.

At the university, Gauss was attracted by philology but repelled by the mathematics courses, and for a time the direction of his future was uncertain. However, at the age of 18 he made a wonderful geometric discovery that caused him to decide in favor of mathematics and gave him great pleasure to the end of his life. The ancient Greeks had known ruler-and-compass constructions for regular polygons of 3, 4, 5, and 15 sides, and for all others obtainable from these by bisecting angles. But this was all, and there the matter rested for 2000 years, until Gauss solved the problem completely. He proved that a regular polygon with n sides is constructible if and only if n is the product of a power of 2 and distinct prime numbers of the form $p_k = 2^{2^k} + 1$. In particular, when $k = 0, 1, 2, 3$, we see that each of the corresponding numbers $p_k = 3, 5, 17, 257$ is prime, so regular polygons with these numbers of sides are constructible.[2]

During these years Gauss was almost overwhelmed by the torrent of ideas which flooded his mind. He began the brief notes of his scientific diary in an effort to record his discoveries, since there were far too many to work out in detail at that time. The first entry, dated March 30, 1796, states the constructibility of the regular polygon with 17 sides, but even earlier than this he was penetrating deeply into several unexplored continents in the theory of numbers. In 1795 he discovered the law of quadratic reciprocity, and as he later wrote, "For a whole year this theorem tormented me and absorbed my greatest efforts, until at last I found a proof."[3] At that time Gauss was unaware

---

[2] Details of some of these constructions are given in H. Tietze, *Famous Problems of Mathematics*, Chapter IX, Graylock Press, New York, 1965.

[3] See D. E. Smith, *A Source Book in Mathematics*, McGraw-Hill, New York, 1929, pp. 112–118. This selection includes a statement of the theorem and the fifth of eight proofs that Gauss found over a period of many years. There are probably more than 50 known today.

that the theorem had already been imperfectly stated without proof by Euler, and correctly stated with an incorrect proof by Legendre. It is the core of the central part of his famous treatise *Disquisitiones Arithmeticae,* which was published in 1801 although completed in 1798.[4] Apart from a few fragmentary results of earlier mathematicians, this great work was wholly original. It is usually considered to mark the true beginning of modern number theory, to which it is related in much the same way as Newton's *Principia* is to physics and astronomy. In the introductory pages Gauss develops his method of congruences for the study of divisibility problems and gives the first proof of the fundamental theorem of arithmetic (also called the unique factorization theorem), which asserts that every integer $n > 1$ can be expressed uniquely as a product of primes. The central part is devoted mainly to quadratic congruences, forms, and residues. The last section presents his complete theory of the cyclotomic (circle-dividing) equation, with its applications to the constructibility of regular polygons. The entire work was a gargantuan feast of pure mathematics, which his successors were able to digest only slowly and with difficulty.

In his *Disquisitiones* Gauss also created the modern rigorous approach to mathematics. He had become thoroughly impatient with the loose writing and sloppy proofs of his predecessors, and resolved that his own works would be beyond criticism in this respect. As he wrote to a friend, "I mean the word proof not in the sense of the lawyers, who set two half proofs equal to a whole one, but in the sense of the mathematician, where $\frac{1}{2}$ proof $= 0$ and it is demanded for proof that every doubt becomes impossible." The *Disquisitiones* was composed in this spirit and in Gauss's mature style, which is terse, rigorous, devoid of motivation, and in many places so carefully polished that it is almost unintelligible. In another letter he said, "You know that I write slowly. This is chiefly because I am never satisfied until I have said as much as possible in a few words, and writing briefly takes far more time than writing at length." One of the effects of this habit is that his publications concealed almost as much as they revealed, for he worked very hard at removing every trace of the train of thought that led him to his discoveries. Abel remarked, "He is like the fox, who effaces his tracks in the sand with his tail." Gauss replied to such criticisms by saying that no self-respecting architect leaves the scaffolding in place after completing his building. Nevertheless, the difficulty of reading his works greatly hindered the diffusion of his ideas; and it seems a perverse economy to save a few words or pages at the cost of many unnecessary hours of struggle and frustration by those who wish to learn.

Gauss's doctoral dissertation (1799) was another milestone in the history of mathematics. After several abortive attempts by earlier mathematicians—

---

[4] There is a translation by Arthur A. Clarke, Yale University Press, 1966.

d'Alembert, Euler, Lagrange, Laplace—the fundamental theorem of algebra was here given its first satisfactory proof. This theorem asserts the existence of a real or complex root for any polynomial equation with real or complex coefficients. Gauss's success inaugurated the age of existence proofs, which ever since have played an important part in pure mathematics. Furthermore, in this proof (he gave four altogether), Gauss appears as the earliest mathematician to use complex numbers and the geometry of the complex plane with complete confidence.[5]

The next period of Gauss's life was heavily weighted toward applied mathematics, and with a few exceptions the great wealth of ideas in his diary and notebooks lay in suspended animation.

In the last decades of the eighteenth century, many astronomers were searching for a new planet between the orbits of Mars and Jupiter, where Bode's law (1772) suggested that there ought to be one. The first and largest of the numerous minor planets known as asteroids was discovered in that region in 1801, and was named Ceres. This discovery ironically coincided with an astonishing publication by the philosopher Hegel, who jeered at astronomers for ignoring philosophy: This science (he said) could have saved them from wasting their efforts by demonstrating that no new planet could possibly exist.[6] Hegel continued his career in a similar vein, and later rose to even greater heights of clumsy obfuscation. Unfortunately the tiny new planet was difficult to see under the best of circumstances, and it was soon lost in the light of the sky near the sun. The sparse observational data made it difficult to calculate the orbit with sufficient accuracy to locate Ceres again after it had moved away from the sun. The astronomers of Europe attempted this task without success for many months. Finally, Gauss was attracted by the challenge, and with the aid of his method of least squares and his unparalleled skill at numerical computation he determined the orbit and told the astronomers where to look with their telescopes, and there it was. He had succeeded in rediscovering Ceres after all the experts had failed.

This achievement brought him fame, an increase in his pension from the Duke, and in 1807 an appointment as professor of astronomy and first director of the new observatory at Göttingen. He carried out his duties with his customary thoroughness, but, as it turned out, he disliked administrative chores, committee meetings, and all the tedious red tape involved in the business of being a professor. He also had little enthusiasm for teaching, which he regarded as a waste of his time and as essentially useless (for different

---

[5] The idea of this proof is very clearly explained by F. Klein, *Elementary Mathematics from an Advanced Standpoint*, Dover, New York, 1945, pp. 101–104.

[6] See the last few pages of "De Orbitis Planetarum," vol. 1 of Georg Wilhelm Hegel, *Sämtliche Werke*, Frommann Verlag, Stuttgart, 1965.

reasons) for both talented and untalented students. However, when teaching was unavoidable, he apparently did it superbly. One of his students was the eminent algebraist Richard Dedekind, for whom Gauss's lectures after the passage of 50 years remained "unforgettable in memory as among the finest which I have ever heard."[7] Gauss had many opportunities to leave Göttingen, but he refused all offers and remained there for the rest of his life, living quietly and simply, traveling rarely, and working with immense energy on a wide variety of problems in mathematics and its applications. Apart from science and his family—he had two wives and six children, two of whom emigrated to America—his main interests were history and world literature, international politics, and public finance. He owned a large library of about 6000 volumes in many languages, including Greek, Latin, English, French, Russian, Danish, and of course German. His acuteness in handling his own financial affairs is shown by the fact that although he started with virtually nothing, he left an estate over a hundred times as great as his average annual income during the last half of his life.

In the first two decades of the nineteenth century Gauss produced a steady stream of works on astronomical subjects, of which the most important was the treatise *Theoria Motus Corporum Coelestium* (1809). This remained the bible of planetary astronomers for over a century. Its methods for dealing with perturbations later led to the discovery of Neptune. Gauss thought of astronomy as his profession and pure mathematics as his recreation, and from time to time he published a few of the fruits of his private research. His great work on the hypergeometric series (1812) belongs to this period. This was a typical Gaussian effort, packed with new ideas in analysis that have kept mathematicians busy ever since.

Around 1820 he was asked by the government of Hanover to supervise a geodetic survey of the kingdom, and various aspects of this task—including extensive field work and many tedious triangulations—occupied him for a number of years. It is natural to suppose that a mind like his would have been wasted on such an assignment, but the great ideas of science are born in many strange ways. These apparently unrewarding labors resulted in one of his deepest and most far-reaching contributions to pure mathematics, without which Einstein's general theory of relativity would have been quite impossible.

Gauss's geodetic work was concerned with the precise measurement of large triangles on the earth's surface. This provided the stimulus that led him to the ideas of his paper *Disquisitiones generales circa superficies curvas* (1827),

---

[7] Dedekind's detailed recollections of this course are given in G. Waldo Dunnington, *Carl Friedrich Gauss: Titan of Science*, Hafner, New York, 1955, pp. 259–61. This book is useful mainly for its many quotations, its bibliography of Gauss's publications, and its list of the courses he offered (but often did not teach) from 1808 to 1854.

in which he founded the intrinsic differential geometry of general curved surfaces.[8] In this work he introduced curvilinear coordinates u and v on a surface; he obtained the fundamental quadratic differential form $ds^2 = E\,du^2 + 2F\,du\,dv + G\,dv^2$ for the element of arc length ds, which makes it possible to determine geodesic curves; and he formulated the concepts of Gaussian curvature and integral curvature.[9] His main specific results were the famous *theorema egregium*, which states that the Gaussian curvature depends only on E, F, and G, and is therefore invariant under bending; and the Gauss–Bonnet theorem on integral curvature for the case of a geodesic triangle, which in its general form is the central fact of modern differential geometry in the large. Apart from his detailed discoveries, the crux of Gauss's insight lies in the word *intrinsic*, for he showed how to study the geometry of a surface by operating only on the surface itself and paying no attention to the surrounding space in which it lies. To make this more concrete, let us imagine an intelligent two-dimensional creature who inhabits a surface but has no awareness of a third dimension or of anything not on the surface. If this creature is capable of moving about, measuring distances along the surface, and determining the shortest path (geodesic) from one point to another, then he is also capable of measuring the Gaussian curvature at any point and of creating a rich geometry on the surface—and this geometry will be Euclidean (flat) if and only if the Gaussian curvature is everywhere zero. When these conceptions are generalized to more than two dimensions, they open the door to Riemannian geometry, tensor analysis, and the ideas of Einstein.

Another great work of this period was his 1831 paper on biquadratic residues. Here he extended some of his early discoveries in number theory with the aid of a new method, his purely algebraic approach to complex numbers. He defined these numbers as ordered pairs of real numbers with suitable definitions for the algebraic operations, and in so doing laid to rest the confusion that still surrounded the subject and prepared the way for the later algebra and geometry of n-dimensional spaces. But this was only incidental to his main purpose, which was to broaden the ideas of number theory into the complex domain. He defined complex integers (now called Gaussian integers) as complex numbers $a + ib$ with a and b ordinary integers; he introduced a new concept of prime numbers, in which 3 remains prime but $5 = (1 + 2i)(1 - 2i)$ does not; and he proved the unique factorization theorem for these integers

---

[8] A translation by A. Hiltebeitel and J. Morehead was published under the title *General Investigations of Curved Surfaces* by the Raven Press, Hewlett, New York, in 1965.

[9] These ideas are explained in nontechnical language in C. Lanczos, *Albert Einstein and the Cosmic World Order*, Interscience-Wiley, New York, 1965, chap. 4.

and primes. The ideas of this paper inaugurated algebraic number theory, which has grown steadily from that day to this.[10]

From the 1830s on, Gauss was increasingly occupied with physics, and he enriched every branch of the subject he touched. In the theory of surface tension, he developed the fundamental idea of conservation of energy and solved the earliest problem in the calculus of variations involving a double integral with variable limits. In optics, he introduced the concept of the focal length of a system of lenses and invented the Gauss wide-angle lens (which is relatively free of chromatic aberration) for telescope and camera objectives. He virtually created the science of geomagnetism, and in collaboration with his friend and colleague Wilhelm Weber he built and operated an iron-free magnetic observatory, founded the Magnetic Union for collecting and publishing observations from many places in the world, and invented the electromagnetic telegraph and the bifilar magnetometer. There are many references to his work in James Clerk Maxwell's famous *Treatise on Electricity and Magnetism* (1873). In his preface, Maxwell says that Gauss "brought his powerful intellect to bear on the theory of magnetism and on the methods of observing it, and he not only added greatly to our knowledge of the theory of attractions, but reconstructed the whole of magnetic science as regards the instruments used, the methods of observation, and the calculation of results, so that his memoirs on Terrestrial Magnetism may be taken as models of physical research by all those who are engaged in the measurement of any of the forces in nature." In 1839 Gauss published his fundamental paper on the general theory of inverse square forces, which established potential theory as a coherent branch of mathematics.[11] As usual, he had been thinking about these matters for many years, and among his discoveries were the divergence theorem (also called Gauss's theorem) of modern vector analysis, the basic mean value theorem for harmonic functions, and the very powerful statement which later became known as "Dirichlet's principle" and was finally proved by Hilbert in 1899.

We have discussed the published portion of Gauss's total achievement, but the unpublished and private part was almost equally impressive. Much of this came to light only after his death, when a great quantity of material from his notebooks and scientific correspondence was carefully analyzed and included in his collected works. His scientific diary has already been mentioned. This little booklet of 19 pages, one of the most precious documents in

---

[10] See E. T. Bell, "Gauss and the Early Development of Algebraic Numbers," *National Mathematics Magazine,* vol. 18 (1944), pp. 188–204, 219–33.

[11] George Green's "Essay on the Application of Mathematical Analysis to the Theories of Electricity and Magnetism" (1828) was neglected and almost completely unknown until it was reprinted in 1846.

the history of mathematics, was unknown until 1898, when it was found among family papers in the possession of one of Gauss's grandsons. It extends from 1796 to 1814 and consists of 146 very concise statements of the results of his investigations, which often occupied him for weeks or months.[12] All of this material makes it abundantly clear that the ideas Gauss conceived and worked out in considerable detail, but kept to himself, would have made him the greatest mathematician of his time if he had published them and done nothing else.

For example, the theory of functions of a complex variable was one of the major accomplishments of nineteenth-century mathematics, and the central facts of this discipline are Cauchy's integral theorem (1827) and the Taylor and Laurent expansions of an analytic function (1831, 1843). In a letter written to his friend Bessel in 1811, Gauss explicitly states Cauchy's theorem and then remarks, "This is a very beautiful theorem whose fairly simple proof I will give on a suitable occasion. It is connected with other beautiful truths which are concerned with series expansions."[13] Thus, many years in advance of those officially credited with these important discoveries, he knew Cauchy's theorem and probably knew both series expansions. However, for some reason the "suitable occasion" for publication did not arise. A possible explanation for this is suggested by his comments in a letter to Wolfgang Bolyai, a close friend from his university years with whom he maintained a lifelong correspondence. "It is not knowledge but the act of learning, not possession but the act of getting there, which grants the greatest enjoyment. When I have clarified and exhausted a subject, then I turn away from it in order to go into darkness again." His was the temperament of an explorer who is reluctant to take the time to write an account of his last expedition when he could be starting another. As it was, Gauss wrote a great deal, but to have published every fundamental discovery he made in a form satisfactory to himself would have required several long lifetimes.

Another prime example is non-Euclidean geometry, which has been compared with the Copernican revolution in astronomy for its impact on the minds of civilized people. From the time of Euclid to the boyhood of Gauss, the postulates of Euclidean geometry were universally regarded as necessities of thought. Yet there was a flaw in the Euclidean structure that had long been a focus of attention: the so-called parallel postulate, stating that through a point not on a line there exists a single line parallel to the given line. This postulate was thought not to be independent of the others, and many had tried

---

[12] See Gauss's *Werke*, vol. X, pp. 483–574, 1917. A translation of the first year of diary entries is given on pp. 487–90 of *The History of Mathematics: A Reader,* ed. J. Fauvel and J. Gray, Macmillan Press Ltd., 1987.

[13] *Werke*, vol. VIII, p. 91, 1900.

without success to prove it as a theorem. We now know that Gauss joined in these efforts at the age of 15, and that he also failed. But he failed with a difference, for he soon came to the shattering conclusion—which had escaped all his predecessors—that the Euclidean form of geometry is not the only one possible. He worked intermittently on these ideas for many years, and by 1820 he was in full possession of the main theorems of non-Euclidean geometry (the name is due to him).[14] But he did not reveal his conclusions, and in 1829 and 1832 Lobachevsky and Johann Bolyai (son of Wolfgang) published their own independent work on the subject. One reason for Gauss's silence in this case is quite simple. The intellectual climate of the time in Germany was totally dominated by the philosophy of Kant, and one of the basic tenets of his system was the idea that Euclidean geometry is the only possible way of thinking about space. Gauss knew that this idea was totally false and that the Kantian system was a structure built on sand. However, he valued his privacy and quiet life, and held his peace in order to avoid wasting his time on disputes with the philosophers. In 1829 he wrote as follows to Bessel: "I shall probably not put my very extensive investigations on this subject [the foundations of geometry] into publishable form for a long time, perhaps not in my lifetime, for I dread the shrieks we would hear from the Boeotians if I were to express myself fully on this matter."[15]

The same thing happened again in the theory of elliptic functions, a very rich field of analysis that was launched primarily by Abel in 1827 and also by Jacobi in 1828–1829. Gauss had published nothing on this subject, and claimed nothing, so the mathematical world was filled with astonishment when it gradually became known that he had found many of the results of Abel and Jacobi before these men were born. Abel was spared this devastating knowledge by his early death in 1829, at the age of 26, but Jacobi was compelled to swallow his disappointment and go on with his work. The facts became known partly through Jacobi himself. His attention was caught by a cryptic passage in the *Disquisitiones* (Article 335), whose meaning can only be understood if one knows something about elliptic functions. He visited Gauss on several occasions to verify his suspicions and tell him about his own most recent discoveries, and each time Gauss pulled 30-year-old manuscripts out of his desk and showed Jacobi what Jacobi had just shown him. The depth of Jacobi's chagrin can readily be imagined. At this point in his life Gauss was indifferent to fame and was actually pleased to be relieved of the burden of preparing the treatise on the subject which he had long planned. After a

---

[14] Everything he is known to have written about the foundations of geometry was published in his *Werke,* vol. VIII, pp. 159–268, 1900.

[15] *Werke,* vol. VIII, p. 200. The Boeotians were a dull-witted tribe of the ancient Greeks.

week's visit with Gauss in 1840, Jacobi wrote to his brother, "Mathematics would be in a very different position if practical astronomy had not diverted this colossal genius from his glorious career."

Such was Gauss, the supreme mathematician. He surpassed the levels of achievement possible for ordinary men of genius in so many ways that one sometimes has the eerie feeling that he belonged to a higher species.

A.26

CAUCHY
(1789–1857)

His [Cauchy's] scientific production was enormous. For long periods he appeared before the Academy once a week to present a new paper, so that the Academy, largely on his account, was obliged to introduce a rule restricting the number of articles a member could request to be published in a year.

Oystein Ore

Augustin-Louis Cauchy was one of the most influential French mathematicians of the nineteenth century. He began his career as a military engineer, but when his health broke down in 1813 he followed his natural inclination and devoted himself wholly to mathematics.

In mathematical productivity Cauchy was surpassed only by Euler, and his collected works fill 27 fat volumes. He made substantial contributions to number theory and determinants; is considered to be the originator of the theory of finite groups; and did extensive work in astronomy, mechanics, optics, and the theory of elasticity.

His greatest achievements, however, lay in the field of analysis. Together with his contemporaries Gauss and Abel, he was a pioneer in the rigorous treatment of limits, continuous functions, derivatives, integrals, and infinite series. Several of the basic tests for the convergence of series are associated with his name. He also provided the first existence proof for solutions of differential equations, gave the first proof of the convergence of a Taylor series (using his form of the remainder), and was the first to feel the need for a careful study of the convergence behavior of Fourier series. However, his most important work was in the theory of functions of a complex variable, which in essence he created and which has continued to be one of the dominant branches of both pure and applied mathematics. In this field, Cauchy's integral theorem and Cauchy's integral formula are fundamental tools without which modern analysis could hardly exist.[1]

Unfortunately, his personality did not harmonize with the fruitful power of his mind. He was an arrogant royalist in politics and a self-righteous, preaching, pious believer in religion—all this in an age of republican skepticism—and most of his fellow scientists disliked him and considered him a smug hypocrite. It might be fairer to put first things first and describe him as a great mathematician who happened also to be a sincere but narrow-minded bigot.[2]

---

[1] For mathematicians, much interesting information about Cauchy's role in the early history of complex analysis is given in E. Neuenschwander, "Studies in the History of Complex Function Theory II: Interactions Among the French School, Riemann, and Weierstrass," *Bulletin of the American Mathematical Society* (*New Series*), vol. 5 (September 1981), pp. 87–105.

[2] There is a recent biography of Cauchy by B. Belhoste, *Augustin-Louis Cauchy* (tr. F. Ragland), Springer-Verlag, 1990.

A.27

ABEL
(1802–1829)

Abel has left mathematicians enough to keep them busy for 500 years.

Charles Hermite

Niels Henrik Abel was one of the foremost mathematicians of the nineteenth century and probably the greatest genius produced by the Scandinavian countries. Along with his older contemporaries Gauss and Cauchy, Abel was one of the pioneers in the development of modern mathematics, which is characterized by its insistence on rigorous proof. His career was a poignant blend of good-humored optimism under the strains of poverty and neglect, modest satisfaction in the many towering achievements of his brief maturity, and patient resignation in the face of an early death.

Abel was one of six children of a poor Norwegian country minister. His great abilities were recognized and encouraged by one of his teachers when he was only 16, and soon he was reading and digesting the works of Newton, Euler, and Lagrange. As a comment on this experience, he inserted the following remark in one of his later mathematical notebooks: "It appears to me that if one wants to make progress in mathematics, one should study the masters and not the pupils." When Abel was only 18, his father died and left the family destitute. They subsisted with the aid of friends and neighbors, and somehow the boy, helped by contributions from several professors, managed to enter the University of Oslo in 1821. His earliest researches were published in 1823, and included his solution of the classic tautochrone problem by means of an integral equation that is now known by his name. This was the first solution of an equation of this kind, and it foreshadowed the extensive development of integral equations in the late nineteenth and early twentieth centuries. He also proved that the general fifth-degree equation $ax^5 + bx^4 + cx^3 + dx^2 + ex + f = 0$ cannot be solved in terms of radicals, as is possible for equations of lower degree, and thus disposed of a problem that had baffled mathematicians for 300 years. He published his proof in a small pamphlet at his own expense.

In his scientific development Abel soon outgrew Norway, and longed to visit France and Germany. With the backing of his friends and professors, he applied to the government, and after the usual red tape and delays, he received a fellowship for a mathematical grand tour of the Continent. He spent most of his first year abroad in Berlin. Here he had the great good fortune to make the acquaintance of August Leopold Crelle, an enthusiastic mathematical amateur who became his close friend, advisor, and protector. In turn, Abel inspired Crelle to launch his famous *Journal für die Reine und Angewandte Mathematik,* which was the world's first periodical devoted wholly to mathematical research. The first three volumes contained 22 contributions by Abel.

Abel's early mathematical training had been exclusively in the older formal tradition of the eighteenth century, as typified by Euler. In Berlin he came under the influence of the new school of thought led by Gauss and Cauchy, which emphasized rigorous deduction as opposed to formal calculation. Except for Gauss's great work on the hypergeometric series, there were hardly any proofs in analysis that would be accepted as valid today. As Abel expressed it in a letter to a friend: "If you disregard the very simplest cases, there is in all of mathematics not a single infinite series whose sum has been rigorously determined. In other words, the most important parts of mathematics stand without a foundation." In this period he wrote his classic study of the binomial series, in which he founded the general theory of convergence and gave the first satisfactory proof of the validity of this series expansion.

Abel had sent to Gauss in Göttingen his pamphlet on the fifth-degree equation, hoping that it would serve as a kind of scientific passport. However, for some reason Gauss put it aside without looking at it, for it was found uncut among his papers after his death 30 years later. Unfortunately for both men, Abel felt that he had been snubbed, and decided to go on to Paris without visiting Gauss.

In Paris he met Cauchy, Legendre, Dirichlet, and others, but these meetings were perfunctory and he was not recognized for what he was. He had already published a number of important articles in Crelle's *Journal,* but the French were hardly aware yet of the existence of this new periodical, and Abel was much too shy to speak of his own work to people he scarcely knew. Soon after his arrival he finished his great *Mémoire sur une Propriété Générale d'une Classe Très Étendue des Fonctions Transcendantes,* which he regarded as his masterpiece. This work contains the discovery about integrals of algebraic functions now known as Abel's theorem, and is the foundation for the later theory of Abelian integrals, Abelian functions, and much of algebraic geometry. Decades later, Hermite is said to have remarked of this *Mémoire,* "Abel has left mathematicians enough to keep them busy for 500 years." Jacobi described Abel's theorem as the greatest discovery in integral calculus of the nineteenth century. Abel submitted his manuscript to the French Academy. He hoped that it would bring him to the notice of the French mathematicians, but he waited in vain until his purse was empty and he was forced to return to Berlin. What happened was this: The manuscript was given to Cauchy and Legendre for examination; Cauchy took it home, mislaid it, and forgot all about it; and it was not published until 1841, when again the manuscript was lost before the proof sheets were read. The original finally turned up in Florence in 1952.[1] In Berlin, Abel finished his first revolutionary article on elliptic functions, a subject he had been working on for several years, and then went back to Norway, deeply in debt.

He had expected on his return to be appointed to a professorship at the University, but once again his hopes were dashed. He lived by tutoring, and for a brief time held a substitute teaching position. During this period he worked incessantly, mainly on the theory of the elliptic functions that he had discovered as the inverses of elliptic integrals. This theory quickly took its place as one of the major fields of nineteenth century analysis, with many applications to number theory, mathematical physics, and algebraic geometry. Meanwhile, Abel's fame had spread to all the mathematical centers of Europe and he stood among the elite of the world's mathematicians, but isolated in Norway he was unaware of it. By early 1829 the tuberculosis he had contracted on his journey had progressed to the point where he was unable to work, and in the spring of that year he died, at the age of 26. As an ironic postscript, shortly after his death Crelle wrote that his efforts had been successful, and that Abel would be appointed to the chair of mathematics in Berlin.

Crelle eulogized Abel in his *Journal* as follows: "All of Abel's works carry the imprint of an ingenuity and force of thought which is amazing. One may say that he was able to penetrate all obstacles down to the very foundations of the problem, with a force of thought which appeared irresistible .... He distinguished himself equally by the purity and nobility of his character and by a rare modesty which made his person cherished to the same unusual degree as was his genius."

Mathematicians, however, have their own ways of remembering their great men, and so we speak of Abel's integral equation, Abelian integrals and

---

[1] For the details of this astonishing story, see the fine book by O. Ore, *Niels Henrik Abel: Mathematician Extraordinary.* University of Minnesota Press, 1957.

functions, Abelian groups, Abel's series, Abel's partial summation formula, Abel's limit theorem in the theory of power series, and Abel summability. Few have had their names linked to so many concepts and theorems in modern mathematics, and what he might have accomplished in a normal lifetime is beyond conjecture.

A.28

DIRICHLET
(1805–1859)

The story was told that young Dirichlet had as a constant companion on all his travels, like a devout man with his prayer book, an old, worn copy of the *Disquisitiones Arithmeticae* of Gauss.

Heinrich Tietze

Peter Gustav Lejeune Dirichlet was a German mathematician who made many contributions of lasting value to analysis and number theory. As a young man he was drawn to Paris by the reputations of Cauchy, Fourier, and Legendre, but he was most deeply influenced by his encounter and lifelong contact with Gauss's *Disquisitiones Arithmeticae* (1801). This prodigious but cryptic work contained many of the great master's far-reaching discoveries in number theory, but it was understood by very few mathematicians at that time. As Kummer later said, "Dirichlet was not satisfied to study Gauss's *Disquisitiones* once or several times, but continued throughout his life to keep in close touch with the wealth of deep mathematical thoughts which it contains by perusing it again and again. For this reason the book was never put on the shelf but had an abiding place on the table at which he worked. Dirichlet was the first one who not only fully understood this work, but also made it accessible to others." In later life Dirichlet became a friend and disciple of Gauss, and also a friend and advisor of Riemann, whom he helped in a small way with his doctoral dissertation. In 1855, after lecturing at Berlin for many years, he succeeded Gauss in the professorship at Göttingen.

One of Dirichlet's earliest achievements was a milestone in analysis: In

191

1829 he gave the first satisfactory proof that certain specific types of functions are actually the sums of their Fourier series. Previous work in this field had consisted wholly of the uncritical manipulation of formulas; Dirichlet transformed the subject into genuine mathematics in the modern sense. As a by-product of this research, he also contributed greatly to the correct understanding of the nature of a function, and gave the definition which is now most often used, namely, that y is a function of x when to each value of x in a given interval there corresponds a unique value of y. He added that it does not matter whether y depends on x according to some "formula" or "law" or "mathematical operation," and he emphasized this by giving the example of the function of x which has the value 1 for all rational x's and the value 0 for all irrational x's.

Perhaps his greatest works were two long memoirs of 1837 and 1839 in which he made very remarkable applications of analysis to the theory of numbers. It was in the first of these that he proved his wonderful theorem that there are an infinite number of primes in any arithmetic progression of the form $a + nb$, where a and b are positive integers with no common factor. His discoveries about absolutely convergent series also appeared in 1837. His important convergence test was published posthumously in his *Vorlesungen über Zahlentheorie* (1863). These lectures went through many editions and had a very wide influence.

He was also interested in mathematical physics, and formulated the so-called Dirichlet principle of potential theory, which asserts the existence of harmonic functions (functions that satisfy Laplace's equation) with prescribed boundary values. Riemann—who gave the principle its name—used it with great effect in some of his profoundest researches. Hilbert gave a rigorous proof of Dirichlet's principle in the early twentieth century.

A.29

LIOUVILLE
(1809–1882)

I would rather discover one cause than be King of Persia.

Democritus

Joseph Liouville was a highly respected professor at the Collège de France in Paris and the founder and editor of the *Journal des Mathématiques Pures et Appliquées,* a famous periodical that played an important role in French mathematical life through the latter part of the nineteenth century. His own remarkable achievements as a creative mathematican have only recently received the appreciation they deserve.[1] The fact that his collected works have never been published is an unfortunate and rather surprising oversight.

He was the first to solve a boundary value problem by solving an equivalent integral equation, a method developed by Fredholm and Hilbert in the early 1900s into one of the major fields of modern analysis. His ingenious theory of fractional differentiation answered the long-standing question of what reasonable meaning can be assigned to the symbol $d^n y / dx^n$ when n is not a positive integer. He discovered the fundamental result in complex analysis now known as *Liouville's theorem*—that a bounded entire function is necessarily a constant—and used it as the basis for his own theory of elliptic

---

[1] See Jesper Lützen, *Joseph Liouville 1809–1882: Master of Pure and Applied Mathematics,* Springer Verlag, 1990, xix + 884 pp.

functions. There is also a well-known Liouville theorem in Hamiltonian mechanics, which states that volume integrals are time-invariant in phase space. His theory of integrals of elementary functions was perhaps the most original of all his achievements, for in it he proved that such integrals as

$$\int e^{-x^2}\, dx, \qquad \int \frac{e^x}{x}\, dx, \qquad \int \frac{\sin x}{x}\, dx, \qquad \int \frac{dx}{\ln x},$$

as well as the elliptic integrals of the first and second kinds, cannot be expressed in terms of a finite number of elementary functions.

The fascinating and difficult theory of transcendental numbers is another important branch of mathematics that originated in Liouville's work. The irrationality of π and e (that is, the fact that these numbers are not roots of any linear equation $ax + b = 0$ whose coefficients are integers) had been proved in the eighteenth century by Lambert and Euler. In 1844 Liouville showed that e is also not a root of any quadratic equation with integral coefficients. This led him to conjecture that e is *transcendental*, which means that it does not satisfy any polynomial equation

$$a_n x^n + a_{n-1} x^{n-1} + \cdots + a_1 x + a_0 = 0$$

with integral coefficients. His efforts to prove this failed, but his ideas contributed to Hermite's success in 1873 and then to Lindemann's 1882 proof that π is also transcendental. Lindemann's result showed at last that the age-old problem of squaring the circle by a ruler-and-compass construction is impossible. One of the great mathematical achievements of modern times was Gelfond's 1929 proof that e^π is transcendental, but nothing is yet known about the nature of any of the numbers $\pi + e$, πe, or π^e. Liouville also discovered a sufficient condition for transcendence and used it in 1844 to produce the first examples of real numbers that are provably transcendental. One of these is

$$\sum_{n=1}^{\infty} \frac{1}{10^{n!}} = \frac{1}{10^1} + \frac{1}{10^2} + \frac{1}{10^6} + \cdots = 0.11000100\ldots.$$

His methods here have also led to extensive further work in the twentieth century.[2]

What he accomplished was certainly better than being King of Persia, or being any king or political leader whatsoever. He was a thinker whose work will live as long as people care about beautiful ideas.

---

[2] More details can be found in Section B.18.

A.30

HERMITE
(1822–1901)

Talk with M. Hermite. He never evokes a concrete image, yet you soon perceive that the most abstract entities are to him like living creatures.

Henri Poincaré

Charles Hermite, one of the most eminent French mathematicians of the nineteenth century, was particularly distinguished for the clean elegance and high artistic quality of his work. As a student, he courted disaster by neglecting his routine assigned work to study the classic masters of mathematics; and though he nearly failed his examinations, he became a first-rate creative mathematician himself while still in his early twenties. In 1870 he was appointed to a professorship at the Sorbonne, where he trained a whole generation of well-known French mathematicians, including Picard, Borel, and Poincaré.

The character of his mind is suggested by the remark of Poincaré quoted above. He disliked geometry, but was strongly attracted to number theory and analysis, and his favorite subject was elliptic functions, where these two fields touch in many remarkable ways. Earlier in the century the Norwegian genius Abel had proved that the general equation of the fifth degree cannot be solved by functions involving only rational operations and root extractions. One of Hermite's most surprising achievements (in 1858) was to show that this equation can be solved by elliptic functions.

His 1873 proof of the transcendence of e (Section B.18) was another high point of his career. If he had been willing to dig even deeper into this vein, he could probably have disposed of π as well, but apparently he'd had enough of a good thing. As he wrote to a friend, "I shall risk nothing on an attempt to prove the transcendence of the number π. If others undertake this enterprise, no one will be happier than I at their success, but believe me, my dear friend, it will not fail to cost them some efforts." As it turned out, Lindemann's proof 9 years later rested on extending Hermite's method by using complex integrals to show that no equation of the form

$$a_n e^{b_n} + \cdots + a_2 e^{b_2} + a_1 e^{b_1} + a_0 = 0$$

can be true if the b's are distinct nonzero algebraic numbers and the a's are algebraic numbers not all of which are zero. The transcendence of π now follows from Euler's equation $e^{\pi i} + 1 = 0$, since if π is algebraic, then πi is also.

Several of his purely mathematical discoveries have had unexpected applications many years later to mathematical physics. For example, the Hermitian forms and matrices which he invented in connection with certain problems of number theory turned out to be crucial for Heisenberg's 1925 formulation of quantum mechanics, and Hermite polynomials and Hermite functions are useful in solving Schrödinger's wave equation. The reason is not clear, but it seems to be true that mathematicians do some of their most valuable practical work when thinking about problems that appear to have nothing whatever to do with physical reality.[1]

---

[1] On this theme, see the article by E. P. Wigner, "The Unreasonable Effectiveness of Mathematics in the Natural Sciences," *Communications on Pure and Applied Mathematics*, vol. 13 (1960), pp. 1–14.

A.31

CHEBYSHEV
(1821–1894)

He was the only man ever able to cope with the refractory character and erratic flow of prime numbers and confine the stream of their progression within algebraic limits, building up, if I may so say, banks on either side which that stream, devious and irregular as are its windings, can never overflow. [These "banks" are the constants $\frac{7}{8}$ and $\frac{9}{8}$ in eq. (2) below].

J. J. Sylvester

Pafnuty Lvovich Chebyshev was the most eminent Russian mathematician of the nineteenth century. He was a contemporary of the famous geometer Lobachevsky (1793–1856), but his work had a much deeper influence throughout Western Europe and he is considered the founder of the great school of mathematics that has been flourishing in Russia for the past century.

As a boy he was fascinated by mechanical toys, and apparently was first attracted to mathematics when he saw the importance of geometry for understanding machines. After his student years in Moscow, he became professor of mathematics at the University of St. Petersburg, a position he held until his retirement. His father was a member of the Russian nobility, but after the famine of 1840 the family estates were so diminished that for the rest of his life Chebyshev was forced to live very frugally and he never married. He spent much of his small income on mechanical models and occasional journeys to Western Europe, where he particularly enjoyed seeing windmills, steam engines, and the like.

Chebyshev was a remarkably versatile mathematician with a rare talent for solving difficult problems by using elementary methods. Most of his effort went into pure mathematics, but he also valued practical applications of his subject, as the following remark suggests: "To isolate mathematics from the practical demands of the sciences is to invite the sterility of a cow shut away from the bulls." He worked in many fields, but his most important achievements were in probability, the theory of numbers, and the approximation of functions (to which he was led by his interest in mechanisms).

In probability, he introduced the concepts of mathematical expectation and variance for sums and arithmetic means of random variables, gave a beautifully simple proof of the law of large numbers based on what is now known as Chebyshev's inequality, and worked extensively on the central limit theorem. He is regarded as the intellectual father of a long series of well-known Russian scientists who contributed to the mathematical theory of probability, including A. A. Markov, S. N. Bernstein, A. N. Kolmogorov, A. Y. Khinchin, and others.

In the late 1840's Chebyshev helped to prepare an edition of some of the works of Euler. It appears that this task caused him to turn his attention to the theory of numbers, particularly to the very difficult problem of the distribution of primes. As the reader certainly knows, a prime number is an integer $p > 1$ that has no positive divisors except 1 and p. The first few are easily seen to be 2, 3, 5, 7, 11, 13, 17, 19, 23, 29, 31, 37, 41, 43, ... It is clear that the primes are distributed among all the positive integers in a rather irregular way; for as we move out, they seem to occur less and less frequently, and yet there are many adjoining pairs separated by a single even number. The problem of discovering the law governing their occurrence—and of understanding the reasons for it—is one that has challenged the curiosity of men for hundreds of years. In 1751 Euler expressed his own bafflement in these words: "Mathematicians have tried in vain to this day to discover some order in the sequence of prime numbers, and we have reason to believe that it is a mystery into which the human mind will never penetrate."

Many attempts have been made to find simple formulas for the nth prime and for the exact number of primes among the first n positive integers. All such efforts have failed, and real progress was achieved only when mathematicians started instead to look for information about the average distribution of the primes among the positive integers. It is customary to denote by $\pi(x)$ the number of primes less than or equal to a positive number x. Thus $\pi(1) = 0$, $\pi(2) = 1$, $\pi(3) = 2$, $\pi(\pi) = 2$, $\pi(4) = 2$, and so on. In his early youth Gauss studied $\pi(x)$ empirically, with the aim of finding a simple function that seems to approximate it with a small relative error for large x. On the basis of his observations he conjectured (perhaps at the age of 14 or 15) that $x/\log x$ is a good approximating function, in the sense that

$$\lim_{x \to \infty} \frac{\pi(x)}{x/\log x} = 1. \tag{1}$$

This statement is the famous *prime number theorem*; and as far as anyone knows, Gauss was never able to support his guess with even a fragment of proof.

Chebyshev, unaware of Gauss's conjecture, was the first mathematician to establish any firm conclusions about this question. Around 1850 he proved that

$$\frac{7}{8} < \frac{\pi(x)}{x/\log x} < \frac{9}{8} \tag{2}$$

for all sufficiently large x, and also that if the limit in (1) exists then its value must be 1. As a by-product of this work, he also proved Bertrand's postulate: for every integer $n \geq 1$ there is a prime p such that $n < p \leq 2n$. Chebyshev's efforts did not bring him to a final proof of the prime number theorem (this came in 1896), but they did stimulate many other mathematicians to continue working on the problem. We shall return to this subject in our account of Riemann in the next section, and also in Section B.16.

A.32

RIEMANN
(1826–1866)

... an extraordinary mathematician.

Salomon Bochner

No great mind of the past has exerted a deeper influence on the mathematics of the twentieth century than Bernhard Riemann, the son of a poor country minister in northern Germany. He studied the works of Euler and Legendre while he was still in secondary school, and it is said that he mastered Legendre's treatise on the theory of numbers in less than a week. But he was shy and modest, with little awareness of his own extraordinary abilities, so at the age of 19 he went to the University of Göttingen with the aim of pleasing his father by studying theology and becoming a minister himself. Fortunately, this worthy purpose soon stuck in his throat, and with his father's willing permission he switched to mathematics.

The presence of the legendary Gauss automatically made Göttingen the center of the mathematical world. But Gauss was remote and unapproachable—particularly to beginning students—and after only a year Riemann left this unsatisfying environment and went to the University of Berlin. There he attracted the friendly interest of Dirichlet and Jacobi, and learned a great deal from both men. Two years later he returned to Göttingen, where he obtained his doctor's degree in 1851. During the next 8 years he endured debilitating poverty and created his greatest works. In 1854 he was appointed Privatdozent (unpaid lecturer), which at that time was the necessary first step on the academic ladder. Gauss died in 1855, and Dirichlet was called to Göttingen as his successor. Dirichlet helped Riemann in every way he could, first with a small salary (about one-tenth of that paid to a full professor) and then with a promotion to an assistant professorship. In 1859 he also died, and Riemann was appointed as a full professor to replace him. Riemann's years of poverty were over, but his health was broken. At the age of 39 he died of tuberculosis in Italy, on the last of several trips he undertook in order to escape the cold, wet climate of northern Germany. Riemann had a short life and published comparatively little, but his works permanently altered the course of mathematics in analysis, geometry, and number theory.[1]

His first published paper was his celebrated dissertation of 1851 on the general theory of functions of a complex variable.[2] Riemann's fundamental aim here was to free the concept of an analytic function from any dependence on explicit expressions such as power series, and to concentrate instead on general principles and geometric ideas. He founded his theory on what are now called the Cauchy–Riemann equations, created the ingenious device of Riemann surfaces for clarifying the nature of multiple-valued functions, and was led to the Riemann mapping theorem. Gauss was rarely enthusiastic about the mathematical achievements of his contemporaries, but in his official report to the faculty he warmly praised Riemann's work: "The dissertation submitted by Herr Riemann offers convincing evidence of the author's thorough and penetrating investigations in those parts of the subject treated in the dissertation, of a creative, active, truly mathematical mind, and of a gloriously fertile originality."

Riemann later applied these ideas to the study of hypergeometric and Abelian functions. In his work on Abelian functions he relied on a remarkable combination of geometric reasoning and physical insight, the latter in the form of Dirichlet's principle from potential theory. He used Riemann surfaces to build a bridge between analysis and geometry which made it possible to give geometric expression to the deepest analytic properties of functions. His powerful intuition often enabled him to discover such properties—for instance, his version of the Riemann–Roch theorem—by simply thinking about possible configurations of closed surfaces and performing imaginary physical experiments on these surfaces. Riemann's geometric methods in complex analysis constituted the true beginning of topology, a rich field of geometry concerned with those properties of figures that are unchanged by continuous deformations.

---

[1] His *Gesammelte Mathematische Werke* (reprinted by Dover in 1953) occupy only a single volume, of which two-thirds consists of posthumously published material. Of the nine papers Riemann published himself, only five deal with pure mathematics.

[2] "Grundlagen für eine allgemeine Theorie der Functionen einer veränderlichen complexen Grösse," in *Werke*, pp. 3–43.

In 1854 he was required to submit a probationary essay in order to be admitted to the position of Privatdozent, and his response was another pregnant work whose influence is indelibly stamped on the mathematics of our own time.[3] The problem he set himself was to analyze Dirichlet's conditions (1829) for the representability of a function by its Fourier series. One of these conditions was that the function must be integrable. But what does this mean? Dirichlet had used Cauchy's definition of integrability, which applies only to functions that are continuous or have at most a finite number of points of discontinuity. Certain functions that arise in number theory suggested to Riemann that this definition should be broadened. He developed the concept of the Riemann integral as it now appears in most textbooks on calculus, established necessary and sufficient conditions for the existence of such an integral, and generalized Dirichlet's criteria for the validity of Fourier expansions. Cantor's famous theory of sets was directly inspired by a problem raised in this paper, and these ideas led in turn to the concept of the Lebesgue integral and even more general types of integration. Riemann's pioneering investigations were therefore the first steps in another new branch of mathematics, the theory of functions of a real variable.

The Riemann rearrangement theorem in the theory of infinite series was an incidental result in the paper just described. He was familiar with Dirichlet's example showing that the sum of a conditionally convergent series can be changed by altering the order of its terms:

$$1 - \frac{1}{2} + \frac{1}{3} - \frac{1}{4} + \frac{1}{5} - \frac{1}{6} + \frac{1}{7} - \frac{1}{8} + \cdots = \ln 2, \tag{1}$$

$$1 + \frac{1}{3} - \frac{1}{2} + \frac{1}{5} + \frac{1}{7} - \frac{1}{4} + \cdots = \frac{3}{2} \ln 2. \tag{2}$$

It is apparent that these two series have different sums but the same terms; for in (2) the first two positive terms in (1) are followed by the first negative term, then the next two positive terms are followed by the second negative term, and so on. Riemann proved that it is possible to rearrange the terms of any conditionally convergent series in such a manner that the new series will converge to an arbitrary preassigned sum or diverge to ∞ or $-\infty$.

In addition to his probationary essay, Riemann was also required to present a trial lecture to the faculty before he could be appointed to his unpaid lectureship. It was the custom for the candidate to offer three titles, and the head of his department usually accepted the first. However, Riemann rashly listed as his third topic the foundations of geometry, a profound subject on

---

[3] "Ueber die Darstellbarkeit einer Function durch eine trigonometrische Reihe," in *Werke,* pp. 227–64.

which he was unprepared but which Gauss had been turning over in his mind for 60 years. Naturally, Gauss was curious to see how this particular candidate's "gloriously fertile originality" would cope with such a challenge, and to Riemann's dismay he designated this as the subject of the lecture. Riemann quickly tore himself away from his other interests at the time—"my investigations of the connection between electricity, magnetism, light, and gravitation"—and wrote his lecture in the next two months. The result was one of the great classical masterpieces of mathematics, and probably the most important scientific lecture ever given.[4] It is recorded that even Gauss was surprised and enthusiastic.

Riemann's lecture presented in nontechnical language a vast generalization of all known geometries, both Euclidean and non-Euclidean. This field is now called Riemannian geometry, and apart from its great importance in pure mathematics, it turned out 60 years later to be exactly the right framework for Einstein's general theory of relativity. Like most of the great ideas of science, Riemannian geometry is quite easy to understand if we set aside the technical details and concentrate on its essential features. Let us recall the intrinsic differential geometry of curved surfaces which Gauss had discovered 25 years earlier. If a surface imbedded in three-dimensional space is defined parametrically by three functions $x = x(u, v)$, $y = y(u, v)$, and $z = z(u, v)$, then u and v can be interpreted as the coordinates of points on the surface. The distance ds along the surface between two nearby points (u, v) and $(u + du, v + dv)$ is given by Gauss's quadratic differential form

$$ds^2 = E\,du^2 + 2F\,du\,dv + G\,dv^2,$$

where E, F, and G are certain functions of u and v. This differential form makes it possible to calculate the lengths of curves on the surface, to find the geodesic (or shortest) curves, and to compute the Gaussian curvature of the surface at any point—all in total disregard of the surrounding space. Riemann generalized this by discarding the idea of a surrounding Euclidean space and introducing the concept of a continuous n-dimensional manifold of points $(x_1, x_2, \ldots, x_n)$. He then imposed an arbitrarily given distance (or metric) ds between nearby points

$$(x_1, x_2, \ldots, x_n) \qquad \text{and} \qquad (x_1 + dx_1, x_2 + dx_2, \ldots, x_n + dx_n)$$

by means of a quadratic differential form

$$ds^2 = \sum_{i,j=1}^{n} g_{ij}\,dx_i\,dx_j, \tag{3}$$

---

[4] "Ueber die Hypothesen, welche der Geometrie zu Grunde liegen," in *Werke*, pp. 272–86. There is a translation in D. E. Smith, *A Source Book in Mathematics*, McGraw-Hill, New York, 1929.

where the g_{ij}'s are suitable functions of $x_1, x_2, \ldots, x_n$ and different systems of g_{ij}'s define different Riemannian geometries on the manifold under discussion. His next steps were to examine the idea of curvature for these Riemannian manifolds and to investigate the special case of constant curvature. All of this depends on massive computational machinery, which Riemann mercifully omitted from his lecture but included in a posthumous paper on heat conduction. In that paper he explicitly introduced the Riemann curvature tensor, which reduces to the Gaussian curvature when $n = 2$ and whose vanishing he showed to be necessary and sufficient for the given quadratic metric to be equivalent to a Euclidean metric. From this point of view, the curvature tensor measures the deviation of the Riemannian geometry defined by formula (3) from Euclidean geometry. Einstein has summarized these ideas in a single statement: "Riemann's geometry of an n-dimensional space bears the same relation to Euclidean geometry of an n-dimensional space as the general geometry of curved surfaces bears to the geometry of the plane."

The physical significance of geodesics appears in its simplest form as the following consequence of Hamilton's principle in the calculus of variations: If a particle is constrained to move on a curved surface, and if no force acts on it, then it glides along a geodesic. A direct extension of this idea is the heart of the general theory of relativity, which is essentially a theory of gravitation. Einstein conceived the geometry of space as a Riemannian geometry in which the curvature and geodesics are determined by the distribution of matter; in this curved space, planets move in their orbits around the sun by simply coasting along geodesics instead of being pulled into curved paths by a mysterious force of gravity whose nature no one has ever really understood.

In 1859 Riemann published his only work on the theory of numbers, a brief but exceedingly profound paper of less than 10 pages devoted to the prime number theorem.[5] This mighty effort started tidal waves in several branches of pure mathematics, and its influence will probably still be felt a thousand years from now. His starting point was a remarkable identity discovered by Euler over a century earlier: If s is a real number greater than 1, then

$$\sum_{n=1}^{\infty} \frac{1}{n^s} = \prod_{p} \frac{1}{1 - (1/p^s)}, \tag{4}$$

where the expression on the right denotes the product of the numbers $(1 - p^{-s})^{-1}$ for all primes p. To understand how this identity arises, we note

---

[5] "Ueber die Anzahl der Primzahlen unter einer gegebenen Grösse," in *Werke*, pp. 145–53. See the statement of the prime number theorem in Section B.16.

that $1/(1-x) = 1 + x + x^2 + \cdots$ for $|x| < 1$, so for each p we have

$$\frac{1}{1-(1/p^s)} = 1 + \frac{1}{p^s} + \frac{1}{p^{2s}} + \cdots.$$

On multiplying these series for all primes p and recalling that each integer $n > 1$ is uniquely expressible as a product of powers of different primes, we see that

$$\prod_p \frac{1}{1-(1/p^s)} = \prod_p \left(1 + \frac{1}{p^s} + \frac{1}{p^{2s}} + \cdots\right)$$

$$= 1 + \frac{1}{2^s} + \frac{1}{3^s} + \cdots + \frac{1}{n^s} + \cdots$$

$$= \sum_{n=1}^{\infty} \frac{1}{n^s},$$

which is the identity (4). The sum of the series on the left of (4) is evidently a function of the real variable $s > 1$, and the identity establishes a connection between the behavior of the function and properties of the primes. Euler himself exploited this connection in several ways, but Riemann perceived that access to the deeper features of the distribution of primes can be gained only by allowing s to be a complex variable. He denoted the resulting function by $\zeta(s)$, and it has since been known as the Riemann zeta function:

$$\zeta(s) = 1 + \frac{1}{2^s} + \frac{1}{3^s} + \cdots, \qquad s = \sigma + it.$$

In his paper he proved several important properties of this function, and in a sovereign way simply stated a number of others without proof. During the century since his death, many of the finest mathematicians in the world have exerted their strongest efforts and created rich new branches of analysis in attempts to prove these statements. The first success was achieved in 1893 by J. Hadamard, and with one exception every statement has since been settled in the sense Riemann expected.[6] This exception is the famous Riemann hypothesis: that all the zeros of $\zeta(s)$ in the strip $0 \le \sigma \le 1$ lie on the central line $\sigma = \frac{1}{2}$. It stands today as the most important unsolved problem of mathematics, and is probably the most difficult problem that the human mind has ever conceived. In a fragmentary note found among his posthumous papers, Riemann wrote that these theorems "follow from an expression for the function $\zeta(s)$ which I

---

[6] Hadamard's work led him to his 1896 proof of the prime number theorem. See E. C. Titchmarsh, *The Theory of the Riemann Zeta Function*, Oxford University Press, London, 1951, Chapter 3. This treatise has a bibliography of 326 items.

have not yet simplified enough to publish."[7] Writing about this fragment in 1944, Hadamard remarked with justified exasperation, "We still have not the slightest idea of what the expression could be."[8] He adds the further comment: "In general, Riemann's intuition is highly geometrical; but this is not the case for his memoir on prime numbers, the one in which that intuition is the most powerful and mysterious."

---

[7] *Werke*, p. 154.

[8] *The Psychology of Invention in the Mathematical Field*, Dover, New York, 1954, p. 118.

A.33

WEIERSTRASS
(1815–1897)

During the last third of the 19th century the greatest analyst of the time and the world's foremost teacher of advanced mathematics were both the same man: Karl Weierstrass in Berlin. His career was also remarkable in another way—and a consolation to all "late-starters"—for he began the solid part of his professional life at the age of almost 40, when most mathematicians are long past their creative years.

He was an excellent student in school, and his father—himself a customs officer—decided that the obvious thing to do with such a bright boy was send him to the University of Bonn to quality for the higher ranks of the Prussian civil service by studying law and commerce. But Karl had no interest in these subjects. He parried his father's insistence by becoming a champion beer-drinker—in a country of champions—and also a superb fencer, where his long reach, great physical strength and lightning reflexes kept him from acquiring a single scar or losing a drop of blood. He rarely attended lectures, got poor grades, and returned home without a degree. His father was furious; but in the end Karl's misspent youth was very well spent indeed, for it shielded him from law and commerce and saved him for mathematics. In order to earn his living he made a fresh start by becoming a secondary school teacher and teaching mathematics, physics, botany, German, penmanship and gymnastics to the children of several small Prussian towns.

Throughout those 15 years as a teacher Weierstrass lived a double life, which was possible only for a strong man with a will of iron. During the days he was just what everyone saw: an able and caring teacher dedicated to the interests of his pupils, and also an amiable and welcome guest in the homes of the young doctors, lawyers and merchants of the towns where he lived. But during the nights he had other intellectual companions from the past, particularly the great Norwegian mathematician Abel—dead in 1829 at the early age of 26—whose pioneering work he continued and developed. His remarkable research on Abelian functions was carried on for years without the knowledge of another living soul; he didn't discuss it with anyone at all, or submit it for publication in the mathematical journals of the day. He would later say that the works of Abel were never far from his hand, and exclaim with admiration and without a trace of envy: "Abel, the lucky fellow! He has done something everlasting! His ideas will always have a fertilizing influence on our science."

He spoke to no one about his work, but on rare occasions a small clue would appear. One day when he missed the beginning of his morning classes, the school principal hurried to his home and found Weierstrass working away by lamplight with the curtains still drawn, completely oblivious to the fact that a new day had dawned. Most scientists and writers have felt a compulsion at least once in their lives to work all through the night, and they know the state of mind: in the words of the old Armenian proverb, When the spring is flowing, keeping it going!

Weierstrass reminisced about this part of his life many years later when writing to a friend: "The infinite emptiness and boredom of those years would have been unendurable without the hard work that made me a recluse—even if I was rated rather a good fellow by the circle of my friends among the Junkers, lawyers, and young officers of the community . . . . The present offered nothing worth mentioning, and it was not my custom to speak of the future."[1]

All this changed in 1854 when Weierstrass at last published an account of

---

[1] There is a striking resemblance between Weierstrass at this time in his life and the Italian political philosopher Machiavelli (1469–1527), who wrote his great book *The Prince* (1513) during the night hours of his exile to his isolated farm—after prison and torture—from his influential position as Secretary of State near the center of Florentine political power. In one of the most remarkable letters in any language, he wrote to his friend Vettori that he spent his days in country occupations: setting bird-traps, inspecting the work of his woodcutters, gossiping at a nearby inn, having his midday meal, and playing cards at the inn with local workmen. But then:

> When evening comes, I return to the house and go into my study; and at the door I take off my country clothes, all caked with mud and slime, and put on court dress; and, when I am thus decently re-clad, I enter into the ancient mansions of the men of ancient days [the books of the writers of antiquity]. And there I am received by my hosts with all lovingkindness, and I feast myself on that food which alone is my true nourishment, and which I was born for.

his research on Abelian functions in Crelle's internationally known *Journal für die Reine und Angewandte Mathematik* ("Journal for Pure and Applied Mathematics"), whose early volumes were filled with Abel's works. This paper, a masterpiece from an unknown schoolteacher in an obscure small town, caused a sensation among mathematicians and brought immediate concrete recognition. An alert professor at the University of Königsberg knew enough to understand and appreciate what Weierstrass had accomplished, and persuaded his university to award him an honorary doctor's degree. The Ministry of Education granted him a year's leave of absence with pay to continue his research, and the next year he was appointed to the University of Berlin, where he remained the rest of his life.

Despite his late start, Weierstrass was widely recognized as the leading analyst of the world during the remaining 35 years of his active career—the "father of modern analysis." His many fundamental contributions are difficult to date because he published little in his later life, preferring instead to make his ideas known through lectures and the spreading influence of his students. For instance, his famous example of a continuous, nowhere-differentiable function was first discussed in his lectures of 1861, but didn't appear in print until it was published in 1874 by one of his students—with his knowledge and permission.

Weierstrass's great creative talents were evenly divided between his thinking and his teaching. His lectures were carefully prepared and continually revised. Each lecture was a creative adventure for himself and his students alike, for his courses were always important new mathematics in the process of being born. The student notes of his lectures, and copies of these notes, and copies of copies, were passed from hand to hand throughout Europe and even America. Like Gauss he was indifferent to fame, but unlike Gauss he endeared himself to generations of students by the generosity with which he encouraged them to develop and publish, and receive credit for, ideas and theorems that he essentially originated himself. The new Weierstrassian analysis was spread throughout the world by the writings and research of such students and followers as Cantor, Schwarz, Hölder, Mittag-Leffler, Sonia Kovalevsky (Weierstrass's favorite student), Hilbert (the greatest mathematician of the next generation), the physicist Max Planck, and many others.

Weierstrass's lectures of 1879 on the calculus of variations initiated a new era of precise reasoning in this subject. One of his students was Oskar Bolza,

---

Footnote 1 continued

The historian Arnold J. Toynbee writes about these events as follows: "In those magic hours of catharsis when he rose above his vexation of spirit, Machiavelli succeeded in transmuting his 'practical' energies into a series of mighty intellectual works." See Toynbee's *A Study of History*, vol. 3, pp. 299–310. And so it was with Weierstrass.

who later became a professor at the newly founded University of Chicago. Bolza's lectures and students over many years were chiefly responsible for developing the important Chicago school of the calculus of variations, which was still influential in the 1930s. Another student of Weierstrass was Willard Gibbs of Yale University, possibly the greatest of all native-born American physical scientists. Gibbs created the science of physical chemistry almost single-handed, and a large part of the American chemical industry (about one-fourth the total industry of the United States) depends directly on his theoretical discoveries. His contributions to thermodynamics and statistical mechanics were at least as important as those of any other physicist. The so-called Bose–Einstein statistics and Fermi–Dirac statistics of modern quantum mechanics are merely modified versions of the fundamental ideas of Gibbs. In mathematics, the modern form of vector analysis is primarily due to Gibbs, and he discovered the "Gibbs phenomenon" in the theory of Fourier series. He studied analysis under Weierstrass in 1868, and doubtless acquired from him some of the mathematical power and precision that energized his theoretical works on physics and chemistry.[2]

To gain a feeling for the nature of Weierstrass's contributions to mathematics, it may be helpful to recall that his subject—"mathematical analysis," or briefly "analysis"—consists of calculus together with all other subjects that arise from calculus: advanced calculus, differential equations, the calculus of variations, infinite series (power series and Fourier series), real analysis (the theory of functions of a real variable), complex analysis (the theory of functions of a complex variable), and so on. Analysis has been the dominant branch of mathematics for 300 years—the others being geometry, number theory and algebra, with many overlaps, of course—and Weierstrass enriched every part of it. To see how, let us begin where he began, with the elliptic functions discovered by Abel.

If $P(t)$ denotes a polynomial in the variable t, then an integral of the type

$$\int_0^x \frac{dt}{\sqrt{P(t)}}$$

---

[2] Gibbs's many-hundred-page papers of the late 1870's stand in much the same relation to chemical thermodynamics as Newton's *Principia* does to mechanics and astronomy. The importance of his profound scientific work was so little understood or appreciated at Yale that for ten years he taught as a professor without any salary, subsisting on his small inherited income. He was offered a pittance, and accepted it, when there was a slight possibility that he might move to Johns Hopkins. There is a delicious story that when the eminent German physicist Helmholtz visited Yale in 1893 and immediately asked to meet the great scientist Gibbs, the university administrators who were escorting him looked at one another in dismay and embarrassment and said, "Who?" For more information see L. P. Wheeler, *Josiah Willard Gibbs: The History of a Great Mind*, Yale University Press, 1962; E. B. Wilson, "Reminiscences of Gibbs by a Student and Colleague," *Bulletin of the American Mathematical Society*, vol. 37 (1931), pp. 401–16; and C. Truesdell, *An Idiot's Fugitive Essays on Science*, Springer-Verlag, 1984, p. 415.

is called an *elliptic integral* if $P(t)$ is of the third or fourth degree.[3] If $P(t)$ is only a second degree polynomial, the integral can be worked out in terms of the elementary functions of calculus; for example,

$$\int_0^x \frac{dt}{\sqrt{1-t^2}} = \sin^{-1} x.$$

Now, if we write

$$y = \int_0^x \frac{dt}{\sqrt{1-t^2}} = \sin^{-1} x, \tag{1}$$

then by inversion we have

$$x = \sin y. \tag{2}$$

Thus, the inversion of the integral (1)—that is, the interchange of the roles of x and y—produces the sine, which is the basic function of trigonometry. It is much easier and more natural to develop trigonometry by using the direct function $x = \sin y$ as the starting point than by using the inverse function $y = \sin^{-1} x$. In particular, (2) is periodic but (1) is not: $\sin(y + 2\pi) = \sin y$. The point of view we are trying to stress is that trigonometry, that is, the theory of the trigonometric functions, can be thought of as produced by inverting the integral (1). Elliptic integrals

$$y = \int_0^x \frac{dt}{\sqrt{P(t)}}, \qquad P(t) \text{ 3rd or 4th degree}, \tag{3}$$

arise in the problem of finding the circumference of an ellipse—hence their name—and also in many other geometric and physical problems. Many mathematicians of the 18th and early 19th centuries labored over elliptic integrals, trying to calculate them (without success), classifying them, and writing treatises on them. All this activity came to a sudden end in 1827 when Abel thought of inverting the elliptic integrals (3) and thereby discovered the *elliptic functions*

$$x = \phi(y).$$

Of course, this inversion was a natural thing to do *after* Abel had thought of it. His first major discovery about these new higher transcendental functions was their double periodicity: that is, there exist two complex numbers a and b whose ratio is not real, such that $\phi(y + a) = \phi(y)$ and $\phi(y + b) = \phi(y)$. Thus, *the elliptic functions are doubly periodic*, whereas the familiar trigonometric and exponential functions are only singly periodic [$\sin(y + 2\pi) = \sin y$, $e^{(y+2\pi i)} = e^y$, the periods being 2π and $2\pi i$, respectively]. Abel rapidly

---

[3] Strictly speaking, elliptic integrals are those in which the integrand is any rational function of t and $\sqrt{P(t)}$, but this technicality can be ignored in the present discussion.

developed the theory of these new functions, and in a paper written just at the end of his life was beginning to study integrals of the form (3) in which $P(t)$ can have degree higher than 4. Integrals of this kind are now called *Abelian integrals,* and their inverses *Abelian functions.*

Weierstrass took up the study of Abelian functions as his life's work, and in the process rebuilt the whole of complex analysis from the ground up. He independently rediscovered Cauchy's Integral Theorem and Formula—which are the foundations of complex analysis—and found the Laurent Expansion before Laurent; created the concept of uniform convergence for the study of power series, and formulated the Weierstrass M-test to put this concept into effect; gave the definition of a complete analytic function by means of power series used to carry out his concept of analytic continuation; and much else. He was especially interested in entire functions, and proved the important Weierstrass Factor Theorem for these functions, which at last provided a satisfactory context for Euler's infinite product for the sine (see Section B.14). Weierstrass's version of complex analysis was characterized by careful logic and rigorous proof, in contrast to Riemann's geometric and highly intuitive approach to the same ideas.[4]

The quality that came to be known as "Weierstrassian rigor" was particularly visible in his contributions to the foundations of real analysis. He refused to accept any statement as "intuitively (or geometrically) obvious," but instead demanded ironclad proof based on explicit properties of the real numbers. He would have agreed fully with the words of Gauss if he had known them (Gauss wrote them in a letter to a friend): "I mean the word proof not in the sense of the lawyers, who set two half proofs equal to a whole one, but in the sense of the mathematician, where $\frac{1}{2}$ proof $= 0$ and it is demanded for proof that every doubt becomes impossible."

For example, most people are willing to accept the following proposition without detailed proof: If a continuous function defined on an interval is sometimes positive and sometimes negative, then it must have the value 0 at some point. Most people, but not Weierstrass. He wanted a proof that would reveal the way this property of functions depends on the structure of the real number system. He provided such a proof, and the result was the now-standard Weierstrass Intermediate Value Theorem. There is also the Weierstrass Extreme Value Theorem, which asserts that a continuous function defined on a closed and bounded interval is bounded and actually assumes maximum and minimum values. The careful reasoning required for these

---

[4] Mathematicians will find much of interest on this subject in E. Neuenschwander, "Studies in the History of Complex Function Theory II: Interactions Among the French School, Riemann, and Weierstrass," *Bulletin of the American Mathematical Society (New Series),* vol. 5 (September 1981), pp. 87–105.

proofs was founded on a crucial property of the real numbers now known as the Bolzano–Weierstrass Theorem.[5] In order to guarantee the soundness of the whole structure, Weierstrass even undertook the task of providing a full construction of the real numbers beginning with the positive integers, a boring process full of mathematical morality but devoid of excitement, or even interest—except possibly to a logician. Weierstrass was not a man to be satisfied with half-measures.

In 1885 he published the famous theorem now called the Weierstrass Approximation Theorem, which asserts that any function continuous on a closed and bounded interval can be uniformly approximated on that interval by polynomials to any given degree of accuracy. This theorem has since found a place in the theoretical foundations of virtually every part of analysis. It has also been given a far-reaching generalization, with many applications, by the modern American mathematician M. H. Stone.[6]

The Weierstrassian program is usually described as "the arithmetization of analysis." Its basic purpose was to confine mathematical proofs to precise reasoning depending only on numbers, by increasing awareness of the fallibility of geometric intuition. Why was this considered necessary?

Almost all mathematicians from Newton to the time of Weierstrass pictured the continuous motions and continuous functions they worked with by means of continuous curves drawn by a continuous sweep of the hand holding a pen or pencil. It seemed obvious to these mathematicians that such a

---

[5] Bernard Bolzano (1781–1848), a major precursor of the revolution in analysis of the late 1800s, was a Bohemian priest who was accused of heresy and in 1820 was dismissed from his position as professor of comparative religion at the University of Prague. He was also a severe critic of the philosophy of Kant. In 1817 he published a paper in which he proved the continuity of polynomial functions and attempted a proof of the Intermediate Value Theorem. This latter proof was not satisfactory because there did not exist at that time a clear concept of real numbers on which to base it; however, the attempt demonstrated his exceptional acuteness of mind in realizing that the validity of the theorem cannot be merely referred to the fact that geometrically it is intuitively obvious. His *Paradoxes of the Infinite* (1850), published two years after his death by a student he had befriended, and unnoticed for about twenty years, contained many foreshadowings of the modern theory of sets. He was quite comfortable with Galileo's paradox that there are exactly as many perfect squares as there are positive integers altogether (the correspondence $n \leftrightarrow n^2$ is one-to-one both ways) and gave many other examples of correspondences between the elements of an infinite set and a proper subset of itself. What puzzled Galileo and interested Bolzano was later elevated by Dedekind (a student of Gauss) and his friend Cantor into the definition of an infinite set: a set is infinite if it can be put into one-to-one correspondence with a proper subset of itself; otherwise, it is finite. These ideas were used by Cantor (1845–1918) to create his great and beautiful theory of infinite cardinal numbers, which is an indispensable part of modern mathematics (see pp. 31–43 of George F. Simmons, *Introduction to Topology and Modern Analysis,* McGraw-Hill, 1963). Weierstrass and Bolzano march arm-in-arm together through the history of mathematics, but not nearly enough seems to be known about Bolzano.

[6] See Simmons, *loc. cit.,* pp. 153–62.

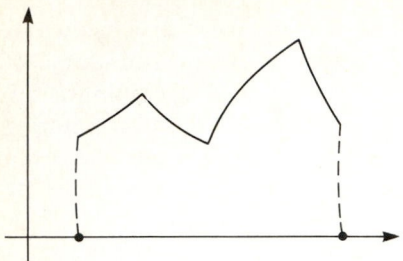

FIGURE A.49

continuous curve must be "differentiable"—that is, have a definite direction or a definite tangent—at every point, or perhaps at every point with a finite number of exceptions (Fig. A.49). The mathematical world was profoundly shocked when Weierstrass produced an example of a continuous function without a derivative at any point at all, a so-called nowhere-differentiable function. Such a function has a graph with many strange features: it has nowhere a tangent, that is, nowhere a direction; its length between any two points is infinite; it has infinitely many infinitely small crinkles; we can plot individual points but we cannot speak of the shape of the curve, and no effort of imagination can picture its graph. A curve with these properties seemed impossible to conceive, and yet there it was! Another even more shattering discovery lay ahead. One of the attributes of a continuous curve had always been its "thinness" or "one-dimensionality." However, in 1890 the Italian mathematician Peano violated common sense even more drastically by constructing a continuous curve that completely fills a square. Since that time many additional space-filling curves have been constructed that are much simpler than Peano's original specimen, but none of them are simple in any reasonable meaning of the word.

The mind quails before these monsters of modern analysis. The English novelist and social critic George Orwell said that "There are some ideas so absurd only an intellectual could believe them." Of course he meant this in a very different sense, because these absurd ideas about continuous curves are true, and they dragged the mathematical world against its will (kicking and screaming) into the era of Weierstrassian rigor.[7]

---

[7] An excellent discussion of these ideas, with pictures, can be found in a lecture by the Austrian mathematician Hans Hahn, "The Crisis in Intuition," pp. 1956–76 of *The World of Mathematics* (ed. James R. Newman), Simon and Schuster, 1956. Weierstrass's example and others are treated in detail in E. C. Titchmarsh, *The Theory of Functions*, 2nd ed., Oxford University Press, 1939, pp. 350–54; and Béla Sz.-Nagy, *Introduction to Real Functions and Orthogonal Expansions*, Oxford University Press, 1965, pp. 101–103.

PART
B

MEMORABLE MATHEMATICS

It is difficult to give an idea of the vast scope of modern mathematics. The word "scope" is not the best; I have in mind an expanse swarming with beautiful details, not the uniform expanse of a bare plain, but a region of a beautiful country, first seen from a distance, but worthy of being surveyed from one end to the other and studied even in its smallest details: its valleys, streams, rocks, woods and flowers.

Arthur Cayley

B.1

The word *geometry* comes from two Greek words meaning "earth" and "measure," which suggests that the subject had some ancient connection with surveying land. The Greek historian Herodotus, who visited Egypt about 450 B.C., tells us that the annual overflow of the river Nile wiped out many boundaries of farmers' fields, and he speculates that geometry may have arisen from the need of the rulers to re-establish these boundaries for the purposes of taxation. The trained surveyors among the Egyptians were called "rope-stretchers," because their main measuring tool was a rope with equally spaced knots.

It would have been quite easy for these surveyors to use such a rope with 11 knots to construct a line *BC* perpendicular to a given line *AB* (Fig. B.1), as follows. If the two ends of the rope are tied together to make an additional knot at *C*, and the closed loop is pulled taut by pegs at *A*, *B* and *C* into a triangle as shown in the figure, then *BC* is visibly perpendicular to *AB*, that is, *ABC* is a right triangle. The word "visibly" here means that the reasoning is based on a simple procedure and works in the following way: If the loop is flipped over and the vertex *A* is placed at *A'* on a sighted straight line with *AB*, and the loop is pulled taut into a triangle again, we find that *BC* seems to be in exactly the same position as before. We conclude that angles *ABC* and *A'BC* are equal, so *ABC* is a right angle (half a straight angle) and *BC* is perpendicular to *AB*. With the advantage of our modern educational background, we know as a theoretical fact that *ABC* is *exactly* a right triangle,

217

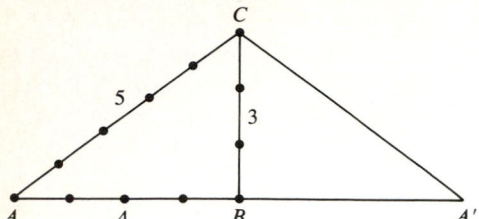

FIGURE B.1

because $3^2 + 4^2 = 5^2$ and we know about the Pythagorean theorem and its converse, but the Egyptians would have known this only as an experimental fact, based on the procedure just described.

There is no documentary evidence to confirm that this is the way Egyptian surveyors actually constructed right angles to aid them in laying out fields, but it could be true. In any event, there is a great gulf between the Egyptians with their ropes and knots, and the Greek geometers with their abstract patterns of lines and proofs of theorems. In Section A.1 we offer some ideas about how the Greeks, in particular Thales, might have crossed this gulf.

Our concern just now is with the Pythagorean theorem itself, which states that if a triangle is a right triangle with hypotenuse c and sides a and b (Fig. B.2), then the square of the hypotenuse equals the sum of the squares of the sides: $c^2 = a^2 + b^2$. Here we have spoken of the square *of* the hypotenuse, meaning the algebraic square of the numerical length of the hypotenuse. Sometimes the theorem is expressed by saying that the square *on* the hypotenuse equals the sum of the squares *on* the sides. In this form we visualize actual geometric squares placed on the hypotenuse and sides (Fig. B.3) and we think of the theorem as stating that the area of the large square equals the sum of the areas of the two smaller squares. Either way we think of

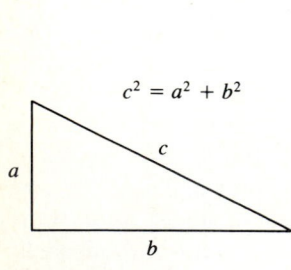

FIGURE B.2

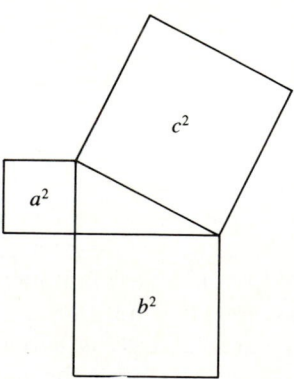

FIGURE B.3

this theorem, it is certainly one of the two or three most important in the whole of mathematics. It is not merely a bizarre item of curiosity at all, but rather an indispensable tool of thought without which geometry, analytic geometry and calculus—and with them modern physical science and engineering—could hardly exist. Even though the Pythagorean theorem is officially part of elementary geometry, this theorem always remains close at hand in our daily work no matter how far we may go in the study of higher mathematics, whether into advanced geometry, or analysis, or even number theory. Our purpose here is to give several of the simpler and more interesting proofs that have been found over the centuries, for with the greatest theorems of mathematics one proof is never enough; we always seek new proofs, new lines of thought, new points of view that may shed additional light on the object of our interest—as when a jeweler examines a cut diamond by turning it this way and that under his lamp.

Proof 1. Perhaps the simplest proof is obtained by arranging four replicas of the triangle in the corners of a square of side $a + b$, as shown in Fig. B.4. The inner figure is easily seen to be a square, and the proof is completed by writing down the fact that the area of the large square equals 4 times the area of the triangle plus the area of the small square:

$$(a + b)^2 = 4(\tfrac{1}{2}ab) + c^2,$$

so

$$a^2 + 2ab + b^2 = 2ab + c^2$$

and therefore

$$a^2 + b^2 = c^2.$$

Proof 2. Another very simple algebraic proof along similar lines can be made by arranging four replicas of the triangle inside a square of side c, as shown in

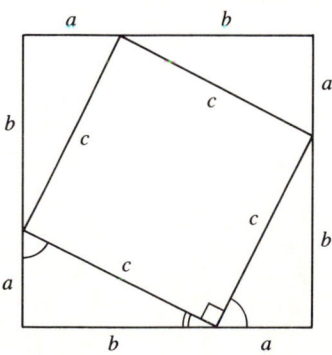

FIGURE B.4

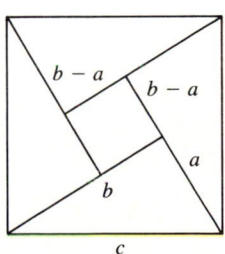

FIGURE B.5

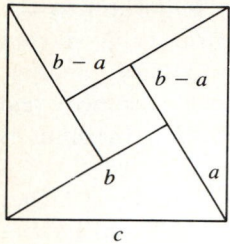 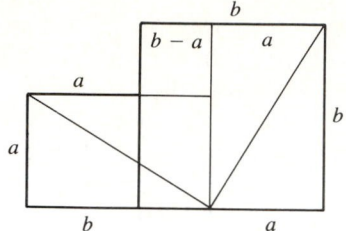

FIGURE B.6

Fig. B.5, so that the square in the middle has side $b - a$. Again the area of the large square equals 4 times the area of the triangle plus the area of the small square:

$$c^2 = 4(\tfrac{1}{2}ab) + (b - a)^2$$
$$= 2ab + b^2 - 2ab + a^2$$
$$= b^2 + a^2.$$

Proof 3. The next proof was discovered by the Indian mathematician Bhaskara in the 12th century. To construct this argument we dissect the square on the left in Fig. B.6 (the same square as in Fig. B.5) and rearrange the pieces as shown on the right, to make up two squares of sides a and b, so that

$$c^2 = a^2 + b^2.$$

It is said that Bhaskara presented his proof by simply drawing Fig. B.6 and speaking the single word "Behold!"

Proof 4. Bhaskara gave a second proof by turning the triangle around as shown in Fig. B.7. If we draw the indicated altitude and use similar triangles, we have

$$\frac{b}{c} = \frac{d}{b} \quad \text{and} \quad \frac{a}{c} = \frac{e}{a}$$

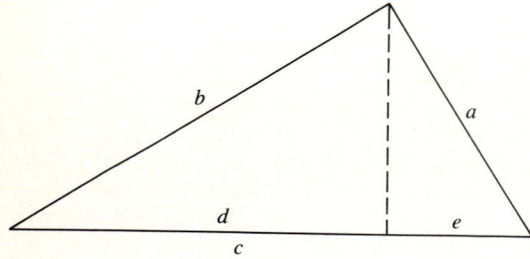

FIGURE B.7

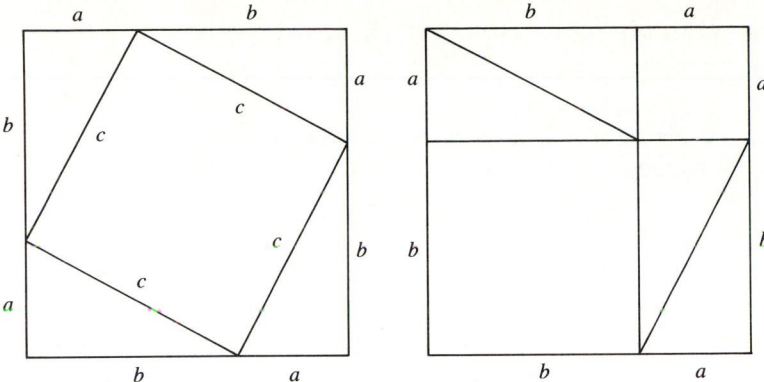

FIGURE B.8

or

$$b^2 = cd \quad \text{and} \quad a^2 = ce.$$

By adding we now obtain

$$b^2 + a^2 = cd + ce = c(d + e) = c^2.$$

This proof was rediscovered in the 17th century by Isaac Newton's friend John Wallis.

Proof 5. For yet another proof, consider the two squares in Fig. B.8, where each has side $a + b$. The first square is dissected into five pieces, of which four are replicas of the given right triangle. The second is dissected into six pieces, of which four are replicas of the given triangle. If we subtract the four triangles from each of the two large squares, what remains tells us that the square on the hypotenuse equals the sum of the squares on the two sides.[1]

The most interesting generalization of the Pythagorean theorem is the theorem of Pappus stated in Section A.8 and illustrated in Fig. A.17 (and proved in the parenthetical remark).

---

[1] This proof is the basis for Aldous Huxley's unforgettable short story, "Young Archimides." In this tragic story the Italian boy Guido, age between 6 and 7—who knows nothing about geometry—*discovers* the Pythagorean theorem for himself by drawing the square on the right of Fig. B.8 on a flagstone with a charred stick, and then moving its triangles around in a second drawing until it becomes the square on the left. Huxley's adult narrator witnesses this extraordinary event without being noticed by Guido, and is led into a lyrical meditation on the role of genius in human history: "This child, I thought, when he grows up, will be to me, intellectually, what a man is to a dog . . . ."

APPENDIX:

THE FORMULAS OF HERON AND BRAHMAGUPTA

In Section A.7 we mentioned a celebrated formula of geometry that rarely finds a place in elementary courses—the so-called "Heron's formula"

$$T = \sqrt{s(s-a)(s-b)(s-c)} \tag{1}$$

for the area T of an arbitrary triangle with vertices A, B, C and sides a, b, c (Fig. B.9), where $s = \frac{1}{2}(a+b+c)$ is the semiperimeter. We also mentioned that it is known from Arabic sources in the Middle Ages that this formula is due to Archimedes.

The most direct approach to establishing this formula is to use the elementary theorem that the area of a triangle is one-half the height times the base. If h is the height from A to the base BC, then we can calculate h by using the Pythagorean theorem, which brings in the segments of the base, and the work can be continued by straightforward but fairly complicated algebraic manipulations (see the Problem). We prefer instead to present an elegant argument more in keeping with the spirit of geometry.

We begin by inscribing a circle with center M and radius r in the triangle ABC. The three triangles AMB, BMC, CMA have areas $\frac{1}{2}rc$, $\frac{1}{2}ra$, $\frac{1}{2}rb$, so

$$T = \tfrac{1}{2}r(c+a+b) = rs. \tag{2}$$

This reduces the problem to expressing r in terms of a, b, c. To accomplish this, we introduce the circle with center N and radius R that is tangent to side AB on the outside and to sides AC and BC extended. We now observe that

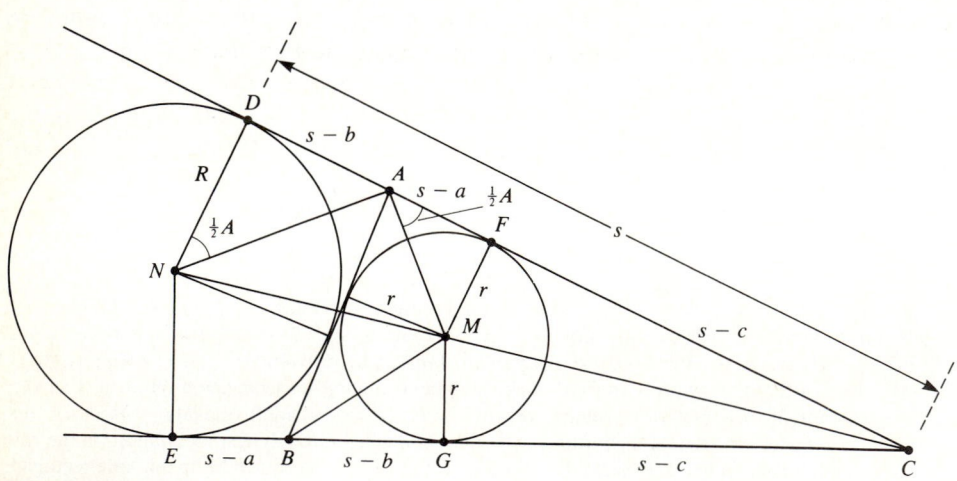

FIGURE B.9

$DC = EC = s$, so $DA = s - b$, $EB = s - a$, and the other tangential segments have the lengths shown in the figure.[2] The rest of the argument depends on the similar triangles *NDC*, *MFC* and *NDA*, *AFM*, from which we obtain

$$\frac{R}{r} = \frac{s}{s-c} \quad \text{and} \quad \frac{R}{s-b} = \frac{s-a}{r}.$$

Solving for R gives

$$R = \frac{rs}{s-c} = \frac{(s-a)(s-b)}{r},$$

so

$$r^2 = \frac{(s-a)(s-b)(s-c)}{s}$$

and (2) becomes

$$T = s\sqrt{\frac{(s-a)(s-b)(s-c)}{s}} = \sqrt{s(s-a)(s-b)(s-c)},$$

as required.

The presence of the factor s under the radical in (1) suggests that this formula for the area of the triangle in Fig. B.10 might be a special case of a more general formula

$$T = \sqrt{(s-a)(s-b)(s-c)(s-d)} \tag{3}$$

for the area of a quadrilateral with sides a, b, c, d, where $s = \frac{1}{2}(a+b+c+d)$ is the semiperimeter of the quadrilateral (Fig. B.11). After all, if the side d shrinks to zero, then the quadrilateral becomes a triangle and (3) collapses to (1), which we know is true.

Unfortunately this conjecture is false, as we see at once by thinking of the quadrilateral as a frame with a joint at each vertex. However, a modified

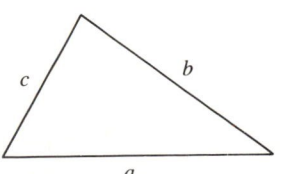

FIGURE B.10

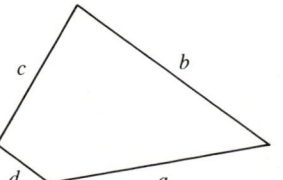

FIGURE B.11

---

[2] Notice that $AF + DA = BG + EB$ and $AF + BG = c$.

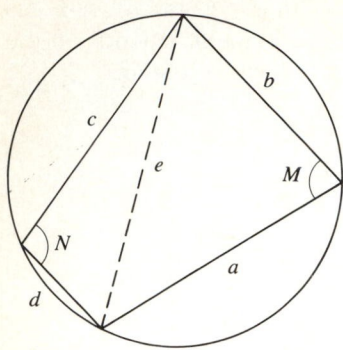

FIGURE B.12

version is true: If the quadrilateral is inscribed in a circle as shown in Fig. B.12, then formula (3) is valid. Under these circumstances (3) is called *Brahmagupta's formula*, after the 7th century Indian mathematician who discovered it.

The proof we give makes use of trigonometry. We begin by inserting the diagonal e in the quadrilateral of Fig. B.12, and also by labeling the opposite angles M and N. We know that $M + N = 180°$, so

$$\cos N = -\cos M \qquad \text{and} \qquad \sin N = \sin M.$$

By the law of cosines,

$$a^2 + b^2 - 2ab \cos M = e^2 = c^2 + d^2 - 2cd \cos N,$$

so

$$2(ab + cd) \cos M = a^2 + b^2 - c^2 - d^2. \tag{4}$$

Since the area T of the quadrilateral is given by

$$T = \frac{1}{2} ab \sin M + \frac{1}{2} cd \sin N = \frac{1}{2} (ab + cd) \sin M,$$

we also have

$$2(ab + cd) \sin M = 4T. \tag{5}$$

By squaring and adding equations (4) and (5), and using the identity $\sin^2 M + \cos^2 M = 1$, we obtain

$$4(ab + cd)^2 = (a^2 + b^2 - c^2 - d^2)^2 + 16T^2,$$

so

$$16T^2 = (2ab + 2cd)^2 - (a^2 + b^2 - c^2 - d^2)^2.$$

By repeatedly factoring differences of two squares in accordance with the identity $x^2 - y^2 = (x + y)(x - y)$, we obtain

$$16T^2 = [2ab + 2cd + a^2 + b^2 - c^2 - d^2] \cdot [2ab + 2cd - a^2 - b^2 + c^2 + d^2]$$
$$= [(a + b)^2 - (c - d)^2] \cdot [(c + d)^2 - (a - b)^2]$$
$$= [a + b + c - d][a + b - c + d] \cdot [c + d + a - b][c + d - a + b]$$
$$= [2s - 2d][2s - 2c][2s - 2b][2s - 2a]$$

or

$$T^2 = (s - a)(s - b)(s - c)(s - d),$$

which proves Brahmagupta's formula (3).

PROBLEM

1. In Fig. B.9, let h be the height of triangle ABC on the base a and establish Heron's formula (1) by carrying out the following steps:
   (a) If d is the projection of side b on the base a, show that $c^2 - (a - d)^2 = b^2 - d^2$, and therefore $d = (a^2 + b^2 - c^2)/2a$.
   (b) Substitute this value for d in $h^2 = b^2 - d^2$, and simplify by using $2s = a + b + c$.
   (c) Substitute this value for h^2 in $T^2 = (\frac{1}{2}ah)^2 = \frac{1}{4}a^2h^2$ and thereby complete the argument.

B.2

MORE ABOUT NUMBERS: IRRATIONALS, PERFECT NUMBERS, AND MERSENNE PRIMES

It has been known for a very long time that $\sqrt{2}$ is an irrational number. The proof which is traditionally ascribed to Pythagoras is one of the earliest intellectual productions of Western civilization that still retains its vigor and interest. It deserves to be included here, both for its own sake and as an introduction to problems of irrationality.

We begin by recalling that the *even* numbers are the integers 0, ± 2, $\pm 4, \ldots$ that can be written in the form $2n$ for some integer n, and the *odd* numbers are the integers ± 1, ± 3, $\pm 5, \ldots$ that can be written in the form $2n + 1$. It is easy to see that the square of an even number is even, since $(2n)^2 = 4n^2 = 2(2n^2)$, and the square of an odd number is odd, since $(2n + 1)^2 = 4n^2 + 4n + 1 = 2(2n^2 + 2n) + 1$. After these preliminaries, we are now ready to prove that $\sqrt{2}$ is not rational. Let us suppose that it is—contrary to what we want to establish—so that $\sqrt{2} = a/b$ for certain positive integers a and b. We may specify that a and b have no common factors >1; for if they have, we can cancel them away until none remain. It will soon be clear that the possibility of making this specification without loss of generality is crucial for the proof. Now, the equation $\sqrt{2} = a/b$ implies that $2 = a^2/b^2$, so $a^2 = 2b^2$ and a^2 is even. This implies that a is also even; for if it were odd, then a^2 would be odd, which it is not. Since a is even, it has the form $a = 2c$ for some integer c, and therefore $4c^2 = 2b^2$ or $b^2 = 2c^2$, so b^2 is even. As before, this implies that b is even. But since a and b are both even, they have 2 as a common factor. This contradicts our earlier specification and shows that it cannot be true that $\sqrt{2}$ is rational. We are therefore driven to the desired conclusion, that $\sqrt{2}$ is irrational.

It is often quite difficult to determine whether a given specific number is rational or not. For instance, the fact that π is irrational was not discovered until 1761. We shall prove this later with the aid of some rather complicated reasoning depending on the calculus of the trigonometric functions (see Section B.17). Unfortunately, no really simple proof is known even to this day.

The Pythagorean argument given here for $\sqrt{2}$ is essentially an argument from elementary number theory, for it depends only on relatively simple properties of positive integers. There are many interesting kinds of positive integers, with a great variety of remarkable properties that have fascinated curious people through the ages. We mention the prime numbers 2, 3, 5, 7, 11, 13, 17, 19, 23, . . . , the squares 1, 4, 9, 16, 25, . . . , and the perfect numbers 6, 28, ...[1]

The primes are the multiplicative building blocks of the positive integers, in the sense that every integer >1 either is itself prime or is expressible as a product of primes. To see this, notice that if $n > 1$ is not a prime, then $n = ab$, where a and b are $<n$; if either a or b is not a prime, it can be factored similarly; and continuing in this way until all factors are not further factorable proves that n can be written as a product of primes. For example, $198 = 2 \cdot 99 = 2 \cdot 3 \cdot 33 = 2 \cdot 3 \cdot 3 \cdot 11$. A fundamental theorem of number theory (called the *unique factorization theorem*) states that this factorization is always unique except for the order of the factors. In particular, a prime factorization of 198 can never involve 5 as a factor, and the factor 2 can never appear more than once.

The unique factorization theorem is deeper than it looks, but to most people it is obviously true. The following facts are much more surprising, and therefore have greater appeal to the imagination.

a. The *four squares theorem*: Every positive integer can be expressed as the sum of not more than four squares.

b. The *two squares theorem*: Every prime number of the form $4n + 1$ can be expressed as the sum of two squares.

c. To formulate our next statement, we consider the geometric progression whose first term is 1 and whose ratio is any number $r \neq 1$:

$$1, r, r^2, \ldots, r^n, \ldots .$$

---

[1] We remind the reader that a *prime number* is an integer $p > 1$ that has no positive factors (or divisors) except 1 and p; equivalently, it cannot be written in the form $p = ab$, where a and b are both positive integers $<p$. A *perfect number* is a positive integer like $6 = 1 + 2 + 3$ that equals the sum of its positive divisors other than itself. Note also that $28 = 1 + 2 + 4 + 7 + 14$. The next perfect numbers after 6 and 28 are 496, 8128, and 33,550,336, as we shall see below.

We recall from elementary algebra that the sum of the first n terms of this progression is given by the formula[2]

$$1 + r + r^2 + \cdots + r^{n-1} = \frac{1 - r^n}{1 - r}. \tag{1}$$

In particular, if $r = 2$, this formula yields

$$1 + 2 + 2^2 + \cdots + 2^{n-1} = 2^n - 1.$$

The *theorem of even perfect numbers* asserts the following: If the sum $2^n - 1$ is prime, then the product $2^{n-1}(2^n - 1)$ of the last term and the sum is an even perfect number; and conversely, every even perfect number is of this form where $2^n - 1$ is prime.

The first part of the theorem in (c) was proved by Euclid about 300 B.C., and the second by Euler in the mid-eighteenth century. These two statements and their proofs constitute a gem of classical number theory with ramifications that continued to attract the attention of serious mathematicians and computer technologists as late as 1979. The details are brief enough for us to present them here.

Let us first notice that for $n = 1, 2, 3, 4, 5, 6, 7$ the corresponding values of $2^n - 1$ are 1, 3, 7, 15, 31, 63, 127. The only primes in this list are 3, 7, 31, 127. According to the theorem, the first four even perfect numbers are therefore $2 \cdot 3 = 6$, $4 \cdot 7 = 28$, $16 \cdot 31 = 496$, $64 \cdot 127 = 8128$. No odd perfect numbers are known, and the question of whether any exist is one of the oldest unsolved problems of mathematics.[3]

To prove the theorem, we shall need some tools. First, a piece of standard notation: If a is a positive integer, the sum of all the divisors of a (including 1 and a itself) is denoted by the symbol $\sigma(a)$, read "sigma of a." For example, $\sigma(1) = 1$, $\sigma(2) = 1 + 2 = 3$, $\sigma(3) = 1 + 3 = 4$, $\sigma(4) = 1 + 2 + 4 = 7$, $\sigma(5) = 1 + 5 = 6$, $\sigma(6) = 1 + 2 + 3 + 6 = 12$. Since a perfect number is one that

---

[2] To prove (1), denote the sum on the left by s,

$$s = 1 + r + r^2 + \cdots + r^{n-1},$$

multiply by r,

$$rs = r + r^2 + r^3 + \cdots + r^n,$$

and subtract, carrying out all possible cancellations, to obtain

$$s - rs = 1 - r^n \quad \text{or} \quad s(1 - r) = 1 - r^n.$$

Since $r \neq 1$, formula (1) follows at once from the last equation.

[3] It is known, however, that there is no odd perfect number containing less than 100 digits.

equals the sum of its divisors other than itself, perfect numbers are precisely those for which $\sigma(a) = 2a$. The only other tool we need is the following fact.

Lemma. *If a and b are positive integers whose greatest common divisor is 1, then* $\sigma(ab) = \sigma(a)\sigma(b)$.

Proof. First, since a and b have no common factor greater than 1, it is easy to see that the factors of the product ab are precisely numbers of the form $a_i b_j$, where a_i is a factor of a and b_j is a factor of b; or equivalently, any divisor d of ab is expressible in the form

$$d = a_i b_j$$

in one and only one way, where a_i is a divisor of a and b_j is a divisor of b. We denote the divisors of a and b by

$$1, a_1, a_2, \ldots, a \qquad \text{and} \qquad 1, b_1, b_2, \ldots, b,$$

so that their sums are

$$\sigma(a) = 1 + a_1 + a_2 + \cdots + a$$

and

$$\sigma(b) = 1 + b_1 + b_2 + \cdots + b.$$

Now let us consider all divisors $d = a_i b_j$ of ab with the same a_i. Their sum is

$$a_i \cdot 1 + a_i b_1 + a_i b_2 + \cdots + a_i b = a_i(1 + b_1 + b_2 + \cdots + b)$$
$$= a_i \sigma(b).$$

Finally, by adding these numbers for all possible a_i's, we obtain the sum of all the divisors of ab:

$$\sigma(ab) = 1 \cdot \sigma(b) + a_1 \sigma(b) + a_2 \sigma(b) + \cdots + a\sigma(b)$$
$$= (1 + a_1 + a_2 + \cdots + a)\sigma(b) = \sigma(a)\sigma(b),$$

and the proof is complete.

As a last preliminary remark, we point out that formula (1) enables us to calculate the value of $\sigma(p^{n-1})$ whenever p is a prime number. Since the divisors of p^{n-1} are $1, p, p^2, \ldots, p^{n-1}$, we have

$$\sigma(p^{n-1}) = 1 + p + p^2 + \cdots + p^{n-1} = \frac{1 - p^n}{1 - p}$$
$$= \frac{p^n - 1}{p - 1}.$$

In particular, when $p = 2$, this gives

$$\sigma(2^{n-1}) = 2^n - 1.$$

We are now ready to prove the theorem of even perfect numbers, which we divide into two separate statements for the sake of clarity.

Theorem 1   (Euclid). *If n is a positive integer for which $2^n - 1$ is prime, then $a = 2^{n-1}(2^n - 1)$ is an even perfect number.*

Proof. Since $2^n - 1$ is prime, n must be at least 2 and a is even. We prove that a is perfect by showing that $\sigma(a) = 2a$. First, $2^n - 1$ is odd, so 2^{n-1} and $2^n - 1$ have no common factor > 1. The lemma therefore tells us that

$$\sigma[2^{n-1}(2^n - 1)] = \sigma(2^{n-1})\sigma(2^n - 1).$$

Next, since $2^n - 1$ is prime, its only divisors are 1 and itself, so

$$\sigma(2^n - 1) = 1 + (2^n - 1) = 2^n.$$

Finally,

$$\begin{aligned}
\sigma(a) &= \sigma[2^{n-1}(2^n - 1)] \\
&= \sigma(2^{n-1})\sigma(2^n - 1) \\
&= (2^n - 1)2^n \\
&= 2[2^{n-1}(2^n - 1)] = 2a,
\end{aligned}$$

so a is perfect.

Theorem 2   (Euler). *If a is an even perfect number, then $a = 2^{n-1}(2^n - 1)$ for some positive integer n such that $2^n - 1$ is prime.*

Proof. Factor the highest possible power of 2 out of a, and in this way write a in the form

$$a = m2^{n-1},$$

where n is at least 2 and m is odd. We shall prove that $m = 2^n - 1$ and that $2^n - 1$ is prime. Since a is perfect and therefore $\sigma(a) = 2a$, we have

$$m2^n = 2a = \sigma(a) = \sigma(m2^{n-1}) = \sigma(m)\sigma(2^{n-1})$$
$$= \sigma(m)(2^n - 1),$$

so

$$\sigma(m) = \frac{m2^n}{2^n - 1}.$$

But $\sigma(m)$ is an integer, so $2^n - 1$ divides $m2^n$; and since $2^n - 1$ and 2^n have no common factor > 1, $2^n - 1$ divides m. We see from this that $m/(2^n - 1)$ is a divisor of m, and this divisor is less than m, since $2^n - 1$ is at least 3. The equation

$$\sigma(m) = \frac{m2^n}{2^n - 1} = m + \frac{m}{2^n - 1}$$

therefore exhibits $\sigma(m)$ as the sum of m and one other divisor of m. This implies that m has two and only two divisors, so it must be prime. Also, it must be true that

$$\frac{m}{2^n - 1} = 1,$$

so $m = 2^n - 1$ and $2^n - 1$ is prime, which completes the proof.

The ideas discussed here raise the natural question: What numbers of the form $2^n - 1$ are prime? The factorization formula

$$a^n - 1 = (a - 1)(a^{n-1} + a^{n-2} + a^{n-3} + \cdots + 1)$$

shows that $2^n - 1$ cannot be prime if n is not; for example,

$$2^6 - 1 = (2^2)^3 - 1 = (2^2 - 1)[(2^2)^2 + 2^2 + 1] = 3 \cdot 21.$$

We may therefore confine our attention to the case in which the exponent n is assumed to be a prime p, and our question becomes: What numbers of the form $2^p - 1$ are prime? Such primes are called *Mersenne primes*, after Father Mersenne, a French scientist–mathematician–priest of the seventeenth century (see the account of Mersenne in Section A.12). Until 1952 only 12 were known, those corresponding to

$$p = 2, 3, 5, 7, 13, 17, 19, 31, 61, 89, 107, 127.$$

The primality of $2^{127} - 1$, a number of 39 digits, was established in 1876.[4] Beginning in 1952, 15 more have been found with the aid of electronic computers, those corresponding to

$$p = 521, 607, 1279, 2203, 2281, 3217, 4253, 4423, 9689, 9941,$$
$$11{,}213, 19{,}937, 21{,}701, 23{,}209, 44{,}497,$$

the last of all in 1979.[5] The largest currently known prime number is therefore

$$2^{44{,}497} - 1.$$

---

[4] The English mathematician G. H. Hardy remarked, "We may be able to recognize directly that 5, or even 17, is prime, but nobody can convince himself that

$$2^{127} - 1$$

is prime except by studying a proof. No one has ever had an imagination so vivid and comprehensive as that."

[5] See D. Slowinski and H. Nelson, "Searching for the 27th Mersenne Prime," *Journal of Recreational Mathematics*, vol. 11(4) (1978–1979), pp. 258–61. This search continues among those who like to play with supercomputers, and by the time the present book is published, additional Mersenne primes will probably be known. (Note added later: Three more Mersenne primes were found in the next few years, those corresponding to $p = 86{,}243$ and $132{,}049$ in 1983, and that corresponding to $p = 216{,}091$ in 1985. Onward and upward for those with a taste for this hobby!)

This has been computed in decimal form and is a number with 13,395 digits. It is so enormous that it exceeds by far the total number of grains of sand that could be packed solidly into the entire visible universe.

PROBLEMS

1. Prove that $\sqrt{3}$ is irrational. Hint: Modify the method of the text and use the fact that every integer is of the form $3n$, $3n + 1$, or $3n + 2$.
2. Prove that $\sqrt{5}$ and $\sqrt{6}$ are irrational by the method of Problem 1. Why doesn't this method work for $\sqrt{4}$?
3. Prove that $\sqrt[3]{2}$ and $\sqrt[3]{3}$ are irrational.
4. Prove that $\sqrt{2} + \sqrt{3}$ is irrational. Hint: Use the fact that $\sqrt{6}$ is irrational.
5. Prove that $\sqrt{2} + \sqrt{3} - \sqrt{6}$ is irrational.
6. Use the unique factorization theorem to prove that if a positive integer m is not the nth power of another integer, then $\sqrt[n]{m}$ is irrational. Notice that this result includes Problems 1, 2, and 3 as special cases.
7. Prove that $\log_{10} 2$ is irrational. For what positive integers n is $\log_{10} n$ irrational?
8. Prove that if a real number x_0 is a root of a polynomial equation

$$x^n + c_{n-1}x^{n-1} + c_{n-2}x^{n-2} + \cdots + c_1 x + c_0 = 0$$

with integral coefficients, then x_0 is either an irrational number or an integer; and if it is an integer, show that it must be a factor of c_0.
9. Use Problem 8 to show that $\sqrt{2} + \sqrt{3}$, $\sqrt{2} + \sqrt[3]{2}$, and $\sqrt{3} + \sqrt[3]{2}$ are irrational.
10. (a) Verify the four squares theorem for all integers from 1 to 50.
    (b) Every prime except 2 (that is, every odd prime) is of the form $4n + 1$ or $4n + 3$. For the integers from 1 to 50, verify that every prime of the first type is expressible as the sum of two squares, and that no prime of the second type is.
11. Calculate the 5th even perfect number.

B.3

ARCHIMEDES'
QUADRATURE
OF THE
PARABOLA

The title of this section is a slightly altered version of the title of a treatise by Archimedes (*On the Quadrature of the Parabola*). The word *quadrature* is an old-fashioned term meaning the act or process of finding areas.

Our purpose here is to examine the procedure by which Archimedes calculated the area of a parabolic segment, that is, the area of the part of the parabola in Fig. B.13 bounded by the arbitrary chord AB and the arc $ADCEB$. There is no convenient way to inscribe regular polygons in this figure, as there is in the case of the circle, so Archimedes used triangles instead. His first approximation was the triangle ABC, where the vertex C is chosen as that point where the tangent to the parabola is parallel to AB. His second approximation was obtained by adding to the triangle ABC the two triangles ACD and BCE, where the vertex D is the point where the tangent is parallel to AC and the vertex E is the point where the tangent is parallel to BC. To obtain his third approximation, he inscribed triangles in the same way in each of the four regions still not included (one such region is that between the arc CE and the chord CE), so his third approximation to the area of the parabolic segment was the sum of the areas of the triangles ABC, ACD, BCE, and the four new triangles. By continuing to exhaust the parabolic segment in this way, he was able to show that its area is exactly four-thirds the area of the first triangle ABC.

The crux of the proof is the fact that the combined area of triangles ACD and BCE is one-fourth the area of triangle ABC, and similarly at each

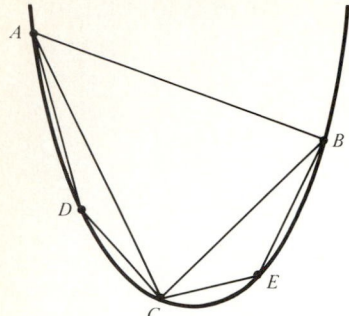

FIGURE B.13

succeeding stage of the process. We shall prove this fact, but first let us see how it enables us to accomplish the quadrature of the parabolic segment. The stated relation between the newly added area and the previous area tells us first that

$$ACD + BCE = \frac{1}{4} ABC;$$

second that

$$\text{the combined area of the next four triangles} = \frac{1}{4} ACD + \frac{1}{4} BCE$$

$$= \frac{1}{4}(ACD + BCE)$$

$$= \frac{1}{4^2} ABC;$$

and so on indefinitely. Thus, by the nature of the exhaustion process, it is clear that

$$\text{the area of the parabolic segment} = ABC + \frac{1}{4} ABC + \frac{1}{4^2} ABC + \cdots$$

$$= ABC\left(1 + \frac{1}{4} + \frac{1}{4^2} + \cdots\right).$$

But from elementary algebra we know the sum of a geometric series,

$$1 + r + r^2 + \cdots = \frac{1}{1 - r},$$

where $-1 < r < 1$.[1] By putting $r = \frac{1}{4}$ we obtain

$$1 + \frac{1}{4} + \frac{1}{4^2} + \cdots = \frac{1}{1 - \frac{1}{4}} = \frac{1}{\frac{3}{4}} = \frac{4}{3},$$

---

[1] This is an immediate consequence of formula (1) in Section B.2.

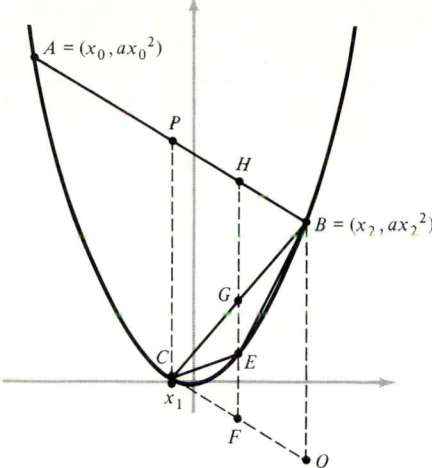

$A = (x_0, ax_0{}^2)$

$B = (x_2, ax_2{}^2)$

FIGURE B.14

and this brings us to our conclusion:

$$\text{the area of the parabolic segment} = \frac{4}{3}ABC,$$

as Archimedes discovered and proved.

We now establish the geometric fact stated above. Archimedes did this by purely geometric reasoning without reference to equations or coordinates, but we shall use analytic geometry instead. Let us suppose that the parabola is the graph of $y = ax^2$, and that A and B have coordinates $A = (x_0, ax_0^2)$ and $B = (x_2, ax_2^2)$, as shown in Fig. B.14. If x_1 is the x-coordinate of C, then by using the formula $dy/dx = 2ax$ we see that the slope of the tangent at C is $2ax_1$. Since the tangent at C is parallel to AB, we have

$$2ax_1 = \frac{ax_0^2 - ax_2^2}{x_0 - x_2} \qquad \text{or} \qquad x_1 = \frac{1}{2}(x_0 + x_2).$$

This tells us that the vertical line through C bisects the chord AB at a point P. It clearly suffices to prove that

$$BCE = \frac{1}{4}BCP. \tag{1}$$

To do this, we begin by completing the parallelogram $CPBQ$. By the same reasoning as before, the vertical line through E bisects the chord BC at a point G, and therefore also bisects the segment BP at H. If we can show that

$$EG = \frac{1}{2}GH, \tag{2}$$

then this will imply that

$$BEG = \frac{1}{2} BGH \qquad \text{and} \qquad CEG = \frac{1}{2} BGH,$$

so that

$$BCE = BGH;$$

and since clearly $BGH = \frac{1}{4}BCP$, this will prove (1). To establish (2), it suffices to show that $FE = \frac{1}{4}FH$, and we do this by showing that $FE = \frac{1}{4}QB$. The calculations are straightforward:

$$
\begin{aligned}
FE &= a[\tfrac{1}{2}(x_1 + x_2)]^2 - [ax_1^2 + 2ax_1 \cdot \tfrac{1}{2}(x_2 - x_1)] \\
&= \tfrac{1}{4}a[(x_1 + x_2)^2 - 4x_1^2 - 4x_1(x_2 - x_1)] \\
&= \tfrac{1}{4}a(x_1^2 - 2x_1x_2 + x_2^2) = \tfrac{1}{4}a(x_1 - x_2)^2;
\end{aligned}
$$

and

$$
\begin{aligned}
QB &= ax_2^2 - [ax_1^2 + 2ax_1(x_2 - x_1)] \\
&= a(x_1^2 - 2x_1x_2 + x_2^2) = a(x_1 - x_2)^2.
\end{aligned}
$$

B.4

THE LUNES OF HIPPOCRATES

According to one tradition, Hippocrates of Chios (ca. 430 B.C.)—not to be confused with his better-known contemporary, the physician Hippocrates of Cos—was originally a merchant whose goods were stolen by pirates.[1] He then went to Athens, where he lived for many years, studied mathematics, and compiled a book on the elements of geometry that strongly influenced Euclid more than a century later.

Hippocrates' discovery was this: The lune (crescent-shaped region) in Fig. B.15 bounded by the circular arcs ADB and AEB (the latter having C as its center) has an area exactly equal to the area of the shaded square whose side is the radius of the circle. (Hippocrates also found the areas of two other kinds of lunes, but we do not discuss these here.)

---

[1] Aristotle, who rarely missed a chance to express his scorn for mathematicians, gives a more demeaning account of Hippocrates' misfortune. "It is well known," he wrote with relish, "that people brilliant in one particular field may be quite foolish in most other things. Thus Hippocrates, though skilled in geometry, was so stupid and spineless that he let a tax collector of Byzantium cheat him out of a fortune." This, from the man who asserted that heavier bodies fall to the ground more rapidly, that men have more teeth than women, and that Greeks desiring sons should have intercourse in the north wind. But everyone makes mistakes, especially people who write a great deal and try to answer many questions; and Aristotle, who is often mistakenly thought of as mainly a philosopher, was really the greatest biologist before Charles Darwin.

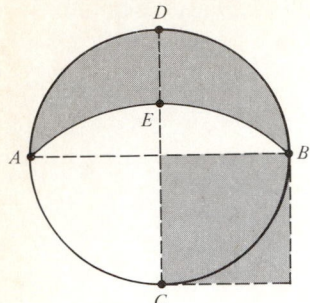

FIGURE B.15

This very surprising theorem seems to be the earliest precise determination of the area of a region bounded by curves. Its proof is simple but ingenious, and depends on the last of the following three geometric facts, each of which implies the next: (a) the areas of two circles are to each other as the squares of the radii (Fig. B.16(a)); (b) sectors of two circles with equal central angles are to each other as the squares of the radii (Fig. B.16(b)); (c) segments of two circles with equal central angles are to each other as the squares of the radii (Fig. B.16(c)). We shall need (c) in the special case of right angles at the center.

The proof of Hippocrates' theorem now proceeds as follows. Redraw the lune as shown in Fig. B.17. The chords joining D with A and B are tangent to the arc AEB and divide the lune into three regions with areas a_1, a_2, a_3. If the radius of the smaller circle is denoted by r, then the Pythagorean theorem tells us that the radius of the larger circle is $\sqrt{2}r$. It is easy to see that a_1 and a_2 are equal segments of the smaller circle and that a_4 is a segment of the larger circle, all with right angles at the center. We now use statement (c) to infer that

$$\frac{a_1}{a_4} = \frac{r^2}{(\sqrt{2}r)^2} = \frac{1}{2}.$$

This yields

$$a_1 = \frac{1}{2}a_4 \quad \text{and} \quad a_2 = \frac{1}{2}a_4,$$

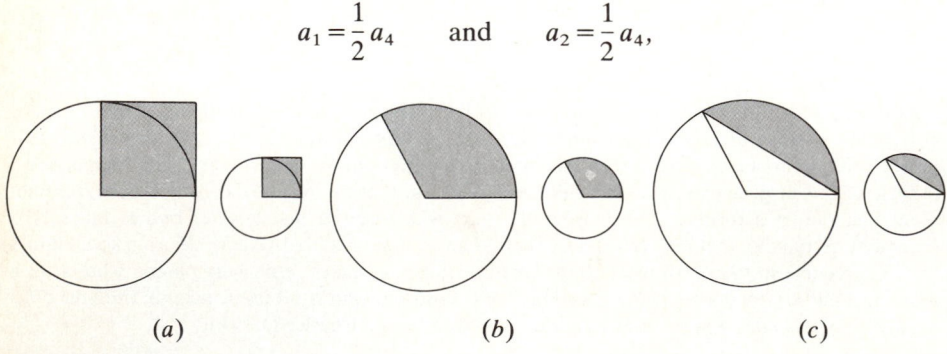

(a)   (b)   (c)

FIGURE B.16

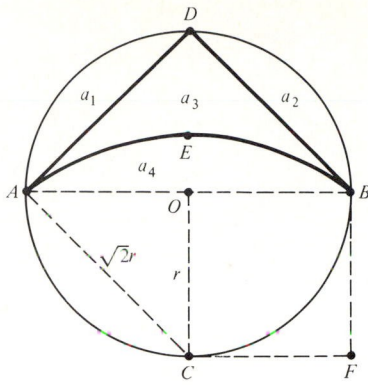

FIGURE B.17

so

$$a_1 + a_2 = a_4.$$

It now follows that

$$\text{area of lune} = a_1 + a_2 + a_3$$

$$= a_4 + a_3$$

$$= \text{area of triangle } ABD$$

$$= r^2 = \text{area of square } OBFC,$$

and the argument is complete.

Hippocrates was a contemporary of Pericles, the great political and cultural leader of Athens during its Golden Age. But nothing Pericles achieved has the enduring quality of this beautiful geometric discovery; even the Parthenon, whose design and construction he supervised, is crumbling away. The reasoning of Hippocrates is a paragon of mathematical proof, untouched by time: In a few elegant steps it converts something easy to understand but difficult to believe into something impossible to doubt.

B.5

FERMAT'S CALCULATION OF $\int_0^b x^n \, dx$ FOR POSITIVE RATIONAL n

To describe Fermat's very ingenious method for calculating this integral, we shall need the following formulas from Sections B.2 and B.3: If $0 < r < 1$, then

$$1 + r + r^2 + \cdots + r^n = \frac{1 - r^{n+1}}{1 - r} \quad \text{and} \quad 1 + r + r^2 + \cdots = \frac{1}{1 - r}. \quad (1)$$

Fermat used upper sums similar to those that are customary in calculus but based instead on a division of the interval $[0, b]$ into an *infinite* number of unequal subintervals, as suggested in Fig. B.18(a). We start with a fixed positive number $r < 1$ (but close to 1) and produce the points of division by moving downward from b by repeatedly multiplying by r, as shown in Fig. B.18(b). The sum of the areas of all the upper rectangles, starting from the right, is an infinite sum depending on r,

$$\begin{aligned}
S_r &= b^n(b - rb) + (rb)^n(rb - r^2 b) + (r^2 b)^n(r^2 b - r^3 b) + \cdots \\
&= b^{n+1}(1 - r) + b^{n+1}r^{n+1}(1 - r) + b^{n+1}r^{2n+2}(1 - r) + \cdots \\
&= b^{n+1}(1 - r)[1 + r^{n+1} + (r^{n+1})^2 + \cdots] \\
&= \frac{b^{n+1}(1 - r)}{1 - r^{n+1}} = \frac{b^{n+1}}{(1 - r^{n+1})/(1 - r)} \\
&= \frac{b^{n+1}}{1 + r + r^2 + \cdots + r^n}.
\end{aligned} \quad (2)$$

Here we used first the second formula of (1) and then the first formula, the

240

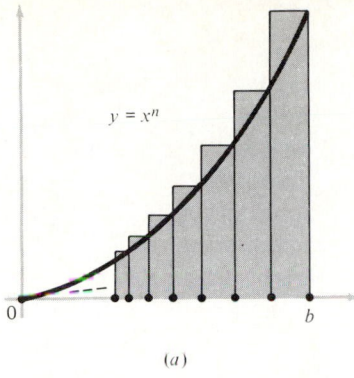
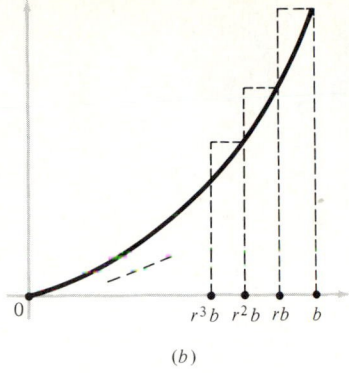

(a)   (b)

FIGURE B.18

latter requiring that n be a positive integer. If we now let $r \to 1$, we see that each of the $n + 1$ terms in the denominator of the last expression also approaches 1, so we have our result:

$$\lim_{r \to 1} S_r = \frac{b^{n+1}}{n+1}, \tag{3}$$

or equivalently,

$$\int_0^b x^n\,dx = \frac{b^{n+1}}{n+1} \tag{4}$$

for every positive integer n.

The only part of this argument in which n must be a positive integer is the last step in arriving at (2). If we assume only that n is a positive rational number p/q, then we can get around this difficulty by making the substitution $s = r^{1/q}$ and calculating as follows:

$$\frac{1-r}{1-r^{n+1}} = \frac{1-s^q}{1-(s^q)^{p/q+1}} = \frac{1-s^q}{1-s^{p+q}}$$

$$= \frac{(1-s^q)/(1-s)}{(1-s^{p+q})/(1-s)} = \frac{1+s+s^2+\cdots+s^{q-1}}{1+s+s^2+\cdots+s^{p+q-1}}.$$

Now, as $r \to 1$, we also have $s \to 1$, and the expression last written tells us that

$$\frac{1-r}{1-r^{n+1}} \to \frac{q}{p+q} = \frac{1}{p/q+1} = \frac{1}{n+1},$$

so (3) and (4) remain valid for any positive rational exponent n.

B.6

HOW
ARCHIMEDES
DISCOVERED
INTEGRATION

Archimedes' discovery of the formula for the volume of a sphere was one of the greatest mathematical achievements of all time. The formula itself was of obvious importance, but even more so was his method of discovering it; for this method amounted to the earliest appearance of the basic idea of integral calculus.

He proved this formula in his treatise *On the Sphere and Cylinder,* by means of a long and rigorous argument of classic perfection. Unfortunately, however, this argument was the kind that compels belief but provides little insight. It was like a great work of architecture whose architect has removed all the scaffolding, burnt the plans, and concealed his private thoughts from which the overall concept emerged. Mathematicians have always been aware—from his formal treatises—of what Archimedes discovered. However, his method of making discoveries remained a mystery until the year 1906, when the Danish scholar Heiberg uncovered a lost manuscript dealing with exactly this question.[1]

In this manuscript Archimedes described to his friend Eratosthenes how he "investigated some problems in mathematics by means of mechanics."[2] The

---

[1] See the *Method* in *The Works of Archimedes*, T. L. Heath (ed.), Dover (no date).

[2] *Method*, p. 13.

most wonderful of these investigations was his discovery of the volume of a sphere. To understand this work, it is necessary to know a little about the level of knowledge from which he started.

As Archimedes states, it was Democritus two centuries earlier who discovered that the volume of a cone is one-third the volume of a cylinder with the same height and the same base. Nothing is definitely known about Democritus' method, but it is believed that he succeeded by considering first a three-sided pyramid (tetrahedron), then an arbitrary pyramid, and finally a cone as the limit of inscribed pyramids.[3]

Also, the Greeks knew a little analytic geometry, but without our notation. They were acquainted with the idea that a locus in a plane can be studied by considering the distances from a moving point to two perpendicular lines; and if the sum of the squares of these distances is constant, then they knew that the locus is a circle. In our notation, this condition amounts to the equation $x^2 + y^2 = a^2$.

Further, Archimedes himself virtually created Greek mechanics. It is well-known that he discovered the law of floating bodies. He also discovered the principle of the lever and many facts about centers of gravity.

We are now ready to follow Archimedes in his search for the volume of a sphere. He considered the sphere to be generated by revolving a circle about its diameter. In modern notation, we start with the circle

$$x^2 + y^2 = 2ax, \tag{1}$$

which has radius a and is tangent to the y-axis at the origin. This circle is shown on the left in Fig. B.19, which is almost identical with Archimedes' original figure. Equation (1) contains the term y^2, and since πy^2 is the area of the variable cross section of the sphere x units to the right of the origin, it is natural to multiply through by π and write (1) in the form

$$\pi x^2 + \pi y^2 = \pi 2ax. \tag{2}$$

This leads us to interpret πx^2 as the area of the variable cross section of the cone generated by revolving the line $y = x$ about the x-axis. This in turn suggests that we seek a similar interpretation for the term $\pi 2ax$ on the right side of (2). If we persist in this search, we might perhaps think of multiplying by $2a$ and thus rewriting (2) as

$$2a(\pi x^2 + \pi y^2) = x\pi(2a)^2. \tag{3}$$

The motivation for this change clearly lies in the fact that $\pi(2a)^2$ is the area of the cross section of the cylinder with the same height and base as the cone.

---

[3] See Section A.3.

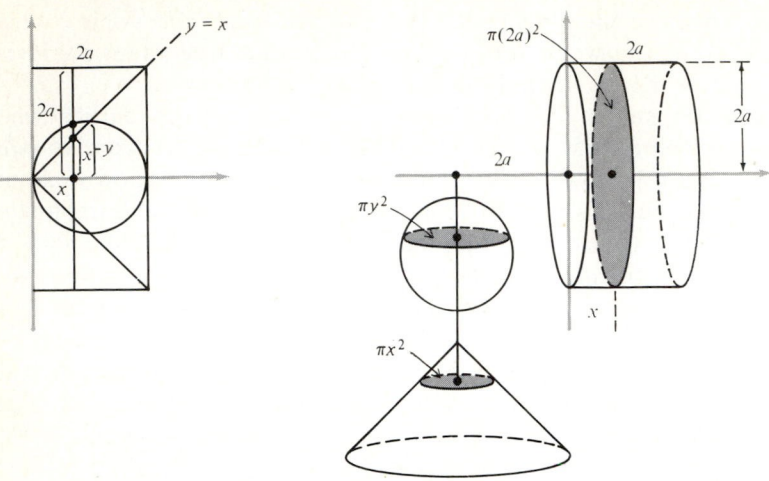

FIGURE B.19
Archimedes' balancing argument.

We therefore have on the left in Fig. B.19 three circular disks of areas πy^2, πx^2, and $\pi(2a)^2$ which are the intersections of a single plane with three solids of revolution. This plane is perpendicular to the x-axis at a distance x units to the right of the origin, and the solids are the sphere, the cone, and the cylinder, whose intersections with the xy-plane are shown in the figure.

On the left-hand side of equation (3), the sum of the first two areas is multiplied by $2a$, and on the right-hand side, the third area is multiplied by x. This observation led Archimedes to the following great idea, as shown on the right in Fig. B.19. He left the disk with radius $2a$ where it is, in a vertical position x units to the right of the origin; and he moved the disks with radii y and x to a point $2a$ units to the left of the origin, where he hung them horizontally with their centers under this point, suspended by a weightless string. The purpose of this maneuver can be understood only if we think of the x-axis as a lever and the origin as its fulcrum or balancing point. It can now be seen that equation (3) deals with moments. (A *moment* is the product of the suspended weight and the length of the lever arm.) From this point of view, equation (3) states that the combined moments of the two disks on the left equals the moment of the single disk on the right, and so, by Archimedes' own principle of the lever, this lever is in equilibrium.

We now carry out the final step of the reasoning. As x increases from 0 to $2a$, the three cross sections sweep through their respective solids and fill these solids. Since the three cross sections are in equilibrium throughout this process, the solids themselves are also in equilibrium. Let V denote the volume of the sphere, which was unknown until Archimedes finished this calculation.

If we use Democritus' formula for the volume of the cone, and also the volume of the cylinder and the obvious location of its center of gravity, then the equilibrium of the solids in the positions shown in the figure yields

$$2a[\tfrac{1}{3}\pi(2a)^2(2a) + V] = a\pi(2a)^2(2a). \tag{4}$$

It is now easy to solve (4) for V and obtain

$$V = \tfrac{4}{3}\pi a^3.$$

The ideas discussed here were created by one who has been described—with good reason—as "the greatest genius of the ancient world." But these ideas are, after all, only a beginning. The crux of the reasoning lies in the transition from (3) to (4), from the moving cross sections to the complete solids. With the advantage of historical perspective, we can recognize this transition as the essence of integration, which we know to be a process of great scope and diversity, with innumerable applications in science and mathematics. Archimedes himself had an inkling of the potential value of his ideas: "I am persuaded that this method will be of no little service to mathematics. For I foresee that once it is understood and established, it will be used to discover other theorems which have not yet occurred to me, by other mathematicians, now living or yet unborn."[4]

---

[4] *Method*, p. 14.

B.7

A SIMPLE
APPROACH
TO $E = Mc^2$

Consider a particle of mass m that starts from rest at the origin on the x-axis and moves in the positive direction under the influence of a constant force F. If we write Newton's second law of motion

$$F = ma \qquad (1)$$

in the form

$$a = \frac{1}{m} F,$$

then we can think of the force F as producing the constant acceleration a. If this acceleration continues long enough, then the velocity v of the particle increases beyond all bounds, and in particular beyond c, the velocity of light. But according to Einstein this cannot happen; nothing can travel faster than light.

The way out of this impasse is to realize that Newton's law is actually somewhat more general than (1); it states that when a force F acts on a body of mass m, it produces *momentum* ($= mv$) at a rate equal to the force:

$$F = \frac{d}{dt} (mv). \qquad (2)$$

This equation reduces to (1) when the mass m is constant. But according to Einstein, the mass is not constant. It increases as the velocity v increases, and is determined as a function of v by the formula

$$m = \frac{m_0}{\sqrt{1 - v^2/c^2}}, \tag{3}$$

where m_0 is the so-called *rest mass*. When this expression for m is inserted in (2), we obtain *Einstein's law of motion*,

$$F = m_0 \frac{d}{dt} \left(\frac{v}{\sqrt{1 - v^2/c^2}} \right). \tag{4}$$

It will be convenient to carry out the differentiation in (4), and in this way introduce the acceleration $a = dv/dt$. We have

$$\frac{d}{dt} \left(\frac{v}{\sqrt{1 - v^2 c^2}} \right) = \frac{d}{dv} \left(\frac{v}{\sqrt{1 - v^2/c^2}} \right) \frac{dv}{dt}$$

$$= a \left[\frac{\sqrt{1 - v^2/c^2} - v(\frac{1}{2})(1 - v^2/c^2)^{-1/2}(-2v/c^2)}{1 - v^2/c^2} \right]$$

$$= a \left[\frac{c^2(1 - v^2/c^2) + v^2}{c^2(1 - v^2/c^2)^{3/2}} \right] = \frac{a}{(1 - v^2/c^2)^{3/2}}.$$

This enables us to write (4) in the form

$$F = \frac{m_0 a}{(1 - v^2/c^2)^{3/2}}, \tag{5}$$

which shows how close Einstein's law is to Newton's law (1) when v is much less than c. However, when v is near the velocity of light, as in most phenomena of atomic physics, then the two laws differ considerably, and all the experimental evidence supports Einstein's version.

We now return to our original problem of the particle starting from rest at the origin on the x-axis, with the slight change that the force F is now assumed only to be positive, so that the acceleration a in (5) is also positive and the velocity is increasing. Our purpose is to show that if the energy of the particle at any stage of the process is understood to be the work done on it by F, then this energy E is related to the increase in the mass, which is $M = m - m_0$, by Einstein's famous equation $E = Mc^2$. If we begin by writing

$$a = \frac{dv}{dt} = \frac{dv}{dx} \frac{dx}{dt} = v \frac{dv}{dx},$$

then (5) yields.[1]

$$E = \int_0^x F\,dx = m_0 \int_0^x \frac{a}{(1 - v^2/c^2)^{3/2}}\,dx$$

$$= m_0 \int_0^v \frac{v\,dv}{(1 - v^2/c^2)^{3/2}}$$

$$= m_0\left(-\frac{c^2}{2}\right)\int_0^v \left(1 - \frac{v^2}{c^2}\right)^{-3/2}\left(-\frac{2v\,dv}{c^2}\right)$$

$$= m_0 c^2 \left(1 - \frac{v^2}{c^2}\right)^{-1/2}\Bigg]_0^v$$

$$= m_0 c^2 \left(\frac{1}{\sqrt{1 - v^2/c^2}} - 1\right)$$

$$= c^2\left(\frac{m_0}{\sqrt{1 - v^2/c^2}} - m_0\right) = c^2(m - m_0) = Mc^2.$$

Needless to say, this is not a proof of Einstein's equation in all cases; it merely shows that this equation connects the increase in mass arising from (3) with the increase in energy associated with the greater velocity.

However, it is necessary to add that the crux of Einstein's equation is the much deeper fact that the rest mass m_0 also has energy associated with it, in the amount of $E = m_0 c^2$. This energy can be thought of as the "energy of being" of the particle, in the sense that mass possesses energy just by virtue of existing. The point of view of modern physics is even more direct than this: Matter *is* energy, in a highly concentrated and localized form. It should also be understood that the constant c^2 is so enormous that a small amount of mass is equivalent to a very large amount of energy. Thus, if the mass of a drop of water could be completely converted into energy in a controlled and useful way, then the resulting energy would be enough to lift several heavy trucks as far as the moon. This is the source of the energy that fuels the sun, in the so-called thermonuclear reactions which physicists are even now seeking to tame to our service.

---

[1] We hope the student is not confused in the following integral by the common practice of using a single letter (x) for both the upper limit of integration and the dummy variable of integration.

B.8

At the beginning of Section B.7 we pointed out that Newton's second law of motion for a force F acting on a body of mass m moving with velocity v can be stated as

$$F = \frac{d}{dt}(mv), \qquad (1)$$

and also that this assumes the more familiar form $F = ma$ when m is constant. In particular, if the body is moving under the action of no external force at all, so that $F = 0$, then (1) tells us that the momentum mv is constant.

As an illustration we consider a rocket of mass m, velocity v, and constant exhaust velocity b, and we assume that this rocket is moving in a straight line in outer space, where no external forces are present. The mass m consists of the structural mass of the rocket plus the mass of the fuel it carries, so m decreases as the fuel is burned. The exhaust gases are ejected at high speed from the tail of the rocket, and this propels it forward just as escaping air propels a toy balloon. We shall find the equation of motion.

Suppose that at time t the mass of the rocket (including fuel) is m and it is moving with velocity v, as shown in Fig. B.20, while at time $t + \Delta t$ the mass is $m + \Delta m$ and the velocity is $v + \Delta v$. The mass of fuel burned in this time is $-\Delta m$ (Δm is clearly negative), and the exhaust products, which are therefore of mass $-\Delta m$, are expelled backward at velocity b relative to the rocket, so that this material has actual velocity $v - b$. The fact that the total momentum

249

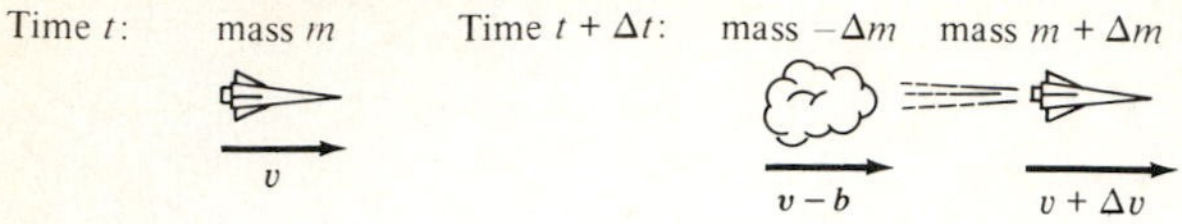

Time t: mass m Time $t + \Delta t$:   mass $-\Delta m$ mass $m + \Delta m$

FIGURE B.20
Rocket acceleration by exhaust expulsion.

of the system is constant means that

$$mv = (m + \Delta m)(v + \Delta v) + (-\Delta m)(v - b).$$

This yields

$$mv = mv + m\,\Delta v + (\Delta m)v + \Delta m\,\Delta v - (\Delta m)v + \Delta m\,b,$$

which reduces to

$$m\,\Delta v = -\Delta m(b + \Delta v). \tag{2}$$

After division by Δt, (2) becomes

$$m\frac{\Delta v}{\Delta t} = -\frac{\Delta m}{\Delta t}(b + \Delta v),$$

and by letting $\Delta t \to 0$ we obtain

$$m\frac{dv}{dt} = -b\frac{dm}{dt}. \tag{3}$$

This is the basic equation of rocket propulsion in outer space.

To illustrate the qualitative conclusions that can be drawn from (3), we use the fact that dv/dt is the acceleration a and write the equation as

$$a = \frac{1}{m}\left[b\left(-\frac{dm}{dt}\right)\right]. \tag{4}$$

Since dm/dt is negative, we see that a is positive, and this means that the velocity is increasing, as we expect. The quantity in brackets here—the product of the exhaust velocity and the rate at which fuel is consumed—is called the *thrust* of the rocket engine. It is clear from (4) that a large acceleration requires a rocket to have a large thrust, and a large thrust is obtained by designing an engine with a large exhaust velocity and a high rate of fuel consumption. Also, if the thrust is constant, then (4) tells us that the acceleration increases as m decreases, that is, as the fuel is burned.

In addition to qualitative inferences of the kind just mentioned, it is also possible to obtain quantitative information from (3).[1] If we write the equation

---

[1] The following material requires the reader to understand the meaning of the formula

$$\int \frac{dm}{m} = \ln m.$$

in the form

$$dv = -b \frac{dm}{m}$$

and integrate from 0 to t, we get

$$v(t) = v(0) - b \ln \frac{m(t)}{m(0)}$$

$$= v(0) + b \ln \frac{m(0)}{m(t)}. \tag{5}$$

As an illustration of the way (5) can be used, suppose the initial mass of the rocket is one-tenth structure and nine-tenths fuel. If the exhaust velocity is $b = 2 \, \text{mi/s}$ and the rocket starts from rest, then its speed at burnout is

$$v = 2 \ln 10 \cong 4.6 \, \text{mi/s}.$$

B.9

To prove Vieta's formula

$$\frac{2}{\pi} = \frac{\sqrt{2}}{2} \cdot \frac{\sqrt{2 + \sqrt{2}}}{2} \cdot \frac{\sqrt{2 + \sqrt{2 + \sqrt{2}}}}{2} \cdots, \tag{1}$$

we need the limit

$$\lim_{\theta \to 0} \frac{\sin \theta}{\theta} = 1, \tag{2}$$

the double-angle formula for the sine in the form

$$\sin \theta = 2 \sin \frac{\theta}{2} \cos \frac{\theta}{2}, \tag{3}$$

and the half-angle formula for the cosine in the form

$$\cos \frac{\theta}{2} = \frac{1}{2} \sqrt{2 + 2 \cos \theta}. \tag{4}$$

By repeatedly applying (3) we obtain

$$1 = \sin \frac{\pi}{2} = 2 \sin \frac{\pi}{4} \cos \frac{\pi}{4}$$

$$= 2^2 \sin \frac{\pi}{8} \cos \frac{\pi}{4} \cos \frac{\pi}{8}$$

$$= 2^3 \sin \frac{\pi}{16} \cos \frac{\pi}{4} \cos \frac{\pi}{8} \cos \frac{\pi}{16}$$

$$= \cdots = 2^{n-1} \sin \frac{\pi}{2^n} \cos \frac{\pi}{4} \cos \frac{\pi}{8} \cdots \cos \frac{\pi}{2^n}.$$

With the aid of (4), this can be written as

$$\frac{2}{\pi} \cdot \frac{\pi/2^n}{\sin \pi/2^n} = \frac{1}{2^{n-1} \sin \pi/2^n}$$

$$= \frac{\sqrt{2}}{2} \cdot \frac{\sqrt{2+\sqrt{2}}}{2} \cdot \frac{\sqrt{2+\sqrt{2+\sqrt{2}}}}{2} \cdots \frac{\sqrt{2+\sqrt{2+\cdots}}}{2}, \qquad (5)$$

where the last factor contains $n - 1$ nested root signs. If we now let $n \to \infty$ and use (2), (5) yields Vieta's formula (1).[1]

---

[1] In the present book, Vieta's formula is only an isolated jewel in the early history of Renaissance mathematics. However, it can be made the first step of a fascinating but demanding journey leading upward to some of the peaks of classical mathematical physics (the kinetic theory of gases, the second law of thermodynamics, etc.). See the little book by M. Kac, *Statistical Independence in Probability, Analysis and Number Theory,* Wiley 1959. As for François Vieta (1540–1603) himself, he was a Frenchman trained in the law who rose to become a royal privy councillor under Henry IV and cultivated mathematics as a hobby. He contributed to the early development of analytic trigonometry and algebra, in particular by his systematic use of letters to represent constants and unknowns.

B.10

$$\pi/4 = 1 - \tfrac{1}{3} + \tfrac{1}{5} - \tfrac{1}{7} + \ldots$$

The formula in question,

$$\frac{\pi}{4} = 1 - \frac{1}{3} + \frac{1}{5} - \frac{1}{7} + \cdots, \tag{1}$$

is easy to derive in a loose, nonrigorous way by simply integrating the infinite geometric series

$$\frac{1}{1+t^2} = 1 - t^2 + t^4 - \cdots$$

term-by-term from 0 to x and then putting $x = 1$. However, how do we know that this term-by-term integration is a valid operation? Our purpose here is to present a fairly simple supporting argument for (1) that cannot be criticized from the point of view of logic.

We begin the proof with the finite formula

$$\frac{1}{1+t^2} = 1 - t^2 + t^4 - t^6 + t^8 - \cdots + t^{4n} - \frac{t^{4n+2}}{1+t^2}, \tag{2}$$

which is valid for all t and can be checked by multiplying through by $1 + t^2$. Now consider a number x such that $0 \le x \le 1$. Integrating (2) over the interval $0 \le t \le x$ gives

$$\tan^{-1} x = \int_0^x \frac{dt}{1+t^2} = x - \frac{x^3}{3} + \frac{x^5}{5} - \frac{x^7}{7} + \cdots + \frac{x^{4n+1}}{4n+1} - R_n(x), \tag{3}$$

where

$$R_n(x) = \int_0^x \frac{t^{4n+2}}{1+t^2}\,dt.$$

It is clear that $1 \le 1 + t^2$, so

$$0 \le R_n(x) \le \int_0^x t^{4n+2}\,dt$$

or

$$0 \le R_n(x) \le \frac{x^{4n+3}}{4n+3}.$$

For the x's under discussion, this shows that

$$0 \le R_n(x) \le \frac{1}{4n+3},$$

so $R_n(x) \to 0$ as $n \to \infty$. This allows us to use (3) to deduce that the formula

$$\tan^{-1} x = x - \frac{x^3}{3} + \frac{x^5}{5} - \frac{x^7}{7} + \cdots \tag{4}$$

is valid for $0 \le x \le 1$. [The right-hand side of (4) is by definition the limit as $n \to \infty$ of the right-hand side of (3) if $R_n(x) \to 0$, which we have just shown to be true for the stated x's.] To obtain Leibniz's formula (1), we now—legally!—set $x = 1$ in (4).

B.11

THE CATENARY, OR CURVE OF A HANGING CHAIN

As a specific example of the use of the methods of integration by trigonometric substitution, we solve the classical problem of determining the exact shape of the curve assumed by a flexible chain of uniform density which is suspended between two points and hangs under its own weight. This curve is called a *catenary*, from the Latin word for chain, *catena*.

Let the y-axis pass through the lowest point of the chain (Fig. B.21), let s be the arc length from this point to a variable point (x, y), and let w_0 be the linear density (weight per unit length) of the chain. We obtain the differential equation of the catenary from the fact that the part of the chain between the lowest point and (x, y) is in static equilibrium under the action of three forces: the tension T_0 at the lowest point; the variable tension T at (x, y), which acts in the direction of the tangent because of the flexibility of the chain; and a downward force $w_0 s$ equal to the weight of the chain between these two points.

Equating the horizontal component of T to T_0 and the vertical component of T to the weight of the chain gives

$$T \cos \theta = T_0 \qquad \text{and} \qquad T \sin \theta = w_0 s,$$

and by dividing we eliminate T and get $\tan \theta = w_0 s / T_0$ or

$$\frac{dy}{dx} = as, \qquad \text{where} \qquad a = \frac{w_0}{T_0}.$$

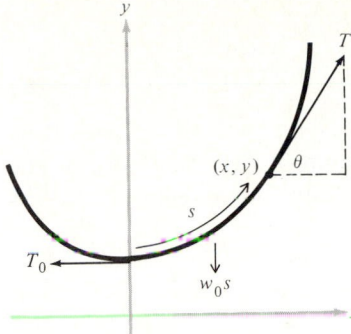

FIGURE B.21

We next eliminate the variable s by differentiating with respect to x,

$$\frac{d^2y}{dx^2} = a\frac{ds}{dx} = a\sqrt{1 + \left(\frac{dy}{dx}\right)^2}. \tag{1}$$

This is the differential equation of the catenary.

We now solve equation (1) by two successive integrations. This process is facilitated by introducing the auxiliary variable $p = dy/dx$, so that (1) becomes

$$\frac{dp}{dx} = a\sqrt{1 + p^2}.$$

On separating variables and integrating, we get

$$\int \frac{dp}{\sqrt{1 + p^2}} = \int a \, dx. \tag{2}$$

To calculate the integral on the left, we make the trigonometric substitution $p = \tan\phi$, so that $dp = \sec^2\phi \, d\phi$ and $\sqrt{1 + p^2} = \sec\phi$. Then

$$\int \frac{dp}{\sqrt{1 + p^2}} = \int \frac{\sec^2\phi \, d\phi}{\sec\phi} = \int \sec\phi \, d\phi$$

$$= \ln(\sec\phi + \tan\phi) = \ln(\sqrt{1 + p^2} + p),$$

so (2) becomes

$$\ln(\sqrt{1 + p^2} + p) = ax + c_1.$$

Since $p = 0$ when $x = 0$, we see that $c_1 = 0$, so

$$\ln(\sqrt{1 + p^2} + p) = ax.$$

It is easy to solve this equation for p, which yields

$$\frac{dy}{dx} = p = \frac{1}{2}(e^{ax} - e^{-ax}),$$

and by integrating we obtain

$$y = \frac{1}{2a}(e^{ax} + e^{-ax}) + c_2.$$

If we now place the origin of the coordinate system in Fig. B.21 at just the right level so that $y = 1/a$ when $x = 0$, then $c_2 = 0$ and our equation takes its final form,

$$y = \frac{1}{2a}(e^{ax} + e^{-ax}). \tag{3}$$

Equation (3) reveals the precise mathematical nature of the catenary and can be used as the basis for further investigations of its properties.[1]

The problem of finding the true shape of the catenary was proposed by James Bernoulli in 1690. Galileo had speculated long before that the curve is a parabola, but Huygens had shown in 1646 (at the age of 17), largely by physical reasoning, that this is not correct, without, however, shedding any light on what the shape might be. Bernoulli's challenge produced quick results, for in 1691 Leibniz, Huygens (now age 62), and James's brother John all published independent solutions of the problem. John Bernoulli was exceedingly pleased that he had been successful in solving the problem, while his brother James, who proposed it, had failed. The taste of victory was still sweet 27 years later, as we see from this passage in a letter John wrote in 1718:

> The efforts of my brother were without success. For my part, I was more fortunate, for I found the skill (I say it without boasting; why should I conceal the truth?) to solve it in full . . . . It is true that it cost me study that robbed me of rest for an entire night. It was a great achievement for those days and for the slight age and experience I then had. The next morning, filled with joy, I ran to my brother, who was still struggling miserably with this Gordian knot without getting anywhere, always thinking like Galileo that the catenary was a parabola. Stop! Stop! I say to him, don't torture yourself any more trying to prove the identity of the catenary with the parabola, since it is entirely false.

---

[1] The hyperbolic cosine defined by $\cosh x = \frac{1}{2}(e^x + e^{-x})$ enables us to write the function (3) in the form

$$y = \frac{1}{a}\cosh ax.$$

This fact is sometimes thought to justify a detailed study of the hyperbolic functions, but the present writer is skeptical.

However, James evened the score by proving in the same year of 1691 that of all possible shapes a chain hanging between two fixed points might have, the catenary has the lowest center of gravity, and therefore the smallest potential energy. This was a very significant discovery, because it was the first hint of the profound idea that in some mysterious way the actual configurations of nature are those that minimize potential energy.

B.12

As an application of integration by parts one can obtain the following reduction formula:

$$\int \sin^n x \, dx = -\frac{1}{n} \sin^{n-1}x \cos x + \frac{n-1}{n} \int \sin^{n-2} x \, dx. \tag{1}$$

This formula leads in an elementary but ingenious way to a very remarkable expression for the number $\pi/2$ as an infinite product,

$$\frac{\pi}{2} = \frac{2}{1} \cdot \frac{2}{3} \cdot \frac{4}{3} \cdot \frac{4}{5} \cdot \frac{6}{5} \cdot \frac{6}{7} \cdots \frac{2n}{2n-1} \cdot \frac{2n}{2n+1} \cdots. \tag{2}$$

This expression was discovered by the English mathematician John Wallis in 1656, and is called *Wallis's product*. Apart from its intrinsic interest, formula (2) underlies other important developments in both pure and applied mathematics, so we prove it here.

If we define I_n by

$$I_n = \int_0^{\pi/2} \sin^n x \, dx,$$

then (1) tells us that

$$I_n = \frac{n-1}{n} I_{n-2}. \tag{3}$$

It is clear that

$$I_0 = \int_0^{\pi/2} dx = \frac{\pi}{2} \quad \text{and} \quad I_1 = \int_0^{\pi/2} \sin x \, dx = 1.$$

We now distinguish the cases of even and odd subscripts, and use (3) to calculate I_{2n} and I_{2n+1}, as follows:

$$I_{2n} = \frac{2n-1}{2n} I_{2n-2} = \frac{2n-1}{2n} \cdot \frac{2n-3}{2n-2} I_{2n-4}$$

$$= \cdots = \frac{2n-1}{2n} \cdot \frac{2n-3}{2n-2} \cdot \frac{2n-5}{2n-4} \cdots \frac{1}{2} I_0$$

$$= \frac{1}{2} \cdot \frac{3}{4} \cdot \frac{5}{6} \cdots \frac{2n-1}{2n} \cdot \frac{\pi}{2}; \tag{4}$$

and

$$I_{2n+1} = \frac{2n}{2n+1} I_{2n-1} = \frac{2n}{2n+1} \cdot \frac{2n-2}{2n-1} I_{2n-3}$$

$$= \cdots = \frac{2n}{2n+1} \cdot \frac{2n-2}{2n-1} \cdot \frac{2n-4}{2n-3} \cdots \frac{2}{3} I_1$$

$$= \frac{2}{3} \cdot \frac{4}{5} \cdot \frac{6}{7} \cdots \frac{2n}{2n+1}. \tag{5}$$

As the next link in the chain of this reasoning, we need the fact that the ratio of these two quantities approaches 1 as $n \to \infty$,

$$\frac{I_{2n}}{I_{2n+1}} \to 1. \tag{6}$$

To establish this, we begin by noticing that on the interval $0 \le x \le \pi/2$ we have $0 \le \sin x \le 1$, and therefore

$$0 \le \sin^{2n+2} x \le \sin^{2n+1} x \le \sin^{2n} x.$$

This implies that

$$0 < \int_0^{\pi/2} \sin^{2n+2} x \, dx \le \int_0^{\pi/2} \sin^{2n+1} x \, dx \le \int_0^{\pi/2} \sin^{2n} x \, dx,$$

or equivalently,

$$0 < I_{2n+2} \le I_{2n+1} \le I_{2n}. \tag{7}$$

If we divide through by I_{2n} and use the fact that by (3) we have

$$\frac{I_{2n+2}}{I_{2n}} = \frac{2n+1}{2n+2},$$

then (7) yields

$$\frac{2n+1}{2n+2} \le \frac{I_{2n+1}}{I_{2n}} \le 1.$$

This implies that

$$\frac{I_{2n+1}}{I_{2n}} \to 1 \qquad \text{as} \qquad n \to \infty,$$

and this is equivalent to (6).

The final steps of the argument are as follows. On dividing (5) by (4), we obtain

$$\frac{I_{2n+1}}{I_{2n}} = \frac{2}{1} \cdot \frac{2}{3} \cdot \frac{4}{3} \cdot \frac{4}{5} \cdot \frac{6}{5} \cdot \frac{6}{7} \cdots \frac{2n}{2n-1} \cdot \frac{2n}{2n+1} \cdot \frac{2}{\pi},$$

so

$$\frac{\pi}{2} = \frac{2}{1} \cdot \frac{2}{3} \cdot \frac{4}{3} \cdot \frac{4}{5} \cdot \frac{6}{5} \cdot \frac{6}{7} \cdots \frac{2n}{2n-1} \cdot \frac{2n}{2n+1} \left(\frac{I_{2n}}{I_{2n+1}} \right).$$

On forming the limit as $n \to \infty$ and using (6), we obtain

$$\frac{\pi}{2} = \lim_{n \to \infty} \frac{2}{1} \cdot \frac{2}{3} \cdot \frac{4}{3} \cdot \frac{4}{5} \cdot \frac{6}{5} \cdot \frac{6}{7} \cdots \frac{2n}{2n-1} \cdot \frac{2n}{2n+1},$$

and this is what (2) means.

We also remark that Wallis's product (2) is equivalent to the formula

$$\left(1 - \frac{1}{2^2}\right)\left(1 - \frac{1}{4^2}\right)\left(1 - \frac{1}{6^2}\right) \cdots = \frac{2}{\pi}. \tag{8}$$

This is easy to see if we write each number in parentheses on the left in factored form. This gives

$$\left(1 - \frac{1}{2}\right)\left(1 + \frac{1}{2}\right)\left(1 - \frac{1}{4}\right)\left(1 + \frac{1}{4}\right)\left(1 - \frac{1}{6}\right)\left(1 + \frac{1}{6}\right) \cdots = \frac{2}{\pi}$$

or

$$\frac{1}{2} \cdot \frac{3}{2} \cdot \frac{3}{4} \cdot \frac{5}{4} \cdot \frac{5}{6} \cdot \frac{7}{6} \cdots = \frac{2}{\pi},$$

which is clearly equivalent to (2). Formula (8) will reappear in Section B.14 as a special case of another even more wonderful formula.[1]

---

[1] Wallis was Savilian Professor of Geometry at Oxford for 54 years, from 1649 until his death in 1703 at the age of 87, and played an important part in forming the climate of thought in which Newton flourished. He introduced negative and fractional exponents as well as the now-standard symbol ∞ for infinity. His infinite product stimulated his friend Lord Brouncker (first president of the Royal Society) to discover the astonishing formula

$$\frac{4}{\pi} = 1 + \cfrac{1^2}{2 + \cfrac{3^2}{2 + \cfrac{5^2}{2 + \cfrac{7^2}{2 + \cfrac{9^2}{2 + \cdots}}}}} ,$$

from which the theory of continued fractions later arose. (No one knows how Brouncker made this discovery, but a proof based on the work of Euler in the next century is given in the chapter on Brouncker in J. L. Coolidge's *The Mathematics of Great Amateurs*, Oxford University Press, 1949.) Among the activities of Wallis's later years was a lively quarrel with the famous philosopher Hobbes, who was under the impression that he had succeeded in squaring the circle and published his erroneous proof. Wallis promptly refuted it, but Hobbes was both arrogant and too ignorant to understand the refutation, and defended himself with a barrage of additional errors, as if a question about the validity of a mathematical proof could be settled by rhetoric and denunciation.

B.13

$$\pi/4 = 1 - \tfrac{1}{3} + \tfrac{1}{5} - \tfrac{1}{7} + \cdots$$

The area of the quarter-circle of radius 1 shown in Fig. B.22 is obviously $\pi/4$. We follow Leibniz and calculate this area in a different way. The part that we actually calculate is the area A of the circular segment cut off by the chord OT, because the remainder of the quarter-circle is clearly an isosceles right triangle of area $\tfrac{1}{2}$.

We obtain the stated area A by integrating the sliverlike elements of area OPQ, where the arc PQ is considered to be so small that it is virtually straight. We think of OPQ as a triangle whose base is the segment PQ of length ds and whose height is the perpendicular distance OR from the vertex O to the base PQ extended. The two similar right triangles in the figure tell us that

$$\frac{ds}{dx} = \frac{OS}{OR} \quad \text{or} \quad OR\,ds = OS\,dx,$$

so the area dA of OPQ is

$$dA = \tfrac{1}{2}OR\,ds = \tfrac{1}{2}OS\,dx = \tfrac{1}{2}y\,dx,$$

where y denotes the length of the segment OS. The element of area dA sweeps across the circular segment in question as x increases from 0 to 1, so

$$A = \int dA = \frac{1}{2}\int_0^1 y\,dx;$$

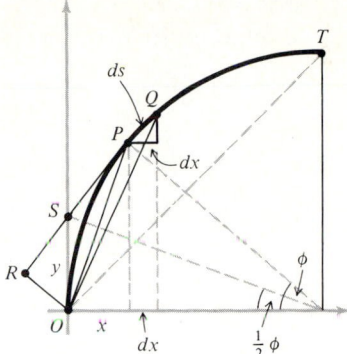

FIGURE B.22

and integrating by parts in order to reverse the roles of x and y gives

$$A = \frac{1}{2}xy \Big]_0^1 - \frac{1}{2}\int_0^1 x\,dy = \frac{1}{2} - \frac{1}{2}\int_0^1 x\,dy, \tag{1}$$

where the limits on the two integrals are understood to be $y = 0$ and $y = 1$. To continue the calculation, we observe that since

$$y = \tan \tfrac{1}{2}\phi \qquad \text{and} \qquad x = 1 - \cos \phi = 2\sin^2 \tfrac{1}{2}\phi,$$

the trigonometric identity

$$\tan^2 \frac{1}{2}\phi = \frac{\sin^2 \tfrac{1}{2}\phi}{\cos^2 \tfrac{1}{2}\phi} = \sin^2 \frac{1}{2}\phi \sec^2 \frac{1}{2}\phi = \sin^2 \frac{1}{2}\phi\left(1 + \tan^2 \frac{1}{2}\phi\right)$$

yields

$$\frac{x}{2} = \frac{y^2}{1 + y^2}.$$

The version of the geometric series given in Section B.10 enables us to write this as

$$\frac{x}{2} = y^2(1 - y^2 + y^4 - y^6 + \cdots) = y^2 - y^4 + y^6 - y^8 + \cdots,$$

so (1) becomes

$$A = \frac{1}{2} - \int_0^1 (y^2 - y^4 + y^6 - y^8 + \cdots)\,dy$$

$$= \frac{1}{2} - \left[\frac{1}{3}y^3 - \frac{1}{5}y^5 + \frac{1}{7}y^7 - \frac{1}{9}y^9 + \cdots\right]_0^1$$

$$= \frac{1}{2} - \left(\frac{1}{3} - \frac{1}{5} + \frac{1}{7} - \frac{1}{9} + \cdots\right)$$

$$= \frac{1}{2} - \frac{1}{3} + \frac{1}{5} - \frac{1}{7} + \frac{1}{9} - \cdots.$$

When $\frac{1}{2}$ is added to this to account for the area of the isosceles right triangle, and the result is equated to the known area $\pi/4$ of the quarter-circle, we have Leibniz's formula

$$\frac{\pi}{4} = 1 - \frac{1}{3} + \frac{1}{5} - \frac{1}{7} + \cdots.$$

Is it any wonder that he took great pleasure and pride in this discovery for the rest of his life?

B.14

One of the most memorable gems of 18th century mathematics is Euler's formula for the sum of the reciprocals of the squares,

$$1 + \frac{1}{4} + \frac{1}{9} + \frac{1}{16} + \cdots = \frac{\pi^2}{6}. \tag{1}$$

Our purpose here is to understand the heuristic reasoning that led Euler to this wonderful discovery.

We begin with some simple algebra. If a and b are $\neq 0$, then it is clear that these numbers are the roots of the equation

$$\left(1 - \frac{x}{a}\right)\left(1 - \frac{x}{b}\right) = 0. \tag{2}$$

This equation can also be written in the form

$$1 - \left(\frac{1}{a} + \frac{1}{b}\right)x + \frac{1}{ab}x^2 = 0, \tag{3}$$

in which it is evident that the negative of the coefficient of x is the sum of the reciprocals of the roots. If we replace x by x^2, and a and b by a^2 and b^2, then (2) and (3) become

$$\left(1 - \frac{x^2}{a^2}\right)\left(1 - \frac{x^2}{b^2}\right) = 0 \tag{4}$$

and

$$1 - \left(\frac{1}{a^2} + \frac{1}{b^2}\right)x^2 + \frac{1}{a^2 b^2}x^4 = 0. \tag{5}$$

The roots of (4) are plainly $\pm a$ and $\pm b$, and (5) is the same equation in polynomial form, from which we see that the negative of the coefficient of x^2 is the sum of the reciprocals of the squares of the positive roots. This pattern persists as we move to equations of higher degree, for

$$\left(1 - \frac{x^2}{a^2}\right)\left(1 - \frac{x^2}{b^2}\right)\left(1 - \frac{x^2}{c^2}\right) = 0$$

(whose roots are obviously $\pm a$, $\pm b$, and $\pm c$) can be written as

$$1 - \left(\frac{1}{a^2} + \frac{1}{b^2} + \frac{1}{c^2}\right)x^2 + \left(\frac{1}{a^2 b^2} + \frac{1}{a^2 c^2} + \frac{1}{b^2 c^2}\right)x^4 - \frac{1}{a^2 b^2 c^2}x^6 = 0,$$

and so on.

Let us now consider the transcendental equation

$$\sin x = 0$$

or

$$x - \frac{x^3}{3!} + \frac{x^5}{5!} - \frac{x^7}{7!} + \cdots = 0.$$

This can be thought of as "a polynomial equation of infinite degree" with an infinite number of roots 0, $\pm\pi$, $\pm 2\pi$, $\pm 3\pi$, ... The root 0 can be removed by dividing by x, which gives the equation

$$\frac{\sin x}{x} = 0$$

or

$$1 - \frac{x^2}{3!} + \frac{x^4}{5!} - \frac{x^6}{7!} + \cdots = 0,$$

with roots $\pm\pi$, $\pm 2\pi$, $\pm 3\pi$, ... In the light of our knowledge of the roots of this equation, the situation described in the previous paragraph suggests that the infinite series

$$\frac{\sin x}{x} = 1 - \frac{x^2}{3!} + \frac{x^4}{5!} - \frac{x^6}{6!} + \cdots$$

can be written as an "infinite product,"

$$\frac{\sin x}{x} = \left(1 - \frac{x^2}{\pi^2}\right)\left(1 - \frac{x^2}{4\pi^2}\right)\left(1 - \frac{x^2}{9\pi^2}\right)\cdots. \tag{6}$$

Further, our analogy also suggests that

$$\frac{1}{\pi^2} + \frac{1}{4\pi^2} + \frac{1}{9\pi^2} + \cdots = \frac{1}{3!},$$

from which Euler's formula (1) follows at once. As an additional observation, it is interesting to note that if we put $x = \pi/2$ in (6), we find that

$$\frac{2}{\pi} = \left(1 - \frac{1}{2^2}\right)\left(1 - \frac{1}{4^2}\right)\left(1 - \frac{1}{6^2}\right)\cdots$$

$$= \left(\frac{1}{2}\cdot\frac{3}{2}\right)\left(\frac{3}{4}\cdot\frac{5}{4}\right)\left(\frac{5}{6}\cdot\frac{7}{6}\right)\cdots,$$

which is equivalent to

$$\frac{\pi}{2} = \left(\frac{2}{1}\cdot\frac{2}{3}\right)\left(\frac{4}{3}\cdot\frac{4}{5}\right)\left(\frac{6}{5}\cdot\frac{6}{7}\right)\cdots.$$

This is *Wallis's product,* which was rigorously proved in Section B.12.

These daring speculations are characteristic of Euler's unique genius, but we hope that no students will suppose that they carry the force of rigorous proof. It will be seen that the crux of the matter is the question of the meaning and validity of (6), which is known as *Euler's infinite product for the sine.* The shortcomings of this discussion invite the construction of a general theory of infinite products, within which formulas like (6) can take their place as firmly established facts. This aim is achieved in more advanced fields of mathematics.

A rigorous elementary proof of Euler's formula (1) which is based on totally different ideas is given in the next section.

B.15

A RIGOROUS
PROOF OF
EULER'S FORMULA
$$\sum_1^\infty \frac{1}{n^2} = \frac{\pi^2}{6}$$

The proof given here is due to D. P. Giesy, and appeared in the *Mathematics Magazine*, vol. 45 (1972), pp. 148–149. Other elementary proofs (and some not so elementary) can be tracked down through Giesy's references.

We begin by defining a function $f_n(x)$ by

$$f_n(x) = \tfrac{1}{2} + \cos x + \cos 2x + \cdots + \cos nx. \tag{1}$$

We will need the closed formula for this function given by

$$f_n(x) = \frac{\sin\left[(2n+1)x/2\right]}{2\sin(x/2)}, \tag{2}$$

where x is not an integral multiple of 2π. To prove this, we use the trigonometric identity $2\cos\theta\sin\phi = \sin(\theta+\phi) - \sin(\theta-\phi)$ to write the following identities (after the first):

$$2 \cdot \frac{1}{2}\sin\frac{1}{2}x = \sin\frac{1}{2}x,$$

$$2\cos x\sin\frac{1}{2}x = \sin\frac{3}{2}x - \sin\frac{1}{2}x,$$

$$2\cos 2x\sin\frac{1}{2}x = \sin\frac{5}{2}x - \sin\frac{3}{2}x,$$

$$\cdots$$

$$2\cos nx\sin\frac{1}{2}x = \sin\frac{(2n+1)x}{2} - \sin\frac{(2n-1)x}{2}.$$

Formula (2) is now easy to obtain by adding these identities and carrying out the obvious cancellations. We now use (1) to define a number E_n, and also to calculate it, as follows:

$$E_n = \int_0^\pi x f_n(x)\, dx = \frac{\pi^2}{4} + \sum_{k=1}^n \left[\frac{(-1)^k - 1}{k^2} \right]. \tag{3}$$

Here each term of the integral after the first is integrated by parts to obtain the stated result. Since the even terms in the sum on the right are zero, (3) can be written in the form

$$\frac{1}{2} E_{2n-1} = \frac{\pi^2}{8} - \sum_{k=1}^n \frac{1}{(2k-1)^2}. \tag{4}$$

Our next purpose is to show that $\lim E_{2n-1} = 0$, because this will establish the formula

$$\sum_{k=1}^\infty \frac{1}{(2k-1)^2} = \frac{\pi^2}{8}, \tag{5}$$

which will be used to prove our final result. We accomplish this as follows. If we use formula (2) and define $g(x)$ by

$$g(x) = \frac{d}{dx} \left[\frac{x/2}{\sin(x/2)} \right],$$

then an integration by parts yields

$$E_{2n-1} = \frac{1}{4n-1} \left[2 + 2 \int_0^\pi g(x) \cos \frac{(4n-1)x}{2}\, dx \right], \tag{6}$$

where we make use in the calculation of the familiar fact that

$$\lim_{x \to 0} \frac{x/2}{\sin(x/2)} = 1.$$

Our desired conclusion that $\lim E_{2n-1} = 0$ will follow from (6) if we can show that $g(x)$ is bounded on the interval of integration. But $g(x)$ is increasing and is therefore bounded on this interval by $g(\pi) = \frac{1}{2}$, and this establishes (5).

To complete our proof of Euler's formula, we divide the positive integers into the evens and odds and use (5) to write

$$\sum_1^\infty \frac{1}{k^2} = \sum_1^\infty \frac{1}{(2k)^2} + \sum_1^\infty \frac{1}{(2k-1)^2} = \frac{1}{4} \sum_1^\infty \frac{1}{k^2} + \frac{\pi^2}{8}.$$

This yields

$$\frac{3}{4} \sum_1^\infty \frac{1}{k^2} = \frac{\pi^2}{8},$$

so

$$\sum_1^\infty \frac{1}{k^2} = \frac{4}{3} \cdot \frac{\pi^2}{8} = \frac{\pi^2}{6}.$$

B.16

THE SEQUENCE
OF PRIMES

Number theory is mainly concerned with properties of the familiar positive integers $1, 2, 3, \ldots$. The notion of a positive integer is perhaps the simplest and clearest of all mathematical concepts, and yet, as we shall see, it is easy to ask elementary questions about these numbers which are incapable of being answered by the deepest resources of modern mathematics. This striking mixture of simplicity and profundity is part of the enduring attraction of the subject.

In Section B.2 we considered several interesting topics in the theory of numbers whose treatment did not depend on complicated mathematical machinery. We continue here with some further topics of this kind, and we also extend the range of our inquiry to include a few ideas that cannot be understood without some knowledge of calculus. We begin by reminding students of several concepts that were briefly discussed in the earlier section.

It is obvious that every positive integer is divisible by 1 and by itself. If an integer $p > 1$ has no positive divisors except 1 and p, it is called a *prime number,* or simply a *prime*; otherwise, it is said to be *composite*. The first few primes are easily seen to be

$$2, 3, 5, 7, 11, 13, 17, 19, 23, 29, 31, 37, 41, 43, \ldots .$$

It is a matter of common experience that every positive integer > 1 either is a prime or can be split into prime factors. Thus, for example, $45 = 3 \cdot 15 = 3 \cdot 3 \cdot 5$, $84 = 2 \cdot 42 = 2 \cdot 2 \cdot 21 = 2 \cdot 2 \cdot 3 \cdot 7$ and $630 = 2 \cdot 315 = 2 \cdot 3 \cdot 105 = 2 \cdot 3 \cdot 3 \cdot 35 =$

$2 \cdot 3 \cdot 3 \cdot 5 \cdot 7$. We also feel that we ought to get the same prime factors no matter what method of factorization is used. These remarks are the content of the *unique factorization theorem*—also called the *fundamental theorem of arithmetic*—which we state formally as follows.

Theorem 1. *Every positive integer > 1 either is a prime or can be expressed as a product of primes, and this expression is unique except for the order of the prime factors.*

Proof. We ask students to consider this in Problem 1.

This statement seems at first sight to be so obviously true that most people are inclined to take it for granted and accept it without proof. Nevertheless, it is far from trivial, and it takes on greater significance when one encounters systems of "integers" and "primes" for which it is false (see Problem 2). The questions that arise in this connection lead into the branch of modern mathematics known as *algebraic number theory*.

As we move out along the sequence of positive integers, we notice that the primes seem to occur less and less frequently. This is quite reasonable; a large number is more likely to be composite than a small one, since it lies beyond more numbers that might qualify as its factors. It is even conceivable that the primes might come to an end, and that all sufficiently large numbers are composite. Euclid's proof that this is not the case has been a model of mathematical elegance for more than two thousand years.

Theorem 2 (Euclid's Theorem). *There are infinitely many primes*

Proof. It suffices to show that if p is any given prime, then there exists a prime $> p$. Let $2, 3, 5, \ldots, p$ be the complete list of primes up to p. If we form the number $N = (2 \cdot 3 \cdot 5 \cdots p) + 1$, then it is clear that $N > p$ and also that N is not divisible by any of the primes $2, 3, 5, \ldots, p$. However, we know that N either is itself a prime or is divisible by some prime $q < N$, and in the latter case the preceding remark implies that $q > p$. Thus, in each case there exists a prime $> p$, and this is what we set out to prove.

Since every prime > 2 is odd, Euclid's theorem is equivalent to the assertion that the arithmetic progression

$$1, 3, 5, \ldots, 2n + 1, \ldots$$

of all odd positive integers contains infinitely many primes. It is therefore natural to wonder about primes in other arithmetic progressions. For example, it is clear that every odd prime lies in one of the two progressions

(a) $1, 5, 9, 13, 17, \ldots, 4n + 1, \ldots$;
(b) $3, 7, 11, 15, 19, \ldots, 4n + 3, \ldots$.

We know that both progressions together contain infinitely many primes, but it is still possible that one of them might contain only a finite number. We can dispose of this possibility for progression (b) by a slight refinement of Euclid's argument.

> **Theorem 3.** *The arithmetic progression* 3, 7, 11, ..., $4n + 3$, ... *contains infinitely many primes.*

> **Proof.** It is clear that the general term of our progression can also be written as $4n - 1$. Just as in Theorem 2, we show that if p is any given prime of this form, then there necessarily exists a larger prime of this form. Let 3, 7, 11, ..., p be the complete list of primes in progression (b) up to p, and form the number $N = 4(3 \cdot 7 \cdot 11 \cdots p) - 1$. It is clear that $N > p$ and also that N is not divisible by 2 or any of the primes 3, 7, 11, ..., p. If N is prime, then it is itself a prime of the form $4n - 1$ which is $> p$. Suppose that N is not prime. Then by the previous remark it is a product of odd primes $< N$ which cannot include any of the primes 3, 7, 11, ..., p. We have noted that every odd prime is of the form $4n + 1$ or $4n - 1$, and since

$$(4m + 1)(4n + 1) = 4(4mn + m + n) + 1,$$

> it is clear that any product of numbers of the form $4n + 1$ is again of this form. These facts imply that in our present situation, that is, where N is a product of odd primes which cannot include any of the primes 3, 7, 11, ..., p, it must be true that at least one of its prime factors is of the form $4n - 1$ and therefore $> p$. We conclude that in each case there exists a prime in progression (b) which is $> p$, and this completes the proof.

It is also true that progression (a) contains infinitely many primes. However, the idea used in the proof of Theorem 3 breaks down in this case and must be replaced by another (see Problem 3). A similar situation arises with the two progressions whose terms are of the form $6n + 1$ and $6n + 5$, for together they clearly contain all primes except 2 and 3, and reasonably elementary methods suffice to show that there are infinitely many primes in each of them.

Let us now look for primes in a general arithmetic progression

$$a, a + b, a + 2b, \ldots, a + nb, \ldots,$$

where a and b are given positive integers. It is easy to see that none are present (except perhaps a itself) if a and b have a common factor > 1. If we exclude this case, then it is natural to conjecture that the progression will contain infinitely many primes. This generalizes our previous statements about primes in special arithmetic progressions, and is the content of a famous theorem proved by the German mathematician Dirichlet in 1837.

> **Theorem 4** **(Dirichlet's Theorem).** *If a and b are positive integers with no*

common factor > 1, *then the arithmetic progression* a, $a + b$, $a + 2b$, ... , $a + nb$, ... *contains infinitely many primes.*

The methods of proof that work in the special cases we have discussed are unable to cope with the general arithmetic progression of this theorem. Dirichlet's proof used ideas and techniques from advanced analysis, and opened up new lines of thought in the theory of numbers which have been richly productive down to the present day. (*Analysis* is the standard term for that part of mathematics consisting of calculus and other subjects that depend on calculus more or less directly, such as differential equations, advanced calculus, etc.)

Many of the most interesting and important facts about primes were discovered by a combination of observation and experiment. In this kind of investigation it is useful to have available a list of all primes up to some prescribed limit N. An obvious method for constructing such a list is to write down in order all the integers from 2 to N and then systematically eliminate the composite numbers. Thus, since 2 is the first prime and every proper multiple of 2 is composite, we strike out 4, 6, 8, etc. The next remaining number is 3, which is prime because it is not a multiple of the only prime smaller than itself, namely 2. Since the proper multiples of 3 are composite, we strike out all of these numbers not already removed as multiples of 2. The next survivor is 5, which is prime because it is not a multiple of 2 or 3, so we strike out all proper multiples of 5 not already removed in the previous steps. And so on. We note that a composite number n must have a prime factor $\leq \sqrt{n}$. This shows that the process is complete when we have eliminated the proper multiples of all primes $\leq \sqrt{N}$.

The procedure described here is called the *sieve of Eratosthenes,* after its discoverer.[1] The result of applying it to the case $N = 100$ is given in the following table. It should be noted that since $\sqrt{100} = 10$, the table is complete by the time all proper multiples of 7 have been struck out. For the sake of emphasis, the primes are printed in bold.

---

[1] The Greek scientist Eratosthenes (276–194 B.C.) was custodian of the famous Library at Alexandria. He wrote on astronomy, geography, chronology, ethics, mathematics, and other subjects. He is remembered mainly for his prime number sieve and for being the first to measure accurately (within 50 miles!) the circumference of the earth. Being informed that at noon on the summer solstice the sun illuminated the bottom of a well at Aswan (i.e., was directly overhead), he measured the angle between the zenith and the sun at noon on the summer solstice at Alexandria and also the distance between Alexandria and Aswan, which lie on roughly the same meridian. A simple calculation then gave the circumference of the earth. He was also a friend of Archimedes (287–212 B.C.)—the greatest intellect of antiquity—and was the recipient of a famous letter (called the *Method*) in which Archimedes revealed his method of making mathematical discoveries. In his old age Eratosthenes went blind, and is said to have committed suicide by starving himself to death.

	2	3	4	5	6	7	8	9	10	11	12	13	14	15	16	17	18	19	20
21	22	23	24	25	26	27	28	29	30	31	32	33	34	35	36	37	38	39	40
41	42	43	44	45	46	47	48	49	50	51	52	53	54	55	56	57	58	59	60
61	62	63	64	65	66	67	68	69	70	71	72	73	74	75	76	77	78	79	80
81	82	83	84	85	86	87	88	89	90	91	92	93	94	95	96	97	98	99	100

Complete tables of primes up to more than 10,000,000 have been compiled by refinements of this process.[2] These tables provide the investigator with an enormous mass of raw data which can be used to formulate and test hypotheses, almost as if the study of prime numbers were a laboratory science. For example, an inspection of our short table shows that there are several chains of 5 consecutive composite numbers, and one of 7. Can the length of such a chain be made as large as we please? The answer to this question is easily seen to be Yes, for if n is a large positive integer, then $n!+2$, $n!+3$, $n!+4, \ldots, n!+n$ is a chain of $n-1$ consecutive composite numbers. On the other hand, the primes tend to cluster together here and there. In our table there are 8 pairs of *twin primes*, that is, primes like 3 and 5 and 29 and 31 which are separated by a single even number. Are there infinitely many such pairs? The longest tables suggest that there are, but no one knows for sure. In 1921 the Norwegian mathematician Viggo Brun was able to generalize the sieve of Eratosthenes to show that the sum of the reciprocals of the twin primes,

$$\frac{1}{3}+\frac{1}{5}+\frac{1}{7}+\frac{1}{11}+\frac{1}{13}+\frac{1}{17}+\frac{1}{19}+\frac{1}{29}+\frac{1}{31}+\cdots,$$

is either finite or convergent. This result is to be contrasted with the fact (which is proved in Section B.19) that the sum of the reciprocals of all the primes diverges.

We have seen that the primes are very irregularly distributed among all the positive integers. The problem of discovering the law that governs their occurrence—and of understanding the reasons for it—is one that has challenged human curiosity for hundreds of years.[3]

Many attempts have been made to find simple formulas for the nth prime p_n and for the exact number of primes among the first n positive integers. All such efforts have failed, and real progress was achieved only when mathematicians started instead to look for information about the *average* distribution of the primes among the positive integers. It is customary to

---

[2] The standard reference is D. N. Lehmer, *List of Prime Numbers from 1 to* 10,006,721, Carnegie Institution of Washington Publication 165, 1914.

[3] In 1751 Euler expressed his own bafflement as follows: "Mathematicians have tried in vain to this day to discover some order in the sequence of prime numbers, and we have reason to believe that it is a mystery into which the human mind will never penetrate." Fortunately, Euler was wrong in this pessimistic forecast.

denote by $\pi(x)$ the number of primes $\leq$ a positive number x. Thus, $\pi(1) = 0$, $\pi(2) = 1$, $\pi(3) = 2$, $\pi(4) = 2$, and $\pi(p_n) = n$ for every positive integer n. In his early youth the great Gauss studied this function by means of tables of primes, with the aim of finding a simple function that approximates $\pi(x)$ with a small relative error for large x. More precisely, he sought a function $f(x)$ with the property that

$$\lim_{x \to \infty} \frac{f(x) - \pi(x)}{\pi(x)} = \lim_{x \to \infty} \left(\frac{f(x)}{\pi(x)} - 1 \right) = 0,$$

that is, such that

$$\lim_{x \to \infty} \frac{\pi(x)}{f(x)} = 1.$$

On the basis of his observations he conjectured (in 1792, at the age of 14 or 15) that both

$$\frac{x}{\ln x} \quad \text{and} \quad \text{li}(x) = \int_2^x \frac{dt}{\ln t}$$

are good approximations. The function $\text{li}(x)$ is known as the *logarithmic integral*. The accompanying table shows how well these functions succeed.

x	$\pi(x)$	$x/\ln x$	$\text{li}(x)$
1,000	168	145	178
10,000	1,229	1,086	1,246
100,000	9,592	8,686	9,630
1,000,000	78,498	72,382	78,628
10,000,000	664,579	620,421	664,918

Even as an adult, Gauss was unable to prove his conjectures. The first solidly established results in this direction were attained around 1850 by the Russian mathematician Chebyshev, who showed that the inequalities

$$\frac{7}{8} < \frac{\pi(x)}{x/\ln x} < \frac{9}{8}$$

are true for all sufficiently large x. He also proved that if the limit

$$\lim_{x \to \infty} \frac{\pi(x)}{x/\ln x}$$

exists, then its value must be 1. The next step—a very long one—was taken by Riemann in 1859, in a brief paper of only 9 pages which is famous for its wealth of profound ideas. Riemann, however, merely sketched his proofs, and omitted some of them altogether, so his work was inconclusive in several respects. The end of this part of the story came in 1896 when Hadamard and de la Vallée Poussin, working independently of each other but building on Riemann's ideas, established the existence of this limit and thereby completed the proof of the *prime number theorem*:

$$\lim_{x \to \infty} \frac{\pi(x)}{x/\ln x} = 1. \tag{1}$$

This relatively simple law is one of the most remarkable facts in the whole of mathematics. If we write it in the form

$$\lim_{x \to \infty} \frac{\pi(x)/x}{1/\ln x} = 1, \tag{2}$$

then it admits the following interesting interpretation in terms of probability. If n is a positive integer, then the ratio $\pi(n)/n$ is the proportion of primes among the integers $1, 2, \ldots, n$, or equivalently, the probability that one of these integers chosen at random will be prime. We can think of (2) as asserting that this probability is approximately $1/\ln n$ for large n.

It follows quite easily from the prime number theorem that the nth prime is approximately $n \ln n$, in the sense that

$$\lim_{n \to \infty} \frac{p_n}{n \ln n} = 1. \tag{3}$$

To prove this, we use the fact that $\pi(p_n) = n$ and infer from (1) that

$$\lim_{n \to \infty} \frac{n}{p_n/\ln p_n} = 1 \quad \text{or} \quad \lim_{n \to \infty} \frac{p_n}{n \ln p_n} = 1. \tag{4}$$

If we now take the logarithm of (4) and use the continuity of the logarithm in the form $\ln \lim = \lim \ln$, we get

$$\lim_{n \to \infty} (\ln p_n - \ln n - \ln \ln p_n) = 0$$

or

$$\lim_{n \to \infty} \ln p_n \left[1 - \frac{\ln n}{\ln p_n} - \frac{\ln \ln p_n}{\ln p_n} \right] = 0.$$

This implies that the bracketed expression must approach 0; and since the third term in it also approaches 0 [recall that $(\ln n)/n \to 0$], we must have

$$\lim_{n \to \infty} \frac{\ln n}{\ln p_n} = 1. \tag{5}$$

With the aid of (5), (4) now yields

$$\lim_{n \to \infty} \frac{p_n}{n \ln n} = \lim_{n \to \infty} \frac{p_n}{n \ln p_n} \cdot \frac{\ln p_n}{\ln n} = 1,$$

which concludes the proof of (3).

It is also interesting to see that the prime number theorem is equivalent to the statement that

$$\lim_{x\to\infty} \frac{\pi(x)}{\text{li}(x)} = 1. \tag{6}$$

To prove this, it suffices to show that

$$\lim_{x\to\infty} \frac{\text{li}(x)}{x/\ln x} = 1; \tag{7}$$

for if this is so, then

$$\lim_{x\to\infty} \frac{\pi(x)}{x/\ln x} = \lim_{x\to\infty} \frac{\pi(x)}{\text{li}(x)} \cdot \frac{\text{li}(x)}{x/\ln x} = \lim_{x\to\infty} \frac{\pi(x)}{\text{li}(x)}.$$

We establish (7) as follows. On integrating li(x) by parts, we get

$$\text{li}(x) = \int_2^x \frac{dt}{\ln t} = \frac{x}{\ln x} - \frac{2}{\ln 2} + \int_2^x \frac{dt}{(\ln t)^2}. \tag{8}$$

Since $1/(\ln t)^2$ is positive and decreasing for $t > 1$, if $x \geq 4$ we have

$$0 < \int_2^x \frac{dt}{(\ln t)^2} = \int_2^{\sqrt{x}} \frac{dt}{(\ln t)^2} + \int_{\sqrt{x}}^x \frac{dt}{(\ln t)^2}$$

$$< \frac{\sqrt{x} - 2}{(\ln 2)^2} + \frac{x - \sqrt{x}}{(\ln \sqrt{x})^2}$$

$$< \frac{\sqrt{x}}{(\ln 2)^2} + \frac{4x}{(\ln x)^2}.$$

This yields

$$0 < \frac{\int_2^x dt/(\ln t)^2}{x/\ln x} < \frac{\ln x}{\sqrt{x}(\ln 2)^2} + \frac{4}{\ln x},$$

so

$$\lim_{x\to\infty} \frac{\int_2^x dt/(\ln t)^2}{x/\ln x} = 0. \tag{9}$$

If (8) is divided by $x/\ln x$, then (7) follows at once from (9), and the proof is complete. This result shows that both of Gauss's youthful conjectures were vindicated when the prime number theorem was finally proved.

The prime number theorem, as a statement involving the logarithm function and a limit, is obviously related to analysis. This is very surprising in view of the fact that the primes are discrete objects that have no apparent connection with the continuous functions and limit processes that are the essence of analysis. Nevertheless, almost all significant work on both Dirichlet's theorem and the prime number theorem has depended on the advanced analytical machinery of infinite series, complex variable theory, Fourier transforms, and the like. Accordingly, this part of mathematics has come to be known as *analytic number theory*.[4]

PROBLEMS

1. (a) If n is an integer >1 which is not a prime, show that it can be expressed as a product of primes. Hint: There exist integers a and b, both of which are >1 and $<n$, such that $n = ab$.

   (b) If $p_1 p_2 \cdots p_m = q_1 q_2 \cdots q_n$, where the p's and q's are primes such that $p_1 \leq p_2 \leq \cdots \leq p_m$ and $q_1 \leq q_2 \leq \cdots \leq q_n$, show that $m = n$ and $p_1 = q_1$, $p_2 = q_2, \ldots, p_n = q_n$. Hint: Assume that a prime which divides a product of positive integers necessarily divides one of the factors (this fact is called *Euclid's lemma*).

2. (This problem is intended to suggest that the unique factorization theorem may not be as "obvious" as it looks.) The arithmetic progression $1, 4, 7, 10, \ldots, 3n + 1, \ldots$ is a system of numbers which—like the positive integers—is closed under multiplication, in the sense that the product of any two numbers in the progression is again in the progression. Let a number $p > 1$ in this progression be called *pseudo-prime* if its only factorization into factors which are both in the progression is $p = 1 \cdot p$.

   (a) Show that every number >1 in the progression either is a pseudo-prime or can be expressed as a product of pseudo-primes.

   (b) List all pseudo-primes ≤ 100.

   (c) Find a number in the progression that can be expressed as a product of pseudo-primes in two different ways.

3. (a) It is asserted in the text that the arithmetic progression $1, 5, 9, 13, 17, \ldots, 4n + 1, \ldots$ contains infinitely many primes. Try to prove this by imitating the proof of Theorem 3, that is, by listing all the primes in the progression up to some given prime p $(5, 13, 17, \ldots, p)$ and considering the number $N = 4(5 \cdot 13 \cdot 17 \cdots p) + 1$. At what point does this attempted proof break down?

---

[4] For additional information on the topics discussed above, see H. M. Edwards, *Riemann's Zeta Function*, Academic Press, 1974, pp. 1–6; and T. M. Apostol, *Introduction to Analytic Number Theory*, Springer-Verlag, 1976, pp. 1–12. For some extremely interesting plausibility discussions that convey an intuitive feeling for the meaning of the prime number theorem, see David Hawkins, "Mathematical Sieves," *Scientific American*, December 1958; and R. Courant and H. Robbins, *What Is Mathematics?* Oxford University Press, 1941, pp. 482–86.

(b) It is known (and was first proved by Euler in 1749) that any odd prime factor of a number of the form $a^2 + 1$ is necessarily of the form $4n + 1$. Use this to prove the assertion in (a) by considering the number

$$M = (2 \cdot 5 \cdot 13 \cdot 17 \cdots p)^2 + 1$$
$$= 4(5 \cdot 13 \cdot 17 \cdots p)^2 + 1.$$

4. Prove that the arithmetic progression $5, 11, 17, \ldots, 6n + 5, \ldots$ contains infinitely many primes. Hint: The general term of this progression can be written $6n - 1$.

5. Prove equation (7) by using L'Hospital's rule.

B.17

MORE ABOUT IRRATIONAL NUMBERS. π IS IRRATIONAL

Readers who have never thought about the matter before may wonder why we care about irrational numbers. In order to understand this, let us assume for a moment that the only numbers we have are the rationals—which, after all, are the only numbers ever used in making scientific measurements. Under these circumstances the symbol $\sqrt{2}$ has no meaning, since there is no rational number whose square is 2 (see Section B.2). One consequence of this is that the circle $x^2 + y^2 = 4$ and the straight line $y = x$ through its center do not intersect; that is, in spite of appearances, there is no point that lies on both, because both curves are discontinuous in the sense of having many missing points, and each threads its way through a gap in the other. This suggests that the system of rational numbers is an inadequate tool for representing the continuous objects of geometry and the continuous motions of physics. In addition, without irrational numbers most sequences and series would not converge and most integrals would not exist; and since it is also true that e and π would be meaningless (we prove below that π is irrational), the enormous and intricate structure of mathematical analysis would collapse into a heap of rubbish so insignificant as to be hardly worth sweeping up. As a practical matter, it is clear that if the irrationals did not exist, it would be necessary to invent them. It was the ancient Greeks who discovered that irrational numbers are indispensable in geometry, and this was one of their more important contributions to civilization.

It is not difficult to prove that e is irrational by assuming the contrary and constructing a number a which is then shown to be a positive integer < 1—an obvious impossibility (for the details see the Appendix). This strategy is also the key to the proofs of the following two theorems, but the details are somewhat more complicated.

We shall need a few properties of the function $f(x)$ defined by

$$f(x) = \frac{x^n(1-x)^n}{n!} = \frac{1}{n!} \sum_{k=n}^{2n} c_k x^k, \tag{1}$$

where the c_k's are certain integers and n is a positive integer to be specified later. First, it is clear that if $0 < x < 1$, then we have

$$0 < f(x) < \frac{1}{n!}. \tag{2}$$

Next, $f(0) = 0$ and $f^{(m)}(0) = 0$ if $m < n$ or $m > 2n$; also, if $n \le m \le 2n$, then

$$f^{(m)}(0) = \frac{m!}{n!} c_m,$$

and this number is an integer. Thus, $f(x)$ and all its derivatives have integral values at $x = 0$. Since $f(1-x) = f(x)$, the same is true at $x = 1$.

Theorem 1. *e^r is irrational for every rational number $r \ne 0$.*

Proof. If $r = p/q$ and e^r is rational, then so is $(e^r)^q = e^p$. Also, if e^{-p} is rational, so is e^p. It therefore suffices to prove that e^p is irrational for every positive integer p.

Assume that $e^p = a/b$ for certain positive integers a and b. We define $f(x)$ by (1) and $F(x)$ by

$$F(x) = p^{2n}f(x) - p^{2n-1}f'(x) + p^{2n-2}f''(x) - \cdots - pf^{(2n-1)}(x) + f^{(2n)}(x), \tag{3}$$

and we observe that $F(0)$ and $F(1)$ are integers. Next,

$$\frac{d}{dx}[e^{px}F(x)] = e^{px}[F'(x) + pF(x)] = p^{2n+1}e^{px}f(x), \tag{4}$$

where the last equality is obtained from a detailed examination of $F'(x) + pF(x)$ based on (3). Equation (4) shows that

$$b\int_0^1 p^{2n+1}e^{px}f(x)\,dx = b\left[e^{px}F(x)\right]_0^1 = aF(1) - bF(0),$$

which is an integer. However, (2) implies that

$$0 < b\int_0^1 p^{2n+1}e^{px}f(x)\,dx < \frac{bp^{2n+1}e^p}{n!} = bpe^p\frac{(p^2)^n}{n!};$$

and since the expression on the right $\to 0$ as $n \to \infty$, it follows that the integer $aF(1) - bF(0)$ has the property that

$$0 < aF(1) - bF(0) < 1$$

if n is large enough. Since there is no positive integer < 1, this contradiction completes the proof.

If we say that a point (x, y) in the plane is a *rational point* whenever both x and y are rational numbers, then this theorem asserts that the curve $y = e^x$ traverses the plane in such a way that it misses all rational points except $(0, 1)$. An equivalent statement is that $y = \ln x$ misses all rational points except $(1, 0)$, so $\ln 2, \ln 3, \ldots$ are all irrational. It can also be proved that $y = \sin x$ misses all rational points except $(0, 0)$, and that $y = \cos x$ misses all rational points except $(0, 1)$.[1] Each of these theorems implies that π is irrational, since $\sin \pi = 0$ and $\cos \pi = -1$. However, we prefer to prove the irrationality of π by the following more direct argument.

Theorem 2. *π is irrational.*

Proof. If π were rational, then π^2 would be rational, so it is sufficient to prove that π^2 is irrational. We assume the contrary, that $\pi^2 = a/b$ for certain positive integers a and b. We again define $f(x)$ by (1), but this time we put

$$F(x) = b^n[\pi^{2n}f(x) - \pi^{2n-2}f''(x) + \pi^{2n-4}f^{(4)}(x) - \cdots + (-1)^n f^{(2n)}(x)], \qquad (5)$$

and again we observe that $F(0)$ and $F(1)$ are integers. A calculation based on (5) shows that

$$\frac{d}{dx}[F'(x) \sin \pi x - \pi F(x) \cos \pi x] = [F''(x) + \pi^2 F(x)] \sin \pi x$$

$$= b^n \pi^{2n+2} f(x) \sin \pi x = \pi^2 a^n f(x) \sin \pi x,$$

so

$$\int_0^1 \pi a^n f(x) \sin \pi x \, dx = \left[\frac{F'(x) \sin \pi x}{\pi} - F(x) \cos \pi x\right]_0^1$$

$$= F(1) + F(0),$$

which is in integer. But (2) implies that

$$0 < \int_0^1 \pi a^n f(x) \sin \pi x \, dx < \frac{\pi a^n}{n!} < 1$$

if n is large enough; and this contradiction—that $F(1) + F(0)$ is a positive integer < 1—concludes the proof.

The underlying method of proof in Theorems 1 and 2 was devised by the French mathematician Hermite in 1873, but the details of the latter argument were first published by Niven in 1947.

---

[1] The details can be found in Chapter II of I. Niven's excellent book, *Irrational Numbers*, Wiley, 1956.

APPENDIX:
A PROOF THAT e IS IRRATIONAL

We use several of the simpler properties of series to prove the theorem of Euler stated in the title.

Our starting point is the fact that

$$e = 1 + 1 + \frac{1}{2!} + \cdots + \frac{1}{n!} + \cdots ,$$

from which it follows that the number

$$e - 1 - 1 - \frac{1}{2!} - \cdots - \frac{1}{n!} = \frac{1}{(n+1)!} + \frac{1}{(n+2)!} + \cdots \tag{1}$$

is positive for every positive integer n. We assume that e is rational, so that $e = p/q$ for certain positive integers p and q, and we deduce a contradiction from this assumption. Let n in (1) be chosen so large that $n > q$, and define a number a by

$$a = n! \left[e - 1 - 1 - \frac{1}{2!} - \cdots - \frac{1}{n!} \right].$$

Since q divides $n!$, a is a positive integer. However, (1) implies that

$$a = n! \left[\frac{1}{(n+1)!} + \frac{1}{(n+2)!} + \cdots \right]$$

$$= \frac{1}{n+1} + \frac{1}{(n+1)(n+2)} + \cdots$$

$$< \frac{1}{n+1} + \frac{1}{(n+1)^2} + \cdots$$

$$= \frac{1}{n+1} \left[1 + \frac{1}{n+1} + \frac{1}{(n+1)^2} + \cdots \right]$$

$$= \frac{1}{n+1} \cdot \frac{1}{1 - 1/(n+1)} = \frac{1}{n}.$$

This contradiction (there is no positive integer $< 1/n$) completes the argument.

B.18

ALGEBRAIC AND TRANSCENDENTAL NUMBERS. e IS TRANSCENDENTAL

In Section B.17 we considered the classification of real numbers into the rationals and the irrationals. We now discuss a similar but much deeper and more substantive distinction, between algebraic and transcendental numbers.

A real number is said to be *algebraic* if it satisfies a polynomial equation of the form

$$a_n x^n + a_{n-1} x^{n-1} + \cdots + a_1 x + a_0 = 0 \tag{1}$$

with integral coefficients, where $a_n \neq 0$; if it satisfies no such equation, it is called *transcendental*.[1] For example, $\sqrt{2}$ and $\sqrt[3]{2 + \sqrt{2}}$ are algebraic because they are roots of $x^2 - 2 = 0$ and $x^6 - 4x^3 + 2 = 0$. Any rational number p/q is algebraic, since it satisfies the first-degree equation $qx - p = 0$. The algebraic numbers can therefore be viewed as a natural generalization of the rational numbers. If an algebraic number satisfies (1) but no such equation of lower degree, it is said to be of *degree n*. Thus, the rationals are of degree 1, and $\sqrt{2}$ is of degree 2 because it is irrational and satisfies a second-degree equation.

---

[1] Complex numbers are classified in exactly the same way, but in most of this section we shall be concerned with real numbers alone.

As their name suggests, algebraic numbers are of great interest and importance for many reasons stemming from both algebra and number theory. We shall briefly explain how they also arise in geometry, and then conclude with some classical applications of analysis to this subject.

GEOMETRIC CONSTRUCTIONS

The ancient Greek mathematicians were fond of geometric problems requiring the construction of figures with specified properties. They often succeeded in solving their problems, but sometimes they failed. Most notably they failed in squaring a circle. By this they meant starting with a given circle and constructing a square of equal area. In this context the word "construct" has a special meaning, for the only instruments allowed in the game are an unmarked ruler and a compass.

Any ruler-and-compass construction starts with certain initial data, consisting of a finite collection of points, lines, and circles in a plane. The construction then proceeds to generate new points, lines, and circles by means of some combination of the following operations: With the ruler we can draw the line determined by two given points; with the compass we can draw the circle whose center is a given point and whose radius is the distance between two given points; and we can find the points of intersection (if any) of given lines and circles. When it is understood in this way, the problem of squaring a circle reduces to that of starting with a segment of length 1 (we can take the radius of the given circle as the unit of measurement) and constructing a segment of length $\sqrt{\pi}$. This problem remained unsolved for more than 2000 years, and was settled only in the nineteenth century by the German mathematician Lindemann, who proved it to be unsolvable; that is, it is impossible to "square the circle."

If we start with a single segment of length 1, then it is quite easy to show that the length of any constructible segment is an algebraic number whose degree is a power of 2.[2] In particular, if a real number is not algebraic at all, then it is certainly not constructible in our present sense. This is the way Lindemann proved that a circle cannot be squared: He demonstrated (in 1882) that π is transcendental and therefore not constructible, and from this it

---

[2] A very clear elementary discussion can be found in Chapter III of R. Courant and H. Robbins, *What Is Mathematics?*, Oxford University Press, 1941. For somewhat deeper treatments based more firmly on the theory of field extensions in modern algebra, see Burton W. Jones, *An Introduction to Modern Algebra*, Macmillan, 1975; I. T. Adamson, *Introduction to Field Theory*, Oliver and Boyd, 1964; Seth Warner, *Modern Algebra*, Prentice-Hall, 1965; or C. R. Hadlock, *Field Theory and Its Classical Problems*, Carus Mathematical Monograph No. 19, Mathematical Association of America, 1978.

follows that $\sqrt{\pi}$ is also not constructible.[3] Lindemann's proof is rather difficult, so we now turn our attention to the simpler but still nontrivial problem of showing that transcendental numbers actually exist.

LIOUVILLE'S THEOREM

The first numbers that were known to be transcendental were exhibited by the French mathematician Liouville in 1851. He produced his examples with the aid of the following theorem, which in a loose interpretation says that an irrational algebraic number cannot be approximated very closely by rational numbers p/q unless the denominators q are allowed to be very large.

Theorem 1. *If α is a real algebraic number of degree $n > 1$, then there exists a constant $c > 0$ such that*

$$\left| \alpha - \frac{p}{q} \right| \geq \frac{c}{q^n}$$

for all rational numbers p/q with $q > 0$.

Proof. The assumption is that α is a real root of a polynomial equation

$$f(x) = a_n x^n + a_{n-1} x^{n-1} + \cdots + a_1 x + a_0 = 0,$$

where the coefficients are integers, $n > 1$, and $a_n \neq 0$; and also that α does not satisfy any such equation of lower degree (and in particular is not rational). Let M be any upper bound for $|f'(x)|$ on the interval $\alpha - 1 \leq x \leq \alpha + 1$. If c is defined to be the smaller of the numbers 1 and $1/M$, we shall prove that c has the required property.

First, if $|\alpha - p/q| \geq 1$, then it is obvious that

$$\left| \alpha - \frac{p}{q} \right| \geq c \geq \frac{c}{q^n},$$

since q is a positive integer.

If $|\alpha - p/q| < 1$, we argue as follows. We begin by observing that p/q cannot be a root of $f(x) = 0$. For if it were, we could factor out $x - p/q$ from $f(x)$ and write the equation as $(x - p/q)g(x) = 0$, where $g(x)$ is a polynomial of degree $n - 1$ with integral coefficients; and α (being irrational) would then satisfy the equation $g(x) = 0$, contrary to the hypothesis. We therefore have $f(p/q) \neq 0$, so

$$\left| f\left(\frac{p}{q} \right) \right| = \frac{|a_n p^n + a_{n-1} p^{n-1} q + \cdots + a_0 q^n|}{q^n} \geq \frac{1}{q^n}.$$

---

[3] Ferdinand Lindemann (1852–1939) was a pupil of Weierstrass and taught at the University of Munich from 1893 until his retirement. There is a bust of him at this institution, and beneath his engraved name is the letter π, framed by a circle and a square in memory of his one great scientific achievement (one like this is enough!). He also tried fruitlessly for many years to prove Fermat's last theorem.

Next, the mean value theorem implies that

$$\left| f\left(\frac{p}{q}\right)\right| = \left| f\left(\frac{p}{q}\right) - f(\alpha)\right| = \left|\frac{p}{q} - \alpha\right| \left| f'(a)\right|$$

for some number a that lies between p/q and α, and hence is in the interval on which $|f'(x)|$ is bounded by M. This yields

$$\frac{1}{q^n} \le \left| f\left(\frac{p}{q}\right)\right| \le \left|\alpha - \frac{p}{q}\right| M,$$

so

$$\left|\alpha - \frac{p}{q}\right| \ge \frac{1}{M} \cdot \frac{1}{q^n} \ge \frac{c}{q^n},$$

and the proof is complete.

As a direct consequence of this theorem, we have the following: If α is an irrational number with the property that for each $n > 1$ there exist rationals p/q making

$$q^n \left|\alpha - \frac{p}{q}\right|$$

as small as we please, then α cannot be algebraic and therefore must be transcendental. This fact enables us to produce many specific examples of transcendental numbers, for instance

$$\alpha = \frac{1}{10^{1!}} + \frac{1}{10^{2!}} + \frac{1}{10^{3!}} + \cdots + \frac{1}{10^{m!}} + \cdots$$

$$= 0.110001000000000000000000100 \ldots.$$

To see this, we first note that α is irrational because this decimal is nonrepeating. Next, if an integer $n > 1$ is given and for any $m > n$ we approximate α by the mth partial sum p/q of the series, then $q = 10^{m!}$ and

$$\left|\alpha - \frac{p}{q}\right| = \frac{1}{10^{(m+1)!}} + \frac{1}{10^{(m+2)!}} + \cdots < \frac{2}{10^{(m+1)!}}.$$

It now follows that

$$q^n \left|\alpha - \frac{p}{q}\right| = (10^{m!})^n \left|\alpha - \frac{p}{q}\right|$$

$$< 2 \cdot \frac{(10^{m!})^n}{10^{(m+1)!}} = 2\left(\frac{10^n}{10^{m+1}}\right)^{m!} \to 0$$

as $m \to \infty$, so α is transcendental. It is clear that we can produce many other examples of this type by considering the sums of the series

$$\frac{a_1}{10^{1!}} + \frac{a_2}{10^{2!}} + \cdots + \frac{a_m}{10^{m!}} + \cdots,$$

where the numerators are any of the digits $1, 2, \ldots, 9$. These irrational numbers are all transcendental, and the essential reason is that each is the limit of a very rapidly convergent sequence of rationals.

This approach makes it possible to exhibit many specimens of transcendental numbers, but it is much harder to prove the transcendence of particular numbers that occur more or less naturally in mathematics. The real beginning here was the work of Hermite, who showed in 1873 that e is transcendental (see below); Lindemann's 1882 proof for π was a fairly direct extension of Hermite's ideas. But subsequent progress was slow. In his famous Paris address of 1900, the great German mathematician David Hilbert proposed a list of 23 unsolved problems which he considered to be the outstanding challenges to the mathematicians of the future. The seventh of these (in part) was to prove that e^{π} and $2^{\sqrt{2}}$ are transcendental. Gelfond succeeded for e^{π} in 1929, and Kuzmin for $2^{\sqrt{2}}$ in 1930. This line of thought reached its climax in 1934–1935, when these facts took their place as special cases of a profound theorem of Gelfond and Schneider, which we state without proof as follows.[4]

Theorem 2. *If α and β are real or complex algebraic numbers such that α is neither 0 nor 1 and β is not a rational real number, then α^{β} is transcendental.*

The transcendence of $2^{\sqrt{2}}$ is an obvious consequence of this theorem, and for students who know something about complex numbers, that of e^{π} follows from the fact that e^{π} is one of the values of $e^{-2i \ln i} = i^{-2i}$, since the number last written has the required form. However, in spite of these advances, there are still many unsolved problems, and nothing is known about the nature of any of the following numbers: $e + \pi$, $e\pi$, π^{e}, 2^{e}, 2^{π}.

THE TRANSCENDENCE OF e

The argument we present is Hilbert's simplified version of Hermite's original proof.

We begin by assuming that e is algebraic of degree n, which means that it satisfies an equation of the form

$$a_n e^n + \cdots + a_2 e^2 + a_1 e + a_0 = 0, \tag{2}$$

---

[4] The main references are the book by Niven mentioned in Section B.17; C. L. Siegel, *Transcendental Numbers*, Princeton University Press, 1949; and A. O. Gelfond, *Transcendental and Algebraic Numbers*, Dover, 1960. See also Hilbert's address, "Mathematical Problems," *Bulletin of the American Mathematical Society*, vol. 8 (1902), pp 437–79. The circumstances surrounding Hilbert's address, which perhaps was the second most important scientific lecture ever given, are interestingly described in C. Reid's biography, *Hilbert*, Springer-Verlag, 1970, pp. 69–84.

where the coefficients are integers and a_n, $a_0 \neq 0$. The strategy of the proof is to reach a contradiction by closely approximating the various powers of e by rationals:

$$e^n = \frac{M_n + \epsilon_n}{M}, \ldots, \qquad e^2 = \frac{M_2 + \epsilon_2}{M}, \qquad e = \frac{M_1 + \epsilon_1}{M}, \qquad (3)$$

where M, M_n, $\ldots$, M_2, M_1 are integers and the errors ϵ_n/M, $\ldots$, ϵ_2/M, ϵ_1/M are very small. If we substitute (3) in (2) and multiply by M, then the result is

$$[a_n M_n + \cdots + a_1 M_1 + a_0 M] + [a_n \epsilon_n + \cdots + a_1 \epsilon_1] = 0. \qquad (4)$$

The first expression in brackets is an integer, and we will choose the M's so that it is not zero. At the same time we will choose the ϵ's to be so small that the second expression in brackets is <1 in absolute value,

$$|a_n \epsilon_n + \cdots + a_1 \epsilon_1| < 1. \qquad (5)$$

This contradiction (a nonzero integer plus a number of absolute value <1 cannot be 0) will then complete the proof.

Students will surely agree that this is a simple and reasonable plan. But the remarkable feature of Hermite's proof is the extremely ingenious way in which he defines the M's and ϵ's by means of an integral whose structure is precisely adapted to its purpose. Hermite's integral defines M:

$$M = \int_0^\infty \frac{x^{p-1}[(x-1)(x-2) \cdots (x-n)]^p e^{-x}}{(p-1)!} \, dx, \qquad (6)$$

where n is the degree of (2) and p is a prime to be specified later. We shall see that M is an integer for every choice of p, and it will be essential to know that p can be taken as large as we please (there are infinitely many primes!). It is clear from (3) that for each $k = 1, 2, \ldots, n$ we must have $M_k + \epsilon_k = e^k M$; and this relation will be satisfied if we define M_k and ϵ_k by breaking the interval of integration of $e^k M$ at the point k:

$$M_k = e^k \int_k^\infty \frac{x^{p-1}[(x-1) \cdots (x-n)]^p e^{-x}}{(p-1)!} \, dx, \qquad (7)$$

$$\epsilon_k = e^k \int_0^k \frac{x^{p-1}[(x-1) \cdots (x-n)]^p e^{-x}}{(p-1)!} \, dx. \qquad (8)$$

We now proceed to the detailed demonstration that for a suitable choice of p, these M's and ϵ's have the desired properties.

We will need the following formula from the elementary theory of the gamma function: For any positive integer m,

$$\int_0^\infty x^m e^{-x} \, dx = m!. \qquad (9)$$

The proof is straightforward, and begins with an integration by parts (with $u = x^m$, $dv = e^{-x}\, dx$):

$$\int_0^\infty x^m e^{-x}\, dx = \left[-x^m e^{-x} \right]_0^\infty + m \int_0^\infty x^{m-1} e^{-x}\, dx$$

$$= m \int_0^\infty x^{m-1} e^{-x}\, dx.$$

The integral on the right is the same as the one on the left except that the exponent of x has been reduced by 1, so continuing in the same way yields

$$\int_0^\infty x^m e^{-x}\, dx = m \int_0^\infty x^{m-1} e^{-x}\, dx = m(m-1) \int_0^\infty x^{m-2} e^{-x}\, dx$$

$$= \cdots = m(m-1)(m-2) \cdots 2 \cdot 1 \int_0^\infty e^{-x}\, dx = m!,$$

since the value of the last integral is 1.

We now turn to the evaluation of the integral (6). If the expression $[(x-1)(x-2) \cdots (x-n)]$ is multiplied out, the result is a polynomial

$$x_n + \cdots \pm n!$$

with integral coefficients; raising this to the pth power yields

$$[(x-1)(x-2) \cdots (x-n)]^p = x^{np} + \cdots \pm (n!)^p,$$

where again the coefficients are integers. This enables us to write (6) in the form

$$M = \frac{\pm(n!)^p}{(p-1)!} \int_0^\infty x^{p-1} e^{-x}\, dx + \sum_{i=1}^{np} \frac{b_i}{(p-1)!} \int_0^\infty x^{p-1+i} e^{-x}\, dx,$$

where the b_i's are integers, and an application of (9) gives

$$M = \pm(n!)^p + \sum_{i=1}^{np} b_i \frac{(p-1+i)!}{(p-1)!},$$

which is an integer. If we now restrict ourselves to primes $p > n$, then the first term here is an integer not divisible by p. However,

$$b_i \frac{(p-1+i)!}{(p-1)!} = b_i p(p+1) \cdots (p-1+i)$$

is divisible by p, so M is plainly an integer which is not divisible by p. We may add the further condition that $p > |a_0|$; and since $a_0 \neq 0$, it follows that the term $a_0 M$ in (4) is not divisible by p (if a prime divides a product, it must divide one of the factors).

We next consider the integral (7) defining M_k:

$$M_k = \int_k^\infty \frac{x^{p-1}[(x-1)\cdots(x-n)]^p e^{-(x-k)}}{(p-1)!}\,dx.$$

If a new variable $y = x - k$ is introduced, this becomes

$$M_k = \int_0^\infty \frac{(y+k)^{p-1}[(y+k-1)\cdots y \cdots (y+k-n)]^p e^{-y}}{(p-1)!}\,dy.$$

Now the expression in brackets contains y in the kth place, so its pth power is a polynomial in y with terms running from y^p to y^{np}. This tells us that

$$(y+k)^{p-1}[(y+k-1)\cdots y \cdots (y+k-n)]^p$$

is a polynomial with integral coefficients whose terms run from y^p to y^{p-1+np}, so

$$M_k = \sum_{i=1}^{np} \frac{c_i}{(p-1)!} \int_0^\infty y^{p-1+i} e^{-y}\,dy = \sum_{i=1}^{np} c_i \frac{(p-1+i)!}{(p-1)!},$$

where the c_i's are integers. By the same reasoning as before, each M_k is an integer divisible by p, so the first bracketed expression in (4) is an integer not divisible by p. We conclude that for the primes p under consideration, the number

$$a_n M_n + \cdots + a_1 M_1 + a_0 M$$

is a nonzero integer, for if it were zero it would be divisible by p.

All that remains is to show that (5) is true if p is sufficiently large; and since n is fixed, it is enough to prove that each $|\epsilon_k|$ can be made as small as we please by taking p to be large enough. To establish this, we note that

$$|\epsilon_k| \le e^k \int_0^k \frac{x^{p-1}\,|(x-1)\cdots(x-n)|^p\, e^{-x}}{(p-1)!}\,dx$$

$$\le e^n \int_0^n \frac{x^{p-1}\,|(x-1)\cdots(x-n)|^p\, e^{-x}}{(p-1)!}\,dx$$

$$\le \frac{e^n n^{p-1}}{(p-1)!} \int_0^n |(x-1)\cdots(x-n)|^p\,dx.$$

If B is an upper bound for $|(x-1)\cdots(x-n)|$ on the interval $0 \le x \le n$, then it follows that

$$|\epsilon_k| \le \frac{e^n n^{p-1}}{(p-1)!} B^p n = e^n \frac{(nB)^p}{(p-1)!};$$

and since the expression on the right $\to 0$ as $p \to \infty$, the proof is complete.

B.19

In Section B.16 we gave Euclid's proof of the fact that there exist infinitely many prime numbers. About 2000 years after the time of Euclid, in 1737, Euler discovered two fundamentally different new proofs, and the methods he used laid the foundations of a new branch of mathematics that is now called *analytic number theory*.

In order to understand Euler's ideas, we begin by recalling that the harmonic series

$$1 + \frac{1}{2} + \frac{1}{3} + \cdots + \frac{1}{n} + \cdots$$

diverges. On the other hand, we know that for any exponent $s > 1$, the series

$$1 + \frac{1}{2^s} + \frac{1}{3^s} + \cdots + \frac{1}{n^s} + \cdots$$

converges, and the so-called *zeta function* is defined to be the sum of this series,

$$\zeta(s) = \sum_{n=1}^{\infty} \frac{1}{n^s} = 1 + \frac{1}{2^s} + \frac{1}{3^s} + \cdots,$$

considered as a function of the variable s.[1]

---

[1] We denote the independent variable by s in order to retain the notation that is customary in the theory of numbers.

Euler's basic discovery was a remarkable identity connecting the zeta function with the prime numbers,

$$\zeta(s) = \prod_p \frac{1}{1 - 1/p^s}, \tag{1}$$

where the expression on the right denotes the product of the numbers $1/(1 - p^{-s})$ for all primes $p = 2, 3, 5, 7, 11, \ldots$, that is, where

$$\prod_p \frac{1}{1 - 1/p^s} = \frac{1}{1 - 1/2^s} \cdot \frac{1}{1 - 1/3^s} \cdot \frac{1}{1 - 1/5^s} \cdot \frac{1}{1 - 1/7^s} \cdots.$$

To see how the identity (1) arises, we recall that the geometric series $1/(1 - x) = 1 + x + x^2 + \cdots$ is valid for $|x| < 1$, so for each prime p we have

$$\frac{1}{1 - 1/p^s} = 1 + \frac{1}{p^s} + \frac{1}{p^{2s}} + \frac{1}{p^{3s}} + \cdots.$$

Without stopping to justify the process, we now multiply these series together for all primes p, remembering that each integer $n > 1$ is uniquely expressible as a product of powers of different primes. This yields

$$\prod_p \frac{1}{1 - 1/p^s} = \prod_p \left(1 + \frac{1}{p^s} + \frac{1}{p^{2s}} + \frac{1}{p^{3s}} + \cdots \right)$$

$$= 1 + \frac{1}{2^s} + \frac{1}{3^s} + \cdots + \frac{1}{n^s} + \cdots$$

$$= \sum_{n=1}^{\infty} \frac{1}{n^s} = \zeta(s),$$

which is the identity (1).

One of Euler's arguments is based on (1) and goes this way. We begin by observing that if there were only a finite number of primes, then the product on the right-hand side of (1) would be an ordinary finite product and would clearly have a finite value for every $s > 0$, even for $s = 1$. However, the value of the left-hand side of (1) for $s = 1$ is the harmonic series,

$$\zeta(1) = 1 + \frac{1}{2} + \frac{1}{3} + \cdots,$$

which diverges to infinity. This argument by contradiction, which can be made into a rigorous proof, shows that there must be an infinite number of primes. Euler's second argument rests on his discovery that *the series of the reciprocals of the primes diverges,*

$$\sum \frac{1}{p_n} = \frac{1}{2} + \frac{1}{3} + \frac{1}{5} + \frac{1}{7} + \frac{1}{11} + \cdots = \infty; \tag{2}$$

for if there were only a finite number of primes, it is obvious that this series couldn't possibly diverge.

The proof of (2) that we give here starts with the geometric series

$$\frac{1}{1-\frac{1}{2}} = 1 + \frac{1}{2} + \frac{1}{2^2} + \cdots,$$

$$\frac{1}{1-\frac{1}{3}} = 1 + \frac{1}{3} + \frac{1}{3^2} + \cdots,$$

$$\frac{1}{1-\frac{1}{5}} = 1 + \frac{1}{5} + \frac{1}{5^2} + \cdots,$$

$$\cdots$$

$$\frac{1}{1-1/p_n} = 1 + \frac{1}{p_n} + \frac{1}{p_n^2} + \cdots.$$

If we multiply these series together by forming a new series whose terms are all possible products of one term selected from each of the series on the right, then this new series converges in any order to the product of the numbers on the left. Since every integer greater than 1 is uniquely expressible as a product of powers of different primes, the product of these series is the series of the reciprocals of all positive integers whose prime factors are $\leq p_n$. In particular, all positive integers $\leq p_n$ have this property, so

$$\frac{1}{1-\frac{1}{2}} \cdot \frac{1}{1-\frac{1}{3}} \cdots \frac{1}{1-1/p_n} \geq \sum_{k=1}^{p_n} \frac{1}{k} > \int_1^{p_n+1} \frac{dx}{x} = \ln\,(p_n+1) > \ln p_n.$$

(It is in the transition here from the sum to the integral that we use the ideas underlying the integral test.) It follows that

$$\left(1-\frac{1}{2}\right)\left(1-\frac{1}{3}\right)\cdots\left(1-\frac{1}{p_n}\right) < \frac{1}{\ln p_n},$$

and taking logarithms of both sides yields

$$\sum_{k=1}^{n} \ln\left(1-\frac{1}{p_k}\right) < -\ln \ln p_n. \tag{3}$$

We next show that

$$-\frac{2}{p_k} < \ln\left(1-\frac{1}{p_k}\right), \tag{4}$$

for when this is applied to (3), we get

$$-2 \sum_{k=1}^{n} \frac{1}{p_k} < -\ln \ln p_n$$

or

$$\sum_{k=1}^{n} \frac{1}{p_k} > \frac{1}{2} \ln \ln p_n,$$

and our conclusion that $\Sigma 1/p_n$ diverges will follow from the fact that $\ln \ln p_n \to \infty$. To establish (4) and complete the argument, it suffices to observe that the line $y = 2x$ lies below the curve $y = \ln(1 + x)$ on the interval $-\frac{1}{2} \le x < 0$ and that every prime is ≥ 2.[2]

---

[2] For other proofs of (2), see I. Niven, *American Mathematical Monthly*, 1971, pp. 272–73; and C. V. Eynden, *American Mathematical Monthly*, 1980, pp. 394–97.

B.20

In this section we derive several formulas discovered by Euler that rank among the most elegant truths in the whole of mathematics. We use the word "derive" instead of "prove" because some of our arguments are rather formal and require more advanced ideas than we can provide here to become fully rigorous in the sense demanded by modern concepts of mathematical proof. However, the mere fact that we are not able to seal every crack in the reasoning seems a flimsy excuse for denying students an opportunity to glimpse some of the wonders that can be found in this part of calculus. For those who wish to dig deeper, full proofs are given in the treatise by K. Knopp, *Theory and Application of Infinite Series,* Hafner, 1951.

THE BERNOULLI NUMBERS

Since

$$\frac{e^x - 1}{x} = 1 + \frac{x}{2!} + \frac{x^2}{3!} + \cdots$$

for $x \neq 0$, and this power series has the value 1 at $x = 0$, the reciprocal function $x/(e^x - 1)$ has a power series expansion valid in some neighbourhood of the origin if the value of this function is defined to be 1 at $x = 0$. We write this series in the form

$$\frac{x}{e^x - 1} = \sum_{n=0}^{\infty} \frac{B_n}{n!} x^n = B_0 + B_1 x + \frac{B_2}{2!} x^2 + \cdots. \tag{1}$$

The numbers B_n defined in this way are called the *Bernoulli numbers*, and it is clear that $B_0 = 1$. A bit of algebra reveals that

$$\frac{x}{e^x - 1} = \frac{x}{2}\left(\frac{e^x + 1}{e^x - 1} - 1\right) = -\frac{x}{2} + \frac{x}{2} \cdot \frac{e^x + 1}{e^x - 1}. \tag{2}$$

A routine check shows that the second term on the right is an even function, so $B_1 = -\frac{1}{2}$ and $B_n = 0$ if n is odd and >1. If we write (1) in the form

$$\left(\frac{B_0}{0!} + \frac{B_1}{1!}x + \frac{B_2}{2!}x^2 + \cdots\right)\left(\frac{1}{1!} + \frac{x}{2!} + \frac{x^2}{3!} + \cdots\right) = 1,$$

then it is clear that the coefficient of x^{n-1} in the product on the left equals zero if $n > 1$. By the rule for multiplying power series, this yields

$$\frac{B_0}{0!} \cdot \frac{1}{n!} + \frac{B_1}{1!} \cdot \frac{1}{(n-1)!} + \frac{B_2}{2!} \cdot \frac{1}{(n-2)!} + \cdots + \frac{B_{n-1}}{(n-1)!} \cdot \frac{1}{1!} = 0,$$

and by multiplying through by $n!$ we obtain

$$\frac{n!}{0!\, n!} B_0 + \frac{n!}{1!\,(n-1)!} B_1 + \frac{n!}{2!\,(n-2)!} B_2 + \cdots + \frac{n!}{(n-1)!\, 1!} B_{n-1} = 0. \tag{3}$$

This equation can also be written more briefly as

$$\binom{n}{0}B_0 + \binom{n}{1}B_1 + \binom{n}{2}B_2 + \cdots + \binom{n}{n-1}B_{n-1} = 0$$

or

$$\sum_{k=0}^{n-1}\binom{n}{k}B_k = 0,$$

where $\binom{n}{k}$ is the binomial coefficient $n!/[k!\,(n-k)!]$. By taking $n = 3, 5, 7, 9, 11, \ldots$ in (3) and doing a little arithmetic, we find that

$$B_2 = \tfrac{1}{6}, \qquad B_4 = -\tfrac{1}{30}, \qquad B_6 = \tfrac{1}{42}, \qquad B_8 = -\tfrac{1}{30}, \qquad B_{10} = \tfrac{5}{66}, \ldots.$$

These calculations can be continued recursively as far as we please, so all the Bernoulli numbers can be considered as known, even though considerable labor may be required to make any particular one of them visibly present. We also point out that it is obvious from (3) and the mode of calculation that every B_n is rational.

THE POWER SERIES FOR THE TANGENT

We now begin to explore the uses of these numbers.

In equation (2) we move the term $-x/2$ to the left and use the fact that

$$\frac{x}{2} \cdot \frac{e^x + 1}{e^x - 1} = \frac{x}{2} \cdot \frac{e^{x/2} + e^{-x/2}}{e^{x/2} - e^{-x/2}}$$

to obtain

$$\frac{x}{2} \cdot \frac{e^{x/2} + e^{-x/2}}{e^{x/2} - e^{-x/2}} = \sum_{n=0}^{\infty} \frac{B_{2n}}{(2n)!} x^{2n}. \tag{4}$$

On the left-hand side of this, we now replace x by $2ix$, which yields

$$\frac{2ix}{2} \cdot \frac{e^{ix} + e^{-ix}}{e^{ix} - e^{-ix}} = x \frac{(e^{ix} + e^{-ix})/2}{(e^{ix} - e^{-ix})/2i} = x \cot x,$$

by means of the well-known formulas expressing $\sin x$ and $\cos x$ in terms of the exponential function. Making the same substitution on the right-hand side of (4) gives

$$\sum_{n=0}^{\infty} \frac{B_{2n}}{(2n)!} (2ix)^{2n} = \sum_{n=0}^{\infty} (-1)^n \frac{2^{2n} B_{2n}}{(2n)!} x^{2n},$$

so

$$x \cot x = \sum_{n=0}^{\infty} (-1)^n \frac{2^{2n} B_{2n}}{(2n)!} x^{2n}. \tag{5}$$

The trigonometric identity $\tan x = \cot x - 2 \cot 2x$ now enables us to use (5) to write

$$\tan x = \sum_{n=0}^{\infty} (-1)^n \frac{2^{2n} B_{2n}}{(2n)!} x^{2n-1} - 2 \sum_{n=0}^{\infty} (-1)^n \frac{2^{2n} B_{2n}}{(2n)!} (2x)^{2n-1}$$

$$= \sum_{n=0}^{\infty} (-1)^n \frac{2^{2n} B_{2n}}{(2n)!} x^{2n-1} - \sum_{n=0}^{\infty} (-1)^n \frac{2^{2n} B_{2n}}{(2n)!} 2^{2n} x^{2n-1}$$

$$= \sum_{n=0}^{\infty} (-1)^n \frac{2^{2n} B_{2n}}{(2n)!} (1 - 2^{2n}) x^{2n-1},$$

so

$$\tan x = \sum_{n=1}^{\infty} (-1)^{n+1} \frac{2^{2n}(2^{2n} - 1) B_{2n}}{(2n)!} x^{2n-1}.$$

This is the full power series for $\tan x$ that is usually encountered several times in truncated form in elementary treatments of infinite series. Based on our knowledge of the Bernoulli numbers, the first few terms of this series are easy to calculate explicitly,

$$\tan x = x + \tfrac{1}{3}x^3 + \tfrac{2}{15}x^5 + \tfrac{17}{315}x^7 + \tfrac{67}{2835}x^9 + \cdots.$$

THE PARTIAL FRACTIONS EXPANSION OF THE COTANGENT

By using entirely different methods, Euler discovered another remarkable expansion of the cotangent: If x is not an integer, then

$$\pi \cot \pi x = \frac{1}{x} + 2x \sum_{n=1}^{\infty} \frac{1}{x^2 - n^2}. \tag{6}$$

We will examine this formula from two very different points of view, and give two derivations.

First, it is quite easy to see that (6) is analogous to the expansion of a rational function in partial fractions. For instance, if we consider the rational function $(2x + 1)/(x^2 - 3x + 2)$ and notice that the denominator has zeros 1 and 2 and can therefore be factored into $(x - 1)(x - 2)$, then this leads to the expansion

$$\frac{2x + 1}{x^2 - 3x + 2} = \frac{2x + 1}{(x - 1)(x - 2)} = \frac{c_1}{x - 1} + \frac{c_2}{x - 2}$$

for certain constants c_1 and c_2. The constant c_1 can now be determined by multiplying through by $x - 1$ and allowing x to approach 1, and similarly for c_2. Formally, (6) can be obtained in much the same way by noticing that $\cot \pi x = \cos \pi x / \sin \pi x$ has a denominator with zeros $0, \pm 1, \pm 2, \ldots,$ and should therefore be expressible in the form

$$\cot \pi x = \frac{a}{x} + \sum_{n=1}^{\infty} \left(\frac{b_n}{x - n} + \frac{c_n}{x + n} \right). \tag{7}$$

From this, the constants a, b_n, and c_n can be found by the procedure suggested (they are all equal to $1/\pi$), and (7) can then be rearranged to yield (6). For reasons that will now be obvious, it is customary to refer to (6) as the *partial fractions expansion of the cotangent*. The main gap in this suggestive but rather tentative derivation is of course the fact that we have no prior guarantee that an expansion of the form (7) is possible.

Another way of approaching (6) is to begin with the infinite product (6) in Section B.14:

$$\frac{\sin x}{x} = \left(1 - \frac{x^2}{\pi^2} \right) \left(1 - \frac{x^2}{4\pi^2} \right) \left(1 - \frac{x^2}{9\pi^2} \right) \cdots \left(1 - \frac{x^2}{n^2 \pi^2} \right) \cdots.$$

If we take the logarithm of both sides to obtain

$$\ln \frac{\sin x}{x} = \sum_{n=1}^{\infty} \ln \left(1 - \frac{x^2}{n^2 \pi^2} \right),$$

and then differentiate, the result is seen to be

$$\cot x - \frac{1}{x} = \sum_{n=1}^{\infty} \frac{-2x}{n^2 \pi^2 - x^2}$$

or

$$\cot x = \frac{1}{x} + 2x \sum_{n=1}^{\infty} \frac{1}{x^2 - n^2 \pi^2};$$

and replacing x by πx and then multiplying through by πx yields

$$\pi x \cot \pi x = 1 + 2x^2 \sum_{n=1}^{\infty} \frac{1}{x^2 - n^2}, \tag{8}$$

which is equivalent to (6).

EULER'S FORMULA FOR $\sum 1/n^{2k}$

We now obtain a major payoff from (5) and (8) by replacing x by πx in (5) and equating the two expressions for $\pi x \cot \pi x$,

$$1 + \sum_{n=1}^{\infty} \frac{-2x^2}{n^2 - x^2} = 1 + \sum_{k=1}^{\infty} (-1)^k \frac{2^{2k} B_{2k}}{(2k)!} (\pi x)^{2k}, \tag{9}$$

where we use k as the index of summation on the right for reasons that will appear in a moment. Each term of the series on the left is easy to expand in a geometric series,

$$\frac{-2x^2}{n^2 - x^2} = -2 \frac{x^2/n^2}{1 - x^2/n^2} = -2 \sum_{k=1}^{\infty} \left(\frac{x^2}{n^2}\right)^k = -2 \sum_{k=1}^{\infty} \frac{x^{2k}}{n^{2k}},$$

so (9) can be written as

$$1 + \sum_{n=1}^{\infty} \left(-2 \sum_{k=1}^{\infty} \frac{x^{2k}}{n^{2k}}\right) = 1 + \sum_{k=1}^{\infty} (-1)^k \frac{2^{2k} B_{2k}}{(2k)!} \pi^{2k} x^{2k}.$$

We now interchange the order of summation on the left and obtain

$$1 + \sum_{k=1}^{\infty} \left(-2 \sum_{n=1}^{\infty} \frac{1}{n^{2k}}\right) x^{2k} = 1 + \sum_{k=1}^{\infty} (-1)^k \frac{2^{2k} B_{2k}}{(2k)!} \pi^{2k} x^{2k},$$

and equating the coefficients of x^{2k} yields

$$\sum_{n=1}^{\infty} \frac{1}{n^{2k}} = (-1)^{k-1} \frac{2^{2k} B_{2k}}{2(2k)!} \pi^{2k}$$

for each positive integer k. In particular, for $k = 1, 2, 3$ we get

$$\sum_{n=1}^{\infty} \frac{1}{n^2} = \frac{\pi^2}{6}, \qquad \sum_{n=1}^{\infty} \frac{1}{n^4} = \frac{\pi^4}{90}, \qquad \sum_{n=1}^{\infty} \frac{1}{n^6} = \frac{\pi^6}{945}.$$

It is very remarkable that for almost 250 years there has been no progress whatever toward finding the exact sum of any one of the series

$$\sum_{n=1}^{\infty} \frac{1}{n^3}, \qquad \sum_{n=1}^{\infty} \frac{1}{n^5}, \qquad \sum_{n=1}^{\infty} \frac{1}{n^7}, \cdots.$$

Perhaps a second Euler is needed for this breakthrough, but none is in sight.

B.21

The *cycloid* is the curve traced out by a point on the circumference of a circle when the circle rolls along a straight line in its own plane, as shown in Fig. B.23. We shall see that this curve has many remarkable geometric and physical properties.

The only convenient way of representing a cycloid is by means of parametric equations. We assume that the rolling circle has radius a and that it rolls along the x-axis, starting from a position in which the center of the circle is on the positive y-axis. The curve is the locus of the point P on the circle which is located at the origin O when the center C is on the y-axis. The angle θ in the figure is the angle through which the radius CP turns as the circle rolls to a new position. If x and y are the coordinates of P, then the rolling of the circle implies that $OB = \text{arc } BP = a\theta$, so $x = OB - AB = OB - PQ = a\theta - a \sin \theta = a(\theta - \sin \theta)$. Also, $y = BC - QC = a - a \cos \theta = a(1 - \cos \theta)$. The cycloid therefore has the parametric representation

$$x = a(\theta - \sin \theta), \qquad y = a(1 - \cos \theta). \tag{1}$$

It is clear from Fig. B.23 that y is a function of x, but it is also clear from equations (1) that it is not possible to find a simple formula for this function. The cycloid is one of many curves for which the parametric equations are much simpler and easier to work with than the rectangular equation.

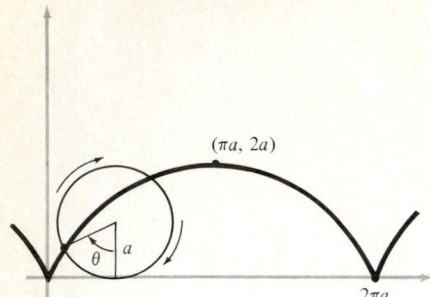 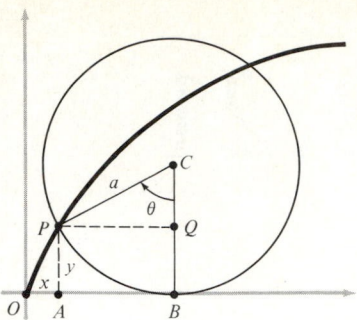

FIGURE B.23
The cycloid.

From equations (1) we have

$$y' = \frac{dy}{dx} = \frac{a \sin \theta \, d\theta}{a(1 - \cos \theta) \, d\theta} = \frac{\sin \theta}{1 - \cos \theta} = \frac{2 \sin \frac{1}{2}\theta \cos \frac{1}{2}\theta}{2 \sin^2 \frac{1}{2}\theta} = \cot \frac{1}{2}\theta. \tag{2}$$

We observe that the derivative y' is not defined for $\theta = 0$, $\pm 2\pi$, $\pm 4\pi$, etc. These values of θ correspond to the points where the cycloid touches the x-axis; these points are called *cusps*. The tangent to the cycloid is vertical at the cusps.

In the following examples we establish the main geometric properties of the cycloid.

Example 1. Show that the area under one arch of the cycloid is three times the area of the rolling circle.

Solution. One arch is traced out as the circle turns through one complete revolution. The usual area integral can therefore be written as follows, using the parameter θ as the variable of integration:

$$A = \int_0^{2\pi a} y \, dx = \int_0^{2\pi} y \frac{dx}{d\theta} \, d\theta = \int_0^{2\pi} a(1 - \cos \theta)a(1 - \cos \theta) \, d\theta$$

$$= a^2 \int_0^{2\pi} (1 - \cos \theta)^2 \, d\theta = a^2 \int_0^{2\pi} (1 - 2 \cos \theta + \cos^2 \theta) \, d\theta$$

$$= a^2 \int_0^{2\pi} (1 + \cos^2 \theta) \, d\theta = a^2 \int_0^{2\pi} d\theta + a^2 \int_0^{2\pi} \tfrac{1}{2}(1 + \cos 2\theta) \, d\theta = 3\pi a^2.$$

Example 2. Show that the length of one arch of the cycloid is four times the diameter of the rolling circle.

Solution. Since $dx = a(1 - \cos \theta) \, d\theta$ and $dy = a \sin d\theta$, the element of arc length ds is given by

$$ds^2 = dx^2 + dy^2 = a^2[(1 - \cos \theta)^2 + \sin^2 \theta) \, d\theta^2$$

$$= 2a^2[1 - \cos \theta] \, d\theta^2 = 4a^2 \sin^2 \tfrac{1}{2}\theta \, d\theta^2,$$

so

$$ds = 2a \sin \tfrac{1}{2}\theta \, d\theta.$$

The length of one arch is therefore

$$L = \int ds = \int_0^{2\pi} 2a \sin \tfrac{1}{2}\theta \, d\theta = -4a \cos \tfrac{1}{2}\theta \, \Big]_0^{2\pi} = 8a.$$

Example 3. Show that the tangent to the cycloid at the point P in Fig. B.23 passes through the top of the rolling circle.

Solution. The point at the top of the circle has coordinates $(a\theta, 2a)$. The slope of the tangent at P is given by (2). The equation of the tangent at P is therefore

$$y - a(1 - \cos \theta) = \frac{\sin \theta}{1 - \cos \theta}(x - a\theta + a \sin \theta).$$

We substitute $x = a\theta$ in this equation and solve for y, which gives

$$y = a(1 - \cos \theta) + \frac{\sin \theta}{1 - \cos \theta} \cdot a \sin \theta = \frac{a(1 - \cos \theta)^2 + a \sin^2 \theta}{1 - \cos \theta} = 2a.$$

This shows that the tangent at P does indeed pass through the point $(a\theta, 2a)$ at the top of the circle.

Galileo seems to have been the first to notice the cycloid and investigate its properties, in the early 1600s. He didn't actually discover any of these properties, but he gave the curve its name and recommended its study to his friends, including Mersenne in Paris. Mersenne informed Descartes and others about it, and in 1638 Descartes found a construction for the tangent which is equivalent to the property given in Example 3. In 1644 Galileo's disciple Torricelli (who invented the barometer) published his discovery of the area under one arch. The length of one arch was discovered in 1658 by the great English architect Christopher Wren.[1] The list of famous men who have worked on the cycloid will be continued, but first we consider some other related curves.

If a circle rolls on the *inside* of a fixed circle, the locus of a point on the rolling circle is called a *hypocycloid*. If a circle rolls on the *outside* of a fixed circle, the locus of a point on the rolling circle is called an *epicycloid*.[2]

We show how to represent a hypocycloid parametrically. Let the fixed circle have radius a and the rolling circle radius b, where $b < a$. Let the fixed

---

[1] Wren was an astronomer and a mathematician—in fact, Savilian Professor of Astronomy at Oxford—before the Great Fire of London in 1666 gave him his opportunity to build St. Paul's Cathedral, as well as dozens of smaller churches throughout the city.

[2] The distinction between these words is easy to remember because the Greek prefix *hypo* means under or beneath, as in "hypodermic," and *epi* means on or above, as in "epicenter."

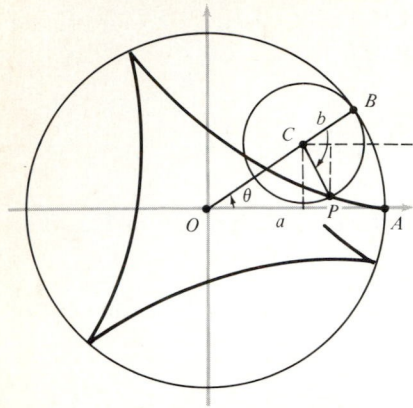

FIGURE B.24
The hypocycloid.

circle have its center at the origin (Fig. B.24), and let the smaller rolling circle start in a position internally tangent to the fixed circle at the point A on the positive x-axis. We consider the point P on the rolling circle that was initially at A. With θ and ϕ as shown in the figure, the rolling of the small circle implies that the arcs AB and BP are equal: $a\theta = b\phi$. We can then see that the coordinates of P are

$$x = (a - b) \cos \theta + b \cos (\phi - \theta),$$
$$y = (a - b) \sin \theta - b \sin (\phi - \theta).$$

But $\phi - \theta = [(a - b)/b]\theta$, so the parametric equations of the hypocycloid are

$$x = (a - b) \cos \theta + b \cos \frac{a - b}{b} \theta,$$

$$y = (a - b) \sin \theta - b \sin \frac{a - b}{b} \theta. \tag{3}$$

The arc length along the fixed circle between successive cusps of the hypocycloid is $2\pi b$. If $2\pi a$ is an integral multiple of $2\pi b$, so that a/b is an integer n, then the hypocycloid has n cusps and the point P returns to A after the smaller circle rolls off its circumference n times on the fixed circle. We leave it to students to decide when P will return to A if a/b is a rational number but not an integer, for example if $a/b = \frac{5}{2}$. A discussion of the case in which a/b is irrational is beyond the scope of this book; it suffices to say that as the smaller circle rolls around and around indefinitely, the cusps of the resulting hypocycloid are evenly and densely distributed on the fixed circle.[3]

---

[3] The curious reader will find additional information in Theorem 439 of G. H. Hardy and E. M. Wright, *Introduction to the Theory of Numbers,* Oxford University Press, 1954; or in Theorem 6.3 of I. Niven, *Irrational Numbers,* Wiley, 1956.

The parametric equations of a hypocycloid of four cusps can be written in a very simple form by using some trigonometric identities. If $a = 4b$, equations (3) become

$$x = 3b \cos \theta + b \cos 3\theta, \qquad y = 3b \sin \theta - b \sin 3\theta.$$

But

$$
\begin{aligned}
\cos 3\theta &= \cos (2\theta + \theta) = \cos 2\theta \cos \theta - \sin 2\theta \sin \theta \\
&= (2 \cos^2 \theta - 1) \cos \theta - 2 \sin^2 \theta \cos \theta \\
&= [2 \cos^2 \theta - 1 - 2(1 - \cos^2 \theta)] \cos \theta \\
&= 4 \cos^3 \theta - 3 \cos \theta,
\end{aligned}
$$

and a similar calculation yields

$$\sin 3\theta = 3 \sin \theta - 4 \sin^3 \theta.$$

Our parametric equations therefore become

$$x = 4b \cos^3 \theta = a \cos^3 \theta, \qquad y = 4b \sin^3 \theta = a \sin^3 \theta. \tag{4}$$

From these equations it is easy to obtain the corresponding rectangular equation,

$$x^{2/3} + y^{2/3} = a^{2/3}. \tag{5}$$

Because of its appearance (Fig. B.25), a hypocycloid of four cusps is often called an *astroid*.

Example 4. Consider the tangent to the astroid at a point P in the first quadrant. Show that the part of this tangent which is cut off by the coordinate axes has constant length, independent of the position of P.

Solution. By equations (4), the slope of the tangent is

$$y' = \frac{dy}{dx} = \frac{3a \sin^2 \theta \cos \theta \, d\theta}{-3a \cos^2 \theta \sin \theta \, d\theta} = -\tan \theta,$$

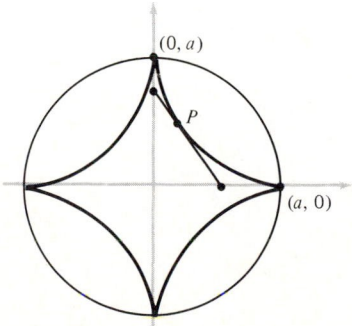

FIGURE B.25
The astroid.

so the equation of the tangent is

$$y - a \sin^3 \theta = -\tan \theta(x - a \cos^3 \theta).$$

We find the x-intercept by putting $y = 0$ and solving for x,

$$x = a \cos^3 \theta + a \sin^2 \theta \cos \theta = a \cos \theta.$$

Similarly, the y-intercept is $y = a \sin \theta$. The length of the part of the tangent cut off by the axes is therefore

$$\sqrt{a^2 \cos^2 \theta + a^2 \sin^2 \theta} = a,$$

which is constant.

We now return to the cycloid discussed earlier, and reflect both it and the y-axis about the x-axis, as shown in Fig. B.26. The parametric equations (1) are still valid, and the resulting curve has several interesting physical properties, which we now describe and analyze.

In 1696 John Bernoulli conceived and solved the now famous *brachistochrone problem*. He published the problem (but not the solution) as a challenge to other mathematicians of the time. The problem is this: Among all smooth curves in a vertical plane that join a given point P_0 to a given lower point P_1 not directly below it, find that particular curve along which a particle will slide down from P_0 to P_1 in the shortest possible time.[4] We can think of the particle as a bead of mass m sliding down an ideal frictionless wire, with the downward force of gravity mg as the only force acting on the bead.

If we assume that the points P_0 and P_1 lie at the origin and at (x_1, y_1) in the first quadrant, as shown in Fig. B.27, then Bernoulli's problem can be stated in mathematical language as follows. The bead is released from rest at P_0, so its initial velocity and initial kinetic energy are zero. The work done by gravity in pulling it down from the origin to an arbitrary point $P = (x, y)$ is

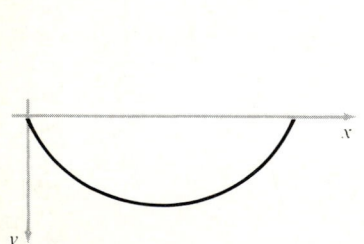

FIGURE B.26

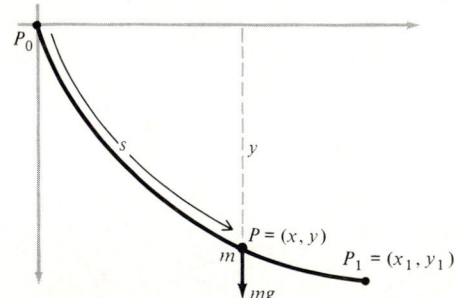

FIGURE B.27

---

[4] The word *brachistochrone* comes from two Greek words meaning shortest time.

mgy. This must equal the increase in the kinetic energy of the bead as it slides down the wire to this point, so

$$\tfrac{1}{2}mv^2 = mgy,$$

and therefore

$$v = \frac{ds}{dt} = \sqrt{2gy}. \tag{6}$$

This can be written as

$$dt = \frac{ds}{\sqrt{2gy}} = \frac{\sqrt{dx^2 + dy^2}}{\sqrt{2gy}} = \frac{\sqrt{1 + (dy/dx)^2}\, dx}{\sqrt{2gy}}. \tag{7}$$

The total time T_1 required for the bead to slide down the wire from P_0 to P_1 will depend on the shape of the wire as specified by its equation $y = f(x)$; it is given by

$$T_1 = \int dt = \int_0^{x_1} \sqrt{\frac{1 + (y')^2}{2gy}}\, dx. \tag{8}$$

The brachistochrone problem therefore amounts to this: to find the particular curve $y = f(x)$ that passes through P_0 and P_1 and minimizes the value of the integral (8).

Since the straight line joining P_0 and P_1 is clearly the shortest path, we might guess that this line also yields the shortest time. However, a moment's consideration of the possibilities will make us more skeptical about this conjecture. There might be an advantage in having the bead slide down more steeply at first, thereby increasing its speed more quickly at the beginning of the motion; for with a faster start, it is reasonable to suppose that the bead might reach P_1 in a shorter time, even though it travels over a longer path. And this is the way it turns out: The brachistochrone curve is an arc of a cycloid through P_0 and P_1 with a cusp at the origin.

Leibniz and Newton, as well as John Bernoulli and his older brother James, solved the problem. John's solution, which is very ingenious but rather specialized in the methods it uses, is given in Section B.22. The cycloid was well known to all these men through the earlier work of the great Dutch scientist Huygens on pendulum clocks (see below). When John found that the cycloid is also the solution of his brachistochrone problem, he was astounded and delighted. He wrote: "With justice we admire Huygens because he first discovered that a heavy particle slides down to the bottom of a cycloid in the same time, no matter where it starts. But you [his readers] will be petrified with astonishment when I say that this very same cycloid, the tautochrone of Huygens, is also the brachistochrone we are seeking."[5]

---

[5] For an English translation of Bernoulli's writings on this subject, see pp. 644–55 of D. E. Smith, *A Source Book in Mathematics*, McGraw-Hill, 1929. Bernoulli's vivid, enthusiastic, personal style is in sharp contrast to the dead, gray, impersonal style of most of the writing in scientific journals nowadays.

Huygens was a profound student of the theory of the pendulum, and in fact was the inventor of the pendulum clock. He was very well aware of the theoretical flaw in such a clock, which is due to the fact that the period of oscillation of a pendulum is not strictly independent of the amplitude of the swing.[6] We can express this flaw in another way by saying that if a bead is released on a frictionless circular wire in a vertical plane, then the time the bead takes to slide down to the bottom will depend on the height of the starting point. Huygens wondered what would happen if the circular wire were replaced by one having the shape of an inverted cycloidal arch. But he did more than merely wonder, for he then went on to make the remarkable discovery referred to in the passage previously quoted, that for a wire of this shape the bead will slide down from any point to the bottom in exactly the same time, no matter where it is released (Fig. B.28). This is the *tautochrone* ("equal time") *property* of the cycloid, and we now prove it by using the formulas given above.

If we write (8) in the equivalent form

$$T_1 = \int \sqrt{\frac{dx^2 + dy^2}{2gy}}$$

and substitute equations (1) into this, we obtain

$$T_1 = \int_0^{\theta_1} \sqrt{\frac{2a^2(1 - \cos \theta)}{2ag(1 - \cos \theta)}} \, d\theta = \theta_1 \sqrt{\frac{a}{g}}$$

as the time required for the bead to slide down a cycloidal wire from P_0 to P_1. The time needed for the bead to reach the bottom of this wire is the value of T_1 when $\theta_1 = \pi$, namely, $\pi\sqrt{a/g}$. Huygens' tautochrone property amounts to the statement that the bead will reach the bottom in exactly the same time if it starts at any intermediate point (x_0, y_0). To prove this, we replace (6) by

$$v = \frac{ds}{dt} = \sqrt{2g(y - y_0)}.$$

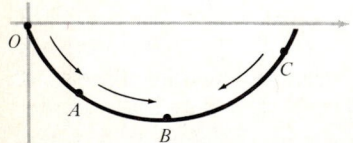

FIGURE B.28
Beads released on the cycloidal wire at O, A, C will reach B in the same amount of time.

---

[6] This fact is often referred to as the "circular error" of pendulum clocks.

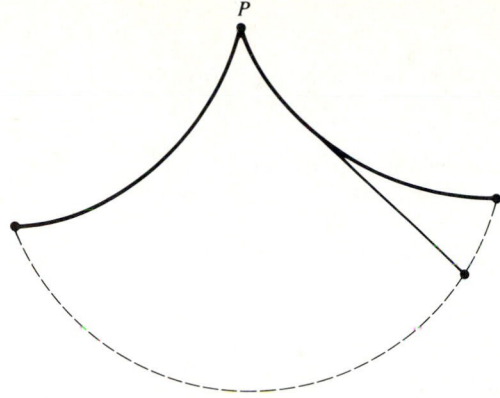

FIGURE B.29
A flexible pendulum constrained by cycloidal jaws swings along another cycloid.

The total time required for the bead to slide down to the bottom is therefore

$$T = \int_{\theta_0}^{\pi} \sqrt{\frac{2a^2(1 - \cos\theta)}{2ag(\cos\theta_0 - \cos\theta)}} \, d\theta = \sqrt{\frac{a}{g}} \int_{\theta_0}^{\pi} \sqrt{\frac{1 - \cos\theta}{\cos\theta_0 - \cos\theta}} \, d\theta$$

$$= \sqrt{\frac{a}{g}} \int_{\theta_0}^{\pi} \frac{\sin\frac{1}{2}\theta \, d\theta}{\sqrt{\cos^2\frac{1}{2}\theta_0 - \cos^2\frac{1}{2}\theta}}, \qquad (9)$$

where the last step makes use of the trigonometric identities $2\sin^2\theta = 1 - \cos 2\theta$ and $\cos 2\theta = \cos^2\theta - \sin^2\theta = 2\cos^2\theta - 1$. If we now use the substitution

$$u = \frac{\cos\frac{1}{2}\theta}{\cos\frac{1}{2}\theta_0}, \qquad du = -\frac{1}{2}\frac{\sin\frac{1}{2}\theta \, d\theta}{\cos\frac{1}{2}\theta_0},$$

then the integral (9) becomes

$$T = -2\sqrt{\frac{a}{g}} \int_{1}^{0} \frac{du}{\sqrt{1 - u^2}} = 2\sqrt{\frac{a}{g}} \sin^{-1}u \Big]_{0}^{1} = \pi\sqrt{\frac{a}{g}}.$$

This shows that T has the same value as before and is therefore independent of the starting point, and the argument is complete.

Once Huygens established the tautochrone property of the cycloid, a further problem presented itself: How could he arrange for a pendulum in a clock to move along a cycloidal, rather than a circular, path? Here he made a further beautiful discovery. If we suspend from the point P at the cusp between two equal inverted cycloidal semiarches a flexible pendulum whose length equals the length of one of the semiarches (Fig. B.29), then the bob will draw up as it swings to the side in such a way that its path is another cycloid.[7]

---

[7] A proof of this statement is given in Section B.23, but to understand this material, one must be acquainted with the concepts of curvature and radius of curvature for curves.

PROBLEMS

1. Find the rectangular equation of the cycloid by eliminating θ from the parametric equations (1). Observe how hopeless it is to try to solve this for y as a simple function of x.

2. Show that for the cycloid (1) the second derivative is given by $y'' = dy'/dx = -a/y^2$. Observe that this fact implies that the cycloid is concave down between the cusps, as shown in Fig. B.23.

3. Use the equation of the normal to the cycloid at P (in Fig. B.23) to show that this normal passes through the point B at the bottom of the rolling circle. Also, obtain this conclusion from the result of Example 3 by using elementary geometry.

4. Assume that the circle in Fig. B.23 rolls to the right along the x-axis at a constant speed, with the center C moving at v_0 units per second. (a) Find the rates of change of the coordinates x and y of the point P. (b) What is the greatest rate of increase of x, and where is P when this is attained? (c) What is the greatest rate of increase of y, and for what value of θ is this attained?

5. If a polygon $ABCD$ rolls (awkwardly) on a straight line $A'D'$, as shown in Fig. B.30, then the point A will trace out in succession several arcs of circles with centers B', C', D'. The tangent to any such arc is evidently perpendicular to the line joining the point of tangency to the corresponding center. Therefore, if the rolling circle that generates a cycloid is thought of as a polygon with an infinite number of sides, then the tangent to the cycloid at any point is the line perpendicular to the line joining the point of tangency to the bottom of the rolling circle. This is Descartes's method for finding the tangent at any point of a cycloid. Verify that it is correct.

6. Find the area inside the astroid (5).

7. Find the total length of the astroid (5).

8. Find the area of the surface generated by revolving the astroid (5) about the x-axis.

9. Show that the hypocycloid of two cusps, with $a = 2b$, is simply the diameter of the fixed circle that lies along the x-axis. In this case, if the center C of the rolling circle moves around with constant angular velocity ω, so that $d\theta/dt = \omega$, show that P moves back and forth on the x-axis with simple harmonic motion of period $2\pi/\omega$ and maximum speed $a\omega$.

10. If the astroid (4) is generated by the small circle rolling around counterclockwise with constant angular velocity ω, find the position of the point P in the first quadrant for which y is increasing most rapidly.

11. The hypocycloid of three cusps, with $a = 3b$, is called a *deltoid*. Sketch this curve, find its parametric equations, and find its total length.

12. Find parametric equations for the epicycloid generated by a circle of radius b

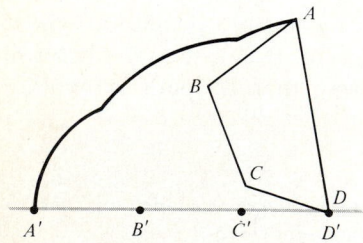

FIGURE B.30

rolling on the outside of a fixed circle of radius a. Use a figure similar to Fig. B.24, where the fixed circle has its center at the origin and the point P is initially at $(a, 0)$.

13. Show that the equations in Problem 12 can be obtained from equations (3) in the text by replacing b by $-b$.

14. The epicycloid of two cusps, with $a = 2b$, is called a *nephroid* (meaning kidney-shaped). Sketch this curve, find its parametric equations, and calculate its total length.

B.22

As explained in Section B.21, we begin with a point P_0 and a lower point P_1, and we seek the shape of the curved wire joining these points down which a bead will slide without friction in the shortest possible time.

We start by considering an apparently unrelated problem in optics. Figure B.31 illustrates a situation in which a ray of light travels from A to P with constant velocity v_1, and then, entering a denser medium, travels from P to B with a smaller velocity v_2. In terms of the notation in the figure, the total time T required for the journey is given by

$$T = \frac{\sqrt{a^2 + x^2}}{v_1} + \frac{\sqrt{b^2 + (c - x)^2}}{v_2}.$$

If we assume that this ray of light is able to select its path from A to B in such a way as to minimize T, then $dT/dx = 0$, and with a little work we see that the minimizing path is characterized by the equation

$$\frac{\sin \alpha_1}{v_1} = \frac{\sin \alpha_2}{v_2}.$$

This is *Snell's law of refraction*. The assumption that light travels from one point to another along the path requiring the shortest time is called *Fermat's principle of least time*. This principle not only provides a rational basis for Snell's law—which is an experimental fact—but also can be applied to find the path of a ray of light through a medium of variable density, where in general

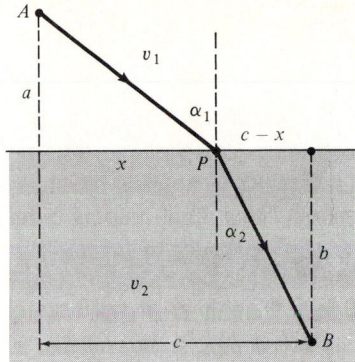

FIGURE B.31
The refraction of light.

light will travel along curves instead of straight lines. In Fig. B.32(a) we have a stratified optical medium. In the individual layers the velocity of light is constant, but the velocity decreases from each layer to the one below it. As the descending ray of light passes from layer to layer, it is refracted more and more toward the vertical, and when Snell's law is applied to the boundaries between the layers, we obtain

$$\frac{\sin \alpha_1}{v_1} = \frac{\sin \alpha_2}{v_2} = \frac{\sin \alpha_3}{v_3} = \frac{\sin \alpha_4}{v_4}.$$

If we next allow these layers to grow thinner and more numerous, then in the limit the velocity of light decreases continuously as the ray descends, and we conclude that

$$\frac{\sin \alpha}{v} = \text{a constant.}$$

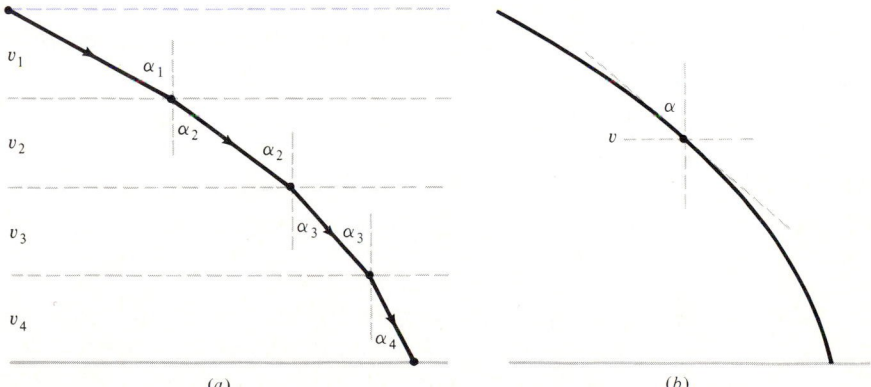

(a)   (b)

FIGURE B.32
Refraction in other optical media.

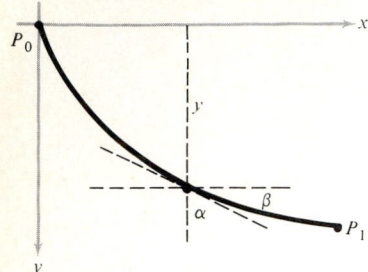

FIGURE B.33

This situation is indicated in Fig. B.32(b); it is approximately what happens to a ray of sunlight falling on the earth as it slows in descending through atmosphere of increasing density.

Returning now to the brachistochrone problem, we introduce a coordinate system as in Fig. B.33 and assume that the bead (like the ray of light) is capable of selecting the path down which it will slide from P_0 to P_1 in the shortest possible time. The argument given above yields

$$\frac{\sin \alpha}{v} = \text{a constant.} \tag{1}$$

If the bead has mass m, so that mg is the downward force that gravity exerts on it, then the fact that the work done by gravity in pulling the bead down the wire equals the increase in the kinetic energy of the bead tells us that $mgy = \frac{1}{2}mv^2$. this gives

$$v = \sqrt{2gy}. \tag{2}$$

From the geometry of the situation we also have

$$\sin \alpha = \cos \beta = \frac{1}{\sec \beta} = \frac{1}{\sqrt{1 + \tan^2 \beta}} = \frac{1}{\sqrt{1 + (y')^2}}. \tag{3}$$

On combining equations (1), (2), and (3)—obtained from optics, mechanics, and calculus—we get

$$y[1 + (y')^2] = c \tag{4}$$

as the differential equation of the brachistochrone.

We now complete our discussion, and discover what curve the brachistochrone actually is, by solving equation (4). When y' is replaced by dy/dx and the variables are separated, (4) becomes

$$dx = \sqrt{\frac{y}{c - y}}\, dy,$$

so

$$x = \int \sqrt{\frac{y}{c - y}}\, dy.$$

We evaluate this integral by starting with the algebraic substituion $u^2 = y/(c - y)$, so that

$$y = \frac{cu^2}{1 + u^2} \quad \text{and} \quad dy = \frac{2cu}{(1 + u^2)^2} \, du.$$

Then

$$x = \int \frac{2cu^2}{(1 + u^2)^2} \, du,$$

and the trigonometric substitution $u = \tan \phi$, $du = \sec^2 \phi \, d\phi$ enables us to write this as

$$x = \int \frac{2c \tan^2 \phi \sec^2 \phi}{(1 + \tan^2 \phi)^2} \, d\phi$$

$$= 2c \int \frac{\tan^2 \phi}{\sec^2 \phi} \, d\phi = 2c \int \sin^2 \phi \, d\phi$$

$$= c \int (1 - \cos 2\phi) \, d\phi = \frac{1}{2} c(2\phi - \sin 2\phi).$$

The constant of integration here is zero because $y = 0$ when $\phi = 0$, and since P_0 is at the origin, we also want to have $x = 0$ when $\phi = 0$. The formula for y gives

$$y = \frac{c \tan^2 \phi}{\sec^2 \phi} = c \sin^2 \phi = \frac{1}{2} c(1 - \cos 2\phi).$$

We now simplify our equations by writing $a = \frac{1}{2}c$ and $\theta = 2\phi$, which yields

$$x = a(\theta - \sin \theta), \qquad y = a(1 - \cos \theta).$$

These are the standard parametric equations of the cycloid with a cusp at the origin. We note that there is a single value of a that makes the first inverted arch of this cycloid pass through the point P_1 in Fig. B.33; for if a is allowed to increase from 0 to ∞, then the arch inflates, sweeps over the first quadrant of the plane, and clearly passes through P_1 for a single suitably chosen value of a.

B.23

Our main purpose here is to establish the property of cycloids stated in the last paragraph of Section B.21. We shall do this by describing as briefly as possible the natural context for this property, within which it appears as a routine example.

Let P be a point moving along a given curve C, and assume that the curvature k is never zero on the part of C we consider, so that the radius of curvature $r = 1/|k|$ exists at every point. Draw the normal to C at P toward the concave side of the curve, and let Q be the point on this normal whose distance from P is r, as shown in Fig. B.34. The point Q is called the *center of curvature* of C corresponding to the point P. As P moves along C, the corresponding center of curvature Q generates a locus called the *evolute* of C. The evolute of a circle is a single point, its center. In general, however, the evolute of a curve is another curve.

We find the center of curvature Q as follows. Suppose C is the graph of a function $y = f(x)$, so that

$$k = \frac{y''}{(1 + y'^2)^{3/2}} \quad \text{and} \quad r = \frac{1}{|k|} = \frac{(1 + y'^2)^{3/2}}{|y''|}.$$

If P is (x, y), let the corresponding point Q be denoted by (X, Y). It is clear from Fig. B.34 that

$$x - X = \pm r \sin \phi, \qquad Y - y = \pm r \cos \phi, \tag{1}$$

318

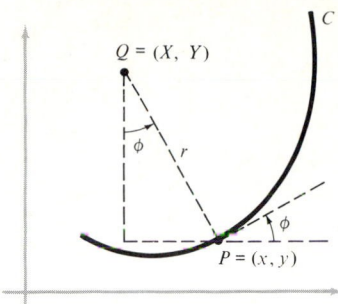

FIGURE B.34
The center of curvature.

where we use the plus or minus signs according as the curve is concave up or down. By recalling the meaning of the sign of the curvature, we see that (1) can be written as

$$x - X = \frac{1}{k}\sin\phi, \qquad Y - y = \frac{1}{k}\cos\phi. \tag{2}$$

Since

$$\sin\phi = \frac{dy}{ds} = \frac{y'}{\sqrt{1+y'^2}} \qquad \text{and} \qquad \cos\phi = \frac{dx}{ds} = \frac{1}{\sqrt{1+y'^2}},$$

equations (2) become

$$X = x - \frac{y'(1+y'^2)}{y''}, \qquad Y = y + \frac{1+y'^2}{y''}. \tag{3}$$

These equations give the coordinates of Q, and are therefore parametric equations for the evolute with the variable x used as the parameter. If the given curve C is defined by parametric equations $x = x(t)$ and $y = y(t)$, then we have

$$k = \frac{x'y'' - y'x''}{(x'^2 + y'^2)^{3/2}}, \qquad \sin\phi = \frac{y'}{\sqrt{x'^2+y'^2}}, \qquad \cos\phi = \frac{x'}{\sqrt{x'^2+y'^2}},$$

where now the primes denote derivatives with respect to t. In this case equations (3) are replaced by

$$X = x - \frac{y'(x'^2 + y'^2)}{x'y'' - y'x''}, \qquad Y = y + \frac{x'(x'^2 + y'^2)}{x'y'' - y'x''}, \tag{4}$$

which are parametric equations for the evolute with t as the parameter.

Example 1. Find the evolute of the parabola $y = x^2$.

Solution. Since $y' = 2x$ and $y'' = 2$, equations (3) yield

$$X = -4x^3, \qquad Y = \frac{6x^2 + 1}{2},$$

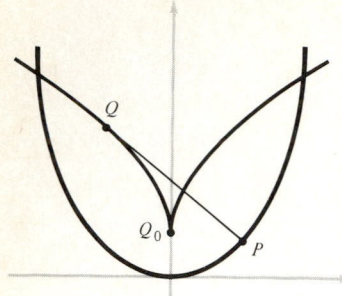

FIGURE B.35
The evolute of a parabola.

after simplification. It is easy to eliminate the parameter x and put the equation of the evolute in the form

$$27X^2 = 16(Y - \tfrac{1}{2})^3.$$

The appearance of this curve in relation to the given parabola is shown in Fig. B.35.

There are two important facts about the relation between a curve and its evolute that are easy to visualize by examining Fig. B.35. First, the line PQ is tangent to the evolute at Q. And second, if x is allowed to increase, so that P moves up the right side of the parabola, then the arc length Q_0Q along the evolute increases at exactly the same rate as the length of the tangent PQ.

These statements can be verified quite easily for the parabola and its evolute by using the formulas given in Example 1. However, instead of pausing to deal with this special case, we shall prove the statements for a general curve, as follows.

It is convenient to simplify the calculations by assuming that $k > 0$ at each point of the given curve C; the case $k < 0$ can be treated in a similar way. This assumption enables us to write equations (2) as

$$X = x - r \sin \phi, \qquad Y = y + r \cos \phi. \tag{5}$$

Since $k = d\phi/ds$ *and* $r = 1/k = ds/d\phi$, we have

$$r \sin \phi = \frac{ds}{d\phi}\frac{dy}{ds} = \frac{dy}{d\phi}, \qquad r \cos \phi = \frac{ds}{d\phi}\frac{dx}{ds} = \frac{dx}{d\phi}. \tag{6}$$

If we now differentiate equations (5) with respect to ϕ and use (6), we obtain

$$\frac{dX}{d\phi} = \frac{dx}{d\phi} - r \cos \phi - \frac{dr}{d\phi}\sin \phi = -\frac{dr}{d\phi}\sin \phi$$

and

$$\frac{dY}{d\phi} = \frac{dy}{d\phi} - r \sin \phi + \frac{dr}{d\phi}\cos \phi = \frac{dr}{d\phi}\cos \phi, \tag{7}$$

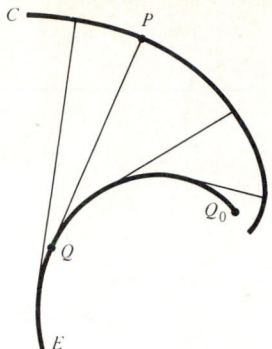

FIGURE B.36
A general evolute.

so

$$\frac{dY}{dX} = -\frac{\cos\phi}{\sin\phi} = -\frac{1}{\tan\phi}. \tag{8}$$

The slopes of the given curve C and its evolute E (Fig. B.36) are $\tan\phi$ and dY/dX. We can therefore conclude from equation (8) that *the normal PQ to the curve at P is tangent to the evolute at the center of curvature Q.* In getting from (7) to (8), it is evidently necessary to assume that $dr/d\phi \neq 0$ on C. That this condition is not superfluous is shown by the example of a circle, whose evolute is a single point.

To establish the second statement, let S be the length of the evolute from a fixed point Q_0 to the variable point Q corresponding to P. Then $dS^2 = dX^2 + dY^2$, and by (7) we have

$$\left(\frac{dS}{d\phi}\right)^2 = \left(\frac{dX}{d\phi}\right)^2 + \left(\frac{dY}{d\phi}\right)^2 = \left(\frac{dr}{d\phi}\right)^2. \tag{9}$$

If the direction of increasing S is chosen so that S and r both increase together, then (9) tells us that

$$\frac{dS}{d\phi} = \frac{dr}{d\phi},$$

and by integrating we obtain

$$S = r + \text{a constant.} \tag{10}$$

It follows from (10) that *the length of an arc of the evolute between any two points Q_1 and Q_2 is equal to the difference between the radii of curvature at the corresponding points P_1 and P_2.*

The results about the geometric relation between a curve C and its evolute E can be expressed in another way. Let a flexible inextensible string be wrapped around the evolute. If the string is held taut and unwound from the evolute, and if in addition the endpoint P of this string starts on the original

FIGURE B.37
The evolute of a cycloid.

curve C, then as we unwind the string, the point P traces out the curve C. This explains the name evolute, for it comes from the Latin *evolvere*, to unwind. Further, the original curve C is called an *involute* of the evolute E.

Example 2. Find the evolute of the cycloid $x = a(\theta - \sin \theta)$, $y = a(1 - \cos \theta)$.

Solution. We calculate

$$x' = a(1 - \cos \theta), \qquad y' = a \sin \theta,$$
$$x'' = a \sin \theta, \qquad y'' = a \cos \theta,$$

and apply formulas (4). After simplification the result is

$$X = a(\theta + \sin \theta),$$
$$Y = -a(1 - \cos \theta).$$

These are parametric equations for another cycloid congruent to the original one, but displaced $2a$ units down and πa units to the right, as shown in Fig. B.37. Since the lower cycloid is the evolute of the upper, the upper is an involute of the lower in the sense of the string property just discussed. This fact provides a complete justification for Huygens' construction of the cycloidal pendulum, described at the end of Section B.21.

B.24

$$\sum_1^\infty \frac{1}{n^2} = \frac{\pi^2}{6}$$

BY DOUBLE INTEGRATION

The geometric series $1/(1-r) = 1 + r + r^2 + \cdots$ enables us to write

$$\int_0^1 \int_0^1 \frac{dx\,dy}{1-xy} = \int_0^1 \int_0^1 (1 + xy + x^2y^2 + \cdots)\,dx\,dy$$

$$= \int_0^1 \left(x + \frac{1}{2}x^2y + \frac{1}{3}x^3y^2 + \cdots\right)\Big]_0^1 dy$$

$$= \int_0^1 \left(1 + \frac{y}{2} + \frac{y^2}{3} + \cdots\right) dy$$

$$= \left(y + \frac{y^2}{2^2} + \frac{y^3}{3^2} + \cdots\right)\Big]_0^1 = 1 + \frac{1}{2^2} + \frac{1}{3^2} + \cdots.$$

The sum of Euler's series $\sum 1/n^2$ is therefore the value of the double integral

$$I = \int_0^1 \int_0^1 \frac{dx\,dy}{1-xy}.$$

We evaluate this integral—and thereby determine the sum of series—by means of a rotation of the coordinate system through the angle $\theta = \pi/4$.

If we rotate the xy-system through an arbitrary angle θ, as shown in Fig. B.38, then the transformation equations are

$$x = u \cos \theta - v \sin \theta,$$

$$y = u \sin \theta + v \cos \theta.$$

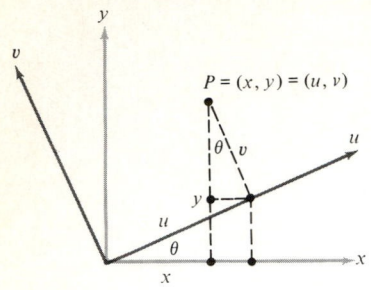

FIGURE B.38

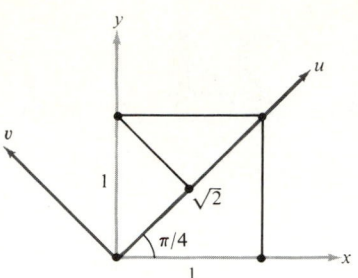

FIGURE B.39

When $\theta = \pi/4$ these equations become

$$x = \tfrac{1}{2}\sqrt{2}(u - v),$$
$$y = \tfrac{1}{2}\sqrt{2}(u + v),$$

so we have

$$xy = \frac{1}{2}(u^2 - v^2) \qquad \text{and} \qquad 1 - xy = \frac{2 - u^2 + v^2}{2}.$$

By inspecting Fig. B.39, we see that the integral I can be written in the form

$$I = 4 \int_0^{\sqrt{2}/2} \int_0^u \frac{dv\,du}{2 - u^2 + v^2} + 4 \int_{\sqrt{2}/2}^{\sqrt{2}} \int_0^{\sqrt{2}-u} \frac{dv\,du}{2 - u^2 + v^2}.$$

If we denote the integrals on the right by I_1 and I_2, then

$$I_1 = 4 \int_0^{\sqrt{2}/2} \left[\int_0^u \frac{dv}{2 - u^2 + v^2} \right] du$$

$$= 4 \int_0^{\sqrt{2}/2} \left[\frac{1}{\sqrt{2 - u^2}} \tan^{-1}\left(\frac{v}{\sqrt{2 - u^2}} \right) \right]_0^u du$$

$$= 4 \int_0^{\sqrt{2}/2} \frac{1}{\sqrt{2 - u^2}} \tan^{-1}\left(\frac{u}{\sqrt{2 - u^2}} \right) du.$$

To continue the calculation, we use the substitution

$$u = \sqrt{2}\sin\theta, \qquad \sqrt{2 - u^2} = \sqrt{2}\cos\theta, \qquad du = \sqrt{2}\cos\theta\,d\theta,$$

$$\tan^{-1}\left(\frac{u}{\sqrt{2 - u^2}} \right) = \tan^{-1}\left(\frac{\sqrt{2}\sin\theta}{\sqrt{2}\cos\theta} \right) = \theta.$$

Then

$$I_1 = 4 \int_0^{\pi/6} \frac{1}{\sqrt{2}\cos\theta} \cdot \theta \cdot \sqrt{2}\cos\theta\,d\theta = 2\theta^2 \Big]_0^{\pi/6} = \frac{\pi^2}{18}.$$

To calculate I_2, we write

$$I_2 = 4 \int_{\sqrt{2}/2}^{\sqrt{2}} \left[\int_0^{\sqrt{2}-u} \frac{dv}{2-u^2+v^2} \right] du$$

$$= 4 \int_{\sqrt{2}/2}^{\sqrt{2}} \left[\frac{1}{\sqrt{2-u^2}} \tan^{-1}\left(\frac{v}{\sqrt{2-u^2}} \right) \right]_0^{\sqrt{2}-u} du$$

$$= 4 \int_{\sqrt{2}/2}^{\sqrt{2}} \frac{1}{\sqrt{2-u^2}} \tan^{-1}\left(\frac{\sqrt{2}-u}{\sqrt{2-u^2}} \right) du.$$

To continue the calculation, we use the same substitution as before, with the additional fact that

$$\tan^{-1}\left(\frac{\sqrt{2}-u}{\sqrt{2-u^2}} \right) = \tan^{-1}\left(\frac{\sqrt{2}-\sqrt{2}\sin\theta}{\sqrt{2}\cos\theta} \right) = \tan^{-1}\left(\frac{1-\sin\theta}{\cos\theta} \right)$$

$$= \tan^{-1}\left(\frac{\cos\theta}{1+\sin\theta} \right) = \tan^{-1}\left(\frac{\sin(\pi/2-\theta)}{1+\cos(\pi/2-\theta)} \right)$$

$$= \tan^{-1}\left(\frac{2\sin\frac{1}{2}(\pi/2-\theta)\cos\frac{1}{2}(\pi/2-\theta)}{2\cos^2\frac{1}{2}(\pi/2-\theta)} \right) = \frac{1}{2}\left(\frac{\pi}{2}-\theta \right).$$

This enables us to write

$$I_2 = 4 \int_{\pi/6}^{\pi/2} \frac{1}{\sqrt{2}\cos\theta} \left(\frac{\pi}{4} - \frac{1}{2}\theta \right) \sqrt{2}\cos\theta\, d\theta = 4\left[\frac{\pi}{4}\theta - \frac{1}{4}\theta^2 \right]_{\pi/6}^{\pi/2}$$

$$= 4\left[\left(\frac{\pi^2}{8} - \frac{\pi^2}{16} \right) - \left(\frac{\pi^2}{24} - \frac{\pi^2}{144} \right) \right] = \frac{\pi^2}{9}.$$

We complete the calculation by putting these results together,

$$\sum_{n=1}^\infty \frac{1}{n^2} = I = I_1 + I_2 = \frac{\pi^2}{18} + \frac{\pi^2}{9} = \frac{\pi^2}{6}.$$

It is interesting to observe that

$$\int_0^1 \int_0^1 \int_0^1 \frac{dx\,dy\,dz}{1-xyz} = \sum_{n=1}^\infty \frac{1}{n^3},$$

so that any person who can evaluate this triple integral will thereby discover the sum of the series on the right—which has remained an unsolved problem since Euler first raised the question in 1736.

B.25

KEPLER'S LAWS AND NEWTON'S LAW OF GRAVITATION

As we know, Isaac Newton conceived the basic ideas of calculus in the years 1665 and 1666 (at age 22 and 23) for the purpose of helping him to understand the movements of the planets against the background of the fixed stars. In order to appreciate what was involved in this achievement, we briefly recall the main stages in the development of astronomical thinking up to his time.

The ancient Greeks constructed an elaborate mathematical model to account for the complicated movements of the sun, moon, and planets as viewed from the earth. A combination of uniform circular motions was used to describe the motion of each body about the earth. It was very natural for them—as it is for all people—to adopt the geocentric point of view that the earth is fixed at the center of the universe and everything else moves around it. Also, they were influenced by the semimystical Pythagorean belief that nothing but motion at constant speed in a perfect circle is worthy of a celestial body.

In this Greek model, each planet P moves uniformly around a small circle (called an *epicycle*) with center C, and at the same time C moves uniformly around a larger circle centered at the earth E, as shown in Fig. B.40. The radius of each circle and the angular speeds of P and C around the centers C and E are chosen to match the observed motion of the planet as closely as possible. This theory of epicycles was given its definitive form in Ptolemy's massive treatise *Almagest* in the second century A.D., and the theory itself is called the *Ptolemaic system*.

The next great step forward was taken by the Polish astronomer

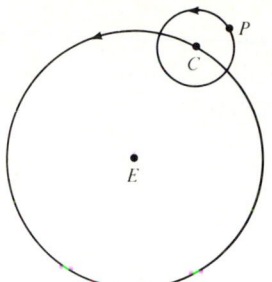

FIGURE B.40
An epicycle.

Copernicus. Shortly before his death in 1543, when he was presumably almost beyond the reach of a wrathful Church, he at last allowed the publication of his heretical book, *On the Revolution of the Celestial Spheres*. This work changed the Ptolemaic point of view by placing the sun, instead of the earth, at the center of each primary circle. Nevertheless, this *heliocentric system* was of much greater cultural than scientific importance. It enlarged the consciousness of many educated Europeans by giving them a better understanding of their place in the scheme of things, but it also kept the clumsy machinery of Ptolemy's circles whose centers move around on other circles.

It was Johannes Kepler (1571–1630) who finally eliminated this jumble of circles. Kepler was the assistant of the wealthy Danish astronomer Tycho Brahe, and when Brahe died in 1601, Kepler inherited the great masses of raw data they had accumulated on the positions of the planets at various times. Kepler worked incessantly on this material for 20 years, and at last succeeded in distilling from it his three beautifully simple laws of planetary motion, which were the climax of thousands of years of purely observational astronomy:

1. The orbit of each planet is an ellipse with the sun at one focus.
2. The line segment joining a planet to the sun sweeps out equal areas in equal times. See Fig. B.41.

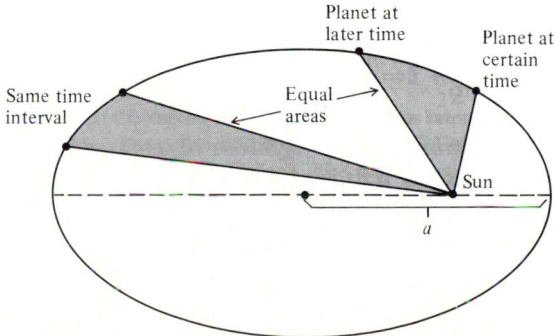

FIGURE B.41
Kepler's second law.

3. The *square* of the period of revolution of a planet is proportional to the *cube* of the semimajor axis of the planet's elliptical orbit. That is, if T is the time required for a planet to make one complete revolution about the sun and a is the semimajor axis shown in the figure, then the ratio T^2/a^3 is the same for all planets in the solar system.

From Kepler's point of view, these were empirical statements that fitted the data, and he had no idea of why they might be true or how they might be related to one another. In short, there was no theory to provide a context within which they could be understood.

Newton created such a theory. In the 1660s he discovered how to derive the inverse square law from Kepler's laws by mathematical reasoning, and also how to derive Kepler's laws from the inverse square law. We recall that *Newton's inverse square law of universal gravitation* states that any two particles of matter in the universe attract each other with a force directed along the line between them and of magnitude

$$G\frac{Mm}{r^2}, \tag{1}$$

where M and m are the masses of the particles, r is the distance between them, and G is a constant of nature called the gravitational constant. With this simple, clean, clear law as the unifying principle of his thinking, Newton published his theory of gravitation in 1687 in his *Principia Mathematica*. In this one book—perhaps the greatest of all scientific treatises—his success in using mathematical methods to explain the most diverse natural phenomena was so profound and far-reaching that he essentially created the sciences of physics and astronomy where only a handful of disconnected observations and simple inferences had existed before. These achievements launched the modern age of science and technology and radically altered the direction of human history.

We now derive Kepler's laws of planetary motion from Newton's law of gravitation, and to this end we discuss the motion of a small particle of mass m (a planet) under the attraction of a fixed large particle of mass M (the sun).

For problems involving a moving particle in which the force acting on it is always directed along the line from the particle to a fixed point, it is usually simplest to resolve the velocity, acceleration, and force into components along and perpendicular to this line. We therefore place the fixed particle M at the origin of a polar coordinate system (Fig. B.42) and express the position vector of the moving particle m in the form

$$\mathbf{R} = r\mathbf{u}_r, \tag{2}$$

where $\mathbf{u}_r$ is the unit vector in the direction of $\mathbf{R}$. It is clear that

$$\mathbf{u}_r = \mathbf{i}\cos\theta + \mathbf{j}\sin\theta, \tag{3}$$

and also that the corresponding unit vector $\mathbf{u}_\theta$, perpendicular to $\mathbf{u}_r$ in the

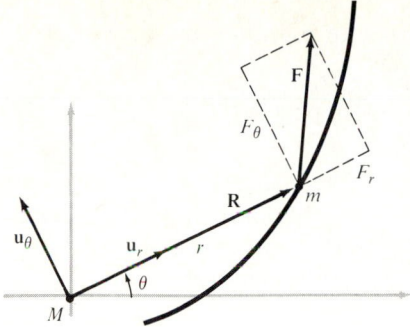

FIGURE B.42

direction of increasing θ, is given by

$$\mathbf{u}_\theta = -\mathbf{i} \sin \theta + \mathbf{j} \cos \theta. \tag{4}$$

It is easy to see by componentwise differentiation that

$$\frac{d\mathbf{u}_r}{d\theta} = \mathbf{u}_\theta \qquad \text{and} \qquad \frac{d\mathbf{u}_\theta}{d\theta} = -\mathbf{u}_r. \tag{5}$$

Thus, differentiating $\mathbf{u}_r$ and $\mathbf{u}_\theta$ with respect to θ has the effect of rotating these vectors 90° in the counterclockwise direction. We shall need the derivatives of $\mathbf{u}_r$ and $\mathbf{u}_\theta$ with respect to the time t. By means of the chain rule we at once obtain the formulas

$$\frac{d\mathbf{u}_r}{dt} = \frac{d\mathbf{u}_r}{d\theta} \frac{d\theta}{dt} = \mathbf{u}_\theta \frac{d\theta}{dt} \qquad \text{and} \qquad \frac{d\mathbf{u}_\theta}{dt} = \frac{d\mathbf{u}_\theta}{d\theta} \frac{d\theta}{dt} = -\mathbf{u}_r \frac{d\theta}{dt}, \tag{6}$$

which are essential for computing the velocity and acceleration vectors $\mathbf{v}$ and $\mathbf{a}$.

Direct calculation from (2) now yields

$$\mathbf{v} = \frac{d\mathbf{R}}{dt} = r \frac{d\mathbf{u}_r}{dt} + \mathbf{u}_r \frac{dr}{dt} = r \frac{d\theta}{dt} \mathbf{u}_\theta + \frac{dr}{dt} \mathbf{u}_r \tag{7}$$

and

$$\mathbf{a} = \frac{d\mathbf{v}}{dt} = \frac{dr}{dt} \frac{d\theta}{dt} \mathbf{u}_\theta + r \frac{d^2\theta}{dt^2} \mathbf{u}_\theta + r \frac{d\theta}{dt} \frac{d\mathbf{u}_\theta}{dt} + \frac{d^2r}{dt^2} \mathbf{u}_r + \frac{dr}{dt} \frac{d\mathbf{u}_r}{dt};$$

and by keeping formulas (6) in mind and rearranging, the latter equation can be written in the form

$$\mathbf{a} = \left(r \frac{d^2\theta}{dt^2} + 2 \frac{dr}{dt} \frac{d\theta}{dt} \right) \mathbf{u}_\theta + \left[\frac{d^2r}{dt^2} - r \left(\frac{d\theta}{dt} \right)^2 \right] \mathbf{u}_r. \tag{8}$$

If the force $\mathbf{F}$ acting on m is written as

$$\mathbf{F} = F_\theta \mathbf{u}_\theta + F_r \mathbf{u}_r, \tag{9}$$

then, from (8) and (9) and Newton's second law of motion $m\mathbf{a} = \mathbf{F}$, we get

$$m\left(r\frac{d^2\theta}{dt^2} + 2\frac{dr}{dt}\frac{d\theta}{dt}\right) = F_\theta \quad \text{and} \quad m\left[\frac{d^2r}{dt^2} - r\left(\frac{d\theta}{dt}\right)^2\right] = F_r. \tag{10}$$

These differential equations govern the motion of the particle m and are called the *equations of motion*; they are valid regardless of the nature of the force $\mathbf{F}$. Our next task is to extract the desired conclusions from these equations by making suitable assumptions about the direction and magnitude of $\mathbf{F}$.

CENTRAL FORCES AND KEPLER'S SECOND LAW

$\mathbf{F}$ is called a *central force* if it has no component perpendicular to $\mathbf{R}$, that is, if $F_\theta = 0$. Under this assumption the first of equations (10) becomes

$$r\frac{d^2\theta}{dt^2} + 2\frac{dr}{dt}\frac{d\theta}{dt} = 0.$$

On multiplying through by r, we obtain

$$r^2\frac{d^2\theta}{dt^2} + 2r\frac{dr}{dt}\frac{d\theta}{dt} = 0$$

or

$$\frac{d}{dt}\left(r^2\frac{d\theta}{dt}\right) = 0,$$

so

$$r^2\frac{d\theta}{dt} = h \tag{11}$$

for some constant h. We shall assume that h is positive, or equivalently that $d\theta/dt$ is positive, which evidently means that m is moving around the origin in a counterclockwise direction.

If $A = A(t)$ is the area swept out by $\mathbf{R}$ from some fixed position of reference, so that $dA = \frac{1}{2}r^2\,d\theta$, then (11) implies that

$$dA = \frac{1}{2}\left(r^2\frac{d\theta}{dt}\right)dt = \tfrac{1}{2}h\,dt.$$

On integrating this from t_1 to t_2, we get

$$A(t_2) - A(t_1) = \tfrac{1}{2}h(t_2 - t_1). \tag{12}$$

The yields Kepler's second law: The line segment joining the sun to a planet sweeps out equal areas in equal intervals of time.

CENTRAL GRAVITATIONAL FORCES AND KEPLER'S FIRST LAW

We now specialize even further, and assume that $\mathbf{F}$ is a central attractive force whose magnitude is given by the inverse square law (1), so that

$$F_r = -G\frac{Mm}{r^2}. \tag{13}$$

If we write (13) in the slightly simpler form

$$F_r = -\frac{km}{r^2}$$

where $k = GM$, then the second of equations (10) becomes

$$\frac{d^2r}{dt^2} - r\left(\frac{d\theta}{dt}\right)^2 = -\frac{k}{r^2}. \tag{14}$$

The next step in this line of thought is difficult to motivate, because it involves considerable technical ingenuity, but we will try. Our purpose is to use the differential equation (14) to obtain the equation of the orbit in the polar form $r = f(\theta)$, so we want to eliminate t from (14) and consider θ as the independent variable. Also, we want r to be the dependent variable, but if (11) is used to put (14) in the form

$$\frac{d^2r}{dt^2} - \frac{h^2}{r^3} = -\frac{k}{r^2}, \tag{15}$$

then the presence of powers of $1/r$ suggests that it might be temporarily convenient to introduce a new dependent variable $z = 1/r$.

To accomplish these various aims, we must first express d^2r/dt^2 in terms of $d^2z/d\theta^2$, by calculating

$$\frac{dr}{dt} = \frac{d}{dt}\left(\frac{1}{z}\right) = -\frac{1}{z^2}\frac{dz}{dt} = -\frac{1}{z^2}\frac{dz}{d\theta}\frac{d\theta}{dt} = -\frac{1}{z^2}\frac{dz}{d\theta}\frac{h}{r^2} = -h\frac{dz}{d\theta}$$

and

$$\frac{d^2r}{dt^2} = -h\frac{d}{dt}\left(\frac{dz}{d\theta}\right) = -h\frac{d}{d\theta}\left(\frac{dz}{d\theta}\right)\frac{d\theta}{dt} = -h\frac{d^2z}{d\theta^2}\frac{h}{r^2} = -h^2z^2\frac{d^2z}{d\theta^2}.$$

When the latter expression is inserted in (15), and r is replaced by $1/z$, we get

$$-h^2z^2\frac{d^2z}{d\theta^2} - h^2z^3 = -kz^2$$

or

$$\frac{d^2z}{d\theta^2} + z = \frac{k}{h^2}. \tag{16}$$

To solve this equation, we observe that, except for the constant term on the right, is the familiar differential equation of simple harmonic motion. To eliminate the constant term, we put

$$w = z - \frac{k}{h^2},$$

so that $d^2w/d\theta^2 = d^2z/d\theta^2$ and (16) becomes

$$\frac{d^2w}{d\theta^2} + w = 0.$$

As we know, the general solution of this familiar equation is

$$w = A \sin \theta + B \cos \theta,$$

or

$$z = A \sin \theta + B \cos \theta + \frac{k}{h^2}. \tag{17}$$

For the sake of simplicity, we now shift the direction of the polar axis in such a way that r is minimal (that is, m is closest to the origin) when $\theta = 0$. This means that z is to be maximal in this direction, so

$$\frac{dz}{d\theta} = 0 \qquad \text{and} \qquad \frac{d^2z}{d\theta^2} < 0$$

when $\theta = 0$. By calculating $dz/d\theta$ and $d^2z/d\theta^2$ from (17), we easily see that these conditions imply that $A = 0$ and $B > 0$. If we now replace z by $1/r$, then (17) can be written as

$$r = \frac{1}{k/h^2 + B \cos \theta} = \frac{h^2/k}{1 + (Bh^2/k) \cos \theta};$$

and if we put $e = Bh^2/k$, then our equation for the orbit becomes

$$r = \frac{h^2/k}{1 + e \cos \theta}, \tag{18}$$

where e is a positive constant.

We recall that (18) is the polar equation of a conic section with focus at the origin and vertical directrix to the right; and furthermore, that this conic section is an ellipse, a parabola, or a hyperbola according as $e < 1$, $e = 1$, or $e > 1$. Since the planets remain in the solar system and do not move infinitely far away from the sun, the ellipse is the only possibility. This yields Kepler's first law: The orbit of each planet is an ellipse with the sun at one focus.[1]

KEPLER'S THIRD LAW

We now restrict ourselves to the case in which m has an elliptic orbit (Fig. B.43) whose polar and rectangular equations are (18) and

$$\frac{x^2}{a^2} + \frac{y^2}{b^2} = 1.$$

We know that $e = c/a$ and $c^2 = a^2 - b^2$, so $e^2 = (a^2 - b^2)/a^2$ and

$$b^2 = a^2(1 - e^2). \tag{19}$$

In astronomy the semimajor axis a of the elliptical orbit is called the *mean distance*, because it is one-half the sum of the least and greatest values of r. These are the values of r corresponding to $\theta = 0$ and $\theta = \pi$ in (18), so by (18) and (19) we have

$$a = \frac{1}{2}\left(\frac{h^2/k}{1+e} + \frac{h^2/k}{1-e}\right) = \frac{h^2}{k(1-e^2)} = \frac{h^2 a^2}{kb^2},$$

which yields

$$b^2 = \frac{h^2 a}{k}. \tag{20}$$

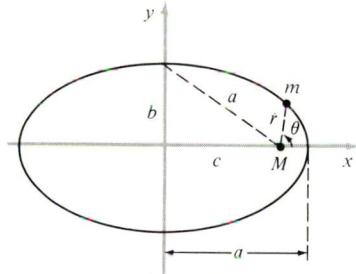

FIGURE B.43

---

[1] In the discussion of equation (17) we have ignored the possibility that r might have a constant value and therefore not be minimal in any direction, so that z has a constant value and is not maximal in any direction. This happens when both $A = 0$ and $B = 0$, so that $z = k/h^2$ and $r = h^2/k$. Under these circumstances we have a circular orbit with radius h^2/k, and this can be included under equation (18) by allowing the possibility that $e = 0$. However, a circular orbit of given radius requires a certain precise orbital speed, which is infinitely unlikely for an actual planet and can be disregarded as a genuine possibility.

If T is the period of m (that is, the time required for one complete revolution in its orbit), then, since the area of the ellipse is πab, it follows from (12) that $\pi ab = \frac{1}{2}hT$, so $T = 2\pi ab/h$. By using (20), we now obtain

$$T^2 = \frac{4\pi^2 a^2 b^2}{h^2} = \left(\frac{4\pi^2}{k}\right)a^3. \tag{21}$$

Since the constant $k = GM$ depends on the central attracting mass M but not on m, (21) holds for all the planets in our solar system and we have Kepler's third law: The squares of the periods of revolution of the planets are proportional to the cubes of their mean distances.

The standard unit of distance among astronomers who work with the solar system is the *astronomical unit*. This is the mean distance from the earth to the sun, which is approximately 93,000,000 miles or 150,000,000 km. Equation (21) takes the more convenient form

$$T^2 = a^3 \tag{22}$$

when time is measured in years and distance in astronomical units. The reason for this, of course, is that 1 year is by definition the period of revolution of the earth in its orbit, so that with these units of measurement $T = 1$ when $a = 1$.

We would like to point out that the mathematical theory discussed in this section is just the beginning of what Newton accomplished, and constitutes only a first approximation to the full story of planetary motion. For instance, we have assumed that only the sun and *one* planet are present. But actually, of course, all the other planets are present as well, and each exerts its own independent gravitational force on the planet under consideration. These additional influences introduce what are called "perturbations" into the idealized elliptical orbit derived here, and the main purpose of the science of celestial mechanics is to take all these complexities into account. One of the great events of nineteenth-century astronomy arose in just this way, namely, the discovery of the planet Neptune by Adams and Leverrier, through their attempts to explain the relatively large deviations of Uranus from its Keplerian orbit.[2]

Also, we have assumed that the sun and planet under discussion are particles, that is, points at which mass is concentrated. In fact, of course, they are extended bodies with substantial dimensions. One of Newton's most remarkable achievements was to prove that the sun and planets behave like particles under the inverse square law of attraction.

Newton's enormous success revived and greatly intensified the almost-forgotten Greek belief that it is possible to understand the universe in a

---

[2] For the details of this dramatic story, see pp. 820–39 of *The World of Mathematics*, ed. James R. Newman, Simon and Schuster, 1956.

rational way. This new confidence in its own intellectual powers permanently altered humanity's perception of itself, and over the past 300 years almost every department of human life has felt its consequences.

PROBLEMS

1. Newton himself did not know the value of the constant of gravitation G. This was determined by means of a classic experiment in 1789 by the English scientist Henry Cavendish. Once G is known, explain how equation (21), written in the form

$$T^2 = \left(\frac{4\pi^2}{GM}\right)a^3,$$

   might be used to calculate the mass of the sun.

2. What is the period of revolution T (in years) of a planet whose mean distance from the sun is
   (a) twice that of the earth?
   (b) three times that of the earth?
   (c) twenty-five times that of the earth?

3. Kepler's first two laws, in the form of equations (11) and (18), imply that m is attracted toward the origin with a force whose magnitude is inversely proportional to the square of r. This was Newton's fundamental discovery, for it caused him to propound his law of gravitation and investigate its consequences. Prove this by assuming (11) and (18) and verifying the following statements:
   (a) $F_\theta = 0$;

   (b) $\dfrac{dr}{dt} = \dfrac{ke}{h}\sin\theta$;

   (c) $\dfrac{d^2r}{dt^2} = \dfrac{ke\cos\theta}{r^2}$;

   (d) $F_r = -\dfrac{mk}{r^2} = -G\dfrac{Mm}{r^2}$.

4. Use formula (17) to show that the speed v of a planet at any point of its orbit is given by

$$v^2 = k\left(\frac{2}{r} - \frac{1}{a}\right).$$

5. Suppose that the earth explodes into fragments which fly off at the same speed in different directions into orbits of their own. Use Kepler's third law and the result of Problem 4 to show that all fragments that do not fall into the sun or escape from the solar system will reunite later at exactly the same point in space.

B.26

EXTENSIONS OF THE COMPLEX NUMBER SYSTEM. ALGEBRAS, QUATERNIONS, AND LAGRANGE'S FOUR SQUARES THEOREM

The history of mathematics shows very clearly that the concept of number—with its gradual growth from the positive integers through several stages to the complex numbers—has always been flexible and adaptive rather than fixed and final. Students have seen enough by the time they reach the end of a calculus course to realize that the extension of the real numbers to the complex numbers is richly productive in its consequences for analysis. Many wonder why we do not extend the complex numbers to an even more inclusive number system, and thereby enrich analysis even further. If two-dimensional vectors, that is, complex numbers, can be multiplied and divided in such a fruitful way, why not do the same for three-dimensional vectors?

This natural question—which was asked but not answered by Gauss and Hamilton—inspired massive developments in modern algebra which reached their logical conclusion fairly recently, in 1958. As a valuable by-product for mathematics as a whole, these developments have fully clarified the unique position that the complex numbers occupy among number systems in general. Our main purpose in this section is to explain the nature of these ideas in terms that can be understood by any reader who is acquainted with the concept of a field and with the elementary theory of vector spaces.[1]

---

[1] The most familiar field is the system of real numbers together with the usual operations of arithmetic, and no harm will be done if the reader simply thinks of this example whenever the word "field" is mentioned below. In this setting, a "scalar" is merely a real number.

Algebras. We are concerned with vector spaces whose vectors can be multiplied in a "reasonable" way. More precisely, an *algebra A* over a field *F*—called the field of *scalars*—is a set of elements called *vectors* together with three operations that can be performed on these vectors: addition, multiplication by elements of *F* (scalar multiplication), and multiplication. It is assumed that the first and second of these operations satisfy the axioms for a vector space; that multiplication is linked to the first two operations by the *distributive laws*

$$a(b + c) = ab + ac \qquad \text{and} \qquad (a + b)c = ac + bc, \tag{1}$$

and by the property that

$$\alpha(ab) = (\alpha a)b = a(\alpha b) \tag{2}$$

for all vectors *a*, *b*, *c* and every scalar α; and that multiplication is *associative* in the sense that

$$a(bc) = (ab)c \tag{3}$$

for all vectors *a*, *b*, *c*. It should be noted particularly that multiplication is not assumed to satisfy the *commutative law* $ab = ba$. These systems are sometimes referred to as "linear algebras," and in the older literature are often called "hypercomplex number systems."

The elementary theory of vector spaces tells us that the algebra *A* has a basis, that is, a set of vectors $e_1, \ldots, e_n$ such that every vector in *A* in uniquely expressible as a linear combination of the form $\alpha_1 e_1 + \cdots + \alpha_n e_n$ for suitable scalars $\alpha_1, \ldots, \alpha_n$ in *F*. The positive integer *n*, which is the number of vectors in the basis, is the same for every basis and is called the *dimension* of the algebra.[2] Since any two vectors can be multiplied, every product $e_i e_j$ of basis vectors is some vector in the algebra. The n^2 products $e_i e_j$ ($i, j = 1, \ldots, n$) comprise a *multiplication table* for the algebra. Once this is known, the product of any two vectors can be calculated by the formula

$$\left(\sum_i \alpha_i e_i \right) \left(\sum_j \beta_j e_j \right) = \sum_i \sum_j \alpha_i \beta_j (e_i e_j). \tag{4}$$

The validity of this formula follows from the assumed properties of the three operations. We can also proceed in the other direction: if $e_1, \ldots, e_n$ is a basis for an *n*-dimensional vector space over *F*, and if a multiplication table is given, then (4) can be used to define a multiplication for all vectors which makes the vector space into an algebra if the associative law $e_i(e_j e_k) = (e_i e_j)e_k$ holds for all the basis vectors.

---

[2] An algebra can have dimension 0 (if it consists of the zero vector alone) or it can be infinite-dimensional. We exclude these cases from the present discussion.

The algebra of complex numbers. Let A be the set of all complex numbers and F the field of real numbers. Then A is a two-dimensional algebra over F with respect to the ordinary addition and multiplication of complex numbers and multiplication of a complex number by a real number. The usual basis is $e_1 = 1$, $e_2 = i$, and the multiplication table for this basis is $1 \cdot 1 = 1$, $1 \cdot i = i$, $i \cdot 1 = i$, $i \cdot i = -1$.

The algebra of quaternions. The quaternions (discovered independently by Gauss in 1819 and Hamilton in 1843) constitute a four-dimensional algebra over the field of real numbers with basis 1, i, j, k and the following multiplication table: $ij = k$, $jk = i$, $ki = j$, $ji = -k$, $kj = -i$, $ik = -j$, $ii = jj = kk = -1$; and $1 \cdot a = a \cdot 1 = a$ for every quaternion a. The elements of the algebra—the quaternions themselves—are therefore vectors of the form

$$a = a_1 \cdot 1 + a_2 \cdot i + a_3 \cdot j + a_4 \cdot k = a_1 + a_2 i + a_3 j + a_4 k \qquad (5)$$

with real coefficients. The reader should notice that this algebra is an extension of the algebra of complex numbers. Addition and scalar multiplication are carried out componentwise, and multiplication by means of (4) and the multiplication table:

$$ab = (a_1 + a_2 i + a_3 j + a_4 k) \cdot (b_1 + b_2 i + b_3 j + b_4 k)$$
$$= a_1 b_1 + a_1 b_2 i + a_1 b_3 j + a_1 b_4 k + a_2 b_1 i + a_2 b_2 ii + a_2 b_3 ij + a_2 b_4 ik$$
$$+ a_3 b_1 j + a_3 b_2 ji + a_3 b_3 jj + a_3 b_4 jk + a_4 b_1 k + a_4 b_2 ki + a_4 b_3 kj + a_4 b_4 kk$$
$$= (a_1 b_1 - a_2 b_2 - a_3 b_3 - a_4 b_4) + (a_1 b_2 + a_2 b_1 + a_3 b_4 - a_4 b_3)i$$
$$+ (a_1 b_3 - a_2 b_4 + a_3 b_1 + a_4 b_2)j + (a_1 b_4 + a_2 b_3 - a_3 b_2 + a_4 b_1)k. \qquad (6)$$

The multiplication of quaternions (like that of real and complex numbers) has the property that division is uniquely possible except by zero: that is, if a and b are given quaternions with $a \neq 0$, then there exist unique quaternions x and y such that $ax = b$ and $ya = b$. We express this fact by saying that the quaternions form a *division algebra*. On the other hand, in contrast to the real and complex numbers, the multiplication of quaternions is not commutative, as we see from the fact that $ij \neq ji$.

To demonstrate that the quaternions form a division algebra, we proceed as follows. By analogy with the complex numbers, the *conjugate* of the quaternion (5) is defined by

$$\bar{a} = a_1 - a_2 i - a_3 j - a_4 k,$$

and it is easy to verify that

$$a\bar{a} = \bar{a}a = a_1^2 + a_2^2 + a_3^2 + a_4^2.$$

The non-negative real number $n(a) = a\bar{a} = \bar{a}a$ is called the *norm* of a, and it is clear that $n(a) \neq 0$ if $a \neq 0$. This enables us to show that every quaternion $a \neq 0$

has an *inverse* given by

$$a^{-1} = \frac{1}{n(a)}\bar{a};$$

for

$$a \cdot a^{-1} = a\left(\frac{1}{n(a)}\bar{a}\right) = \frac{1}{n(a)}(a\bar{a}) = \frac{n(a)}{n(a)} = 1$$

and

$$a^{-1} \cdot a = \left(\frac{1}{n(a)}\bar{a}\right)a = \frac{1}{n(a)}(\bar{a}a) = \frac{n(a)}{n(a)} = 1.$$

Now that we know that every nonzero quaternion has an inverse, it is easy to show that the quaternions form a division algebra. For suppose that a and b are given with $a \neq 0$. To solve $ax = b$ we simply multiply both sides on the left by a^{-1}, which yields $x = a^{-1}b$. To see that this solution is unique, we suppose that x' is also a solution, so that $ax' = b$. Then $ax' = ax$, and by multiplying both sides of this on the left by a^{-1} we obtain $x' = x$. The same type of reasoning tells us that $ya = b$ has the unique solution $y = ba^{-1}$, which in general is not equal to $a^{-1}b$.

It is also worth observing that the norm is *multiplicative* in the sense that

$$n(ab) = n(a)n(b). \tag{7}$$

To prove this, we first remark that an easy computation establishes the relation

$$\overline{ab} = \bar{b}\,\bar{a};$$

and (7) now follows from

$$n(ab) = (ab)(\overline{ab}) = (ab)(\bar{b}\,\bar{a}) = a[b(\bar{b}\,\bar{a})] = a[(b\,\bar{b})\bar{a}]$$
$$= a[n(b)\bar{a}] = n(b)(a\,\bar{a}) = n(b)n(a).$$

If $a = a_1 + a_2 i + a_3 j + a_4 k$ and $b = b_1 + b_2 i + b_3 j + b_4 k$, then by using (6) we see that (7) is equivalent to the curious identity

$$(a_1^2 + a_2^2 + a_3^2 + a_4^2)(b_1^2 + b_2^2 + b_3^2 + b_4^2) = (a_1 b_1 - a_2 b_2 - a_3 b_3 - a_4 b_4)^2$$
$$+ (a_1 b_2 + a_2 b_1 + a_3 b_4 - a_4 b_3)^2$$
$$+ (a_1 b_3 - a_2 b_4 + a_3 b_1 + a_4 b_2)^2$$
$$+ (a_1 b_4 + a_2 b_3 - a_3 b_2 + a_4 b_1)^2. \tag{8}$$

This identity can of course be verified directly by anyone who happens to think of writing it down. Its importance will become clear at the end of this section, in the proof of Lagrange's four squares theorem.

It is hopeless to look for additional division algebras with real scalars, for in 1878 Frobenius proved that the only ones that exist are the real numbers, the complex numbers and the quaternions. We cannot prove this important theorem here.[3] However, we can show quite easily that it is impossible to extend the complex numbers to a three-dimensional algebra (whether division algebra or not) with real scalars. To see this, suppose that 1, i, j form a basis for such an algebra and that 1 and i have their usual properties ($1 \cdot a = a \cdot 1 = a$ for every a and $ii = -1$). Then the product ij must be of the form

$$ij = a_1 + a_2 i + a_3 j \tag{9}$$

for certain real numbers a_1, a_2, a_3. If we multiply each side of (9) by i, we get

$$i(ij) = (ii)j = -j$$

and

$$i(a_1 + a_2 i + a_3 j) = a_1 i - a_2 + a_3 ij = a_1 i - a_2 + a_3(a_1 + a_2 i + a_3 j)$$
$$= a_1 i - a_2 + a_1 a_3 + a_2 a_3 i + a_3^2 j.$$

Equating these results and collecting terms yields

$$(a_1 a_3 - a_2) + (a_1 + a_2 a_3)i + (a_3^2 + 1)j = 0.$$

This implies the contradiction that $a_3^2 = -1$, and the question of Gauss and Hamilton is answered. However, if we are willing to drop the requirement of associativity in the algebras we consider as possible extensions of the complex numbers, then the story continues, as follows.

Nonassociative algebras. A vector space with a multiplication of vectors which satisfies conditions (1) and (2) is called a *nonassociative algebra*.[4]

The most important nonassociative algebras are undoubtedly the *Lie algebras,* in which the following two axioms provide a substitute for the associative law: $a^2 = 0$ and $a(bc) + b(ca) + c(ab) = 0$ for all a, b, c. A simple example of a Lie algebra is the set of vectors in three-dimensional space with cross product as multiplication. Algebraists like Lie algebras for their own sake, but the real significance of these systems lies in their connection with the Lie groups which have proved to be so important in twentieth-century geometry.

Nonassociative division algebras with real scalars have surprising applications to topology, for the existence of such an algebra of dimension n yields certain conclusions about continuous vector fields on the $(n-1)$-dimensional sphere. In 1940, Hopf proved by topological methods that n must be a power of 2. The cases $n = 2$ and $n = 4$ correspond to the algebras of complex numbers

---

[3] For proofs that are accessible to readers with a moderate background in linear algebra, see L. E. Dickson, *Algebras and their Arithmetics*, University of Chicago Press, 1923; A. G. Kurosh, *General Algebra*, Chelsea, 1963, pp. 221–43; or I. N. Herstein, *Topics in Algebra*, Blaisdell, 1964, pp. 326–29.

[4] Since these are merely algebras in which multiplication is not required to be associative, the term "not necessarily associative algebra" would be more accurate.

and quaternions, both of which are associative. For the next case, $n = 8$, there is a nonassociative division algebra discovered by Cayley in 1845—it is an extension of the quaternions—which is now called the *Cayley algebra* or the algebra of *Cayley numbers*.[5] In 1958, by using very refined techniques of algebraic topology, Bott, Milnor and Kervaire proved that the only possible dimensions are 1, 2, 4 and 8. Accordingly, there are only four division algebras (associative or nonassociative) with real scalars: the real numbers, the complex numbers, the quaternions, and the Cayley algebra.

To appreciate the light these discoveries shed on the position of the complex numbers among real division algebras, we should explicitly recognize that with each step up to a higher dimension a desirable property must be sacrificed. In extending the real numbers to the complex numbers we lose the order relation, but the gain more than compensates for this loss. In extending the complex numbers to the quaternions we lose commutativity, and this virtually excludes the quaternions from use in analysis. In the next step, to the Cayley algebra, associativity is lost. And beyond the Cayley algebra there are no real division algebras at all. Thus, for the purposes of analysis, the complex number system is as far as we can reasonably go, and each further step brings us closer to chaos.

Lagrange's four squares theorem. In spite of the fact that quaternions have nothing to offer analysis, they are not altogether devoid of redeeming qualities. We illustrate this point by using the identity (8) to prove one of the central theorems of classical number theory.

As we go out along the sequence of positive integers, it is clear that the perfect squares 1, 4, 9, 16, 25, 36, . . . , occur less and less frequently. Among the numbers that are not squares, there are many that can be expressed as a sum of two squares; thus, $2 = 1 + 1$, $5 = 4 + 1$, $8 = 4 + 4$, and so on. But not every number is a square or a sum of two squares. For instance, $3 = 1 + 1 + 1$ requires three squares, and $7 = 4 + 1 + 1 + 1$ requires four. It is natural to expect that if we go far enough out we will find numbers requiring five squares, six squares, etc.; but this is not true. It is a very remarkable fact that every positive integer can be written as a sum of at most four squares—or exactly four squares if we allow such representations as $1 = 1^2 + 0^2 + 0^2 + 0^2$ and $2 = 1^2 + 1^2 + 0^2 + 0^2$. This statement is known as *Lagrange's four squares theorem* because he was the first to prove it. However, he leaned heavily on preliminary work by Euler, and the proof we give is essentially due to Euler.

The basic tool is a slightly different form of the identity (8). If we replace a_1, a_2, a_3, a_4 by $x_1, -x_2, -x_3, -x_4$ and b_1, b_2, b_3, b_4 by y_1, y_2, y_3, y_4, then (8)

---

[5] A description of this algebra is given on pp. 226–28 of the book by Kurosh.

becomes

$$(x_1^2 + x_2^2 + x_3^2 + x_4^2)(y_1^2 + y_2^2 + y_3^2 + y_4^2) = (x_1 y_1 + x_2 y_2 + x_3 y_3 + x_4 y_4)^2$$
$$+ (x_1 y_2 - x_2 y_1 - x_3 y_4 + x_4 y_3)^2$$
$$+ (x_1 y_3 + x_2 y_4 - x_3 y_1 - x_4 y_2)^2$$
$$+ (x_1 y_4 - x_2 y_3 + x_3 y_2 - x_4 y_1)^2. \quad (10)$$

Our motive in making this change is to obtain an even number of minus signs in each squared expression on the right, for reasons that will appear below. The first thing (10) tells us is that if each of several integers is a sum of four squares, then their product can also be written in this form. But every integer >1 can be factored into a product of primes; and since 1 and 2 can each be written as a sum of four squares, it suffices to prove Lagrange's theorem for prime numbers >2. We establish this in the following two steps.

1. If p is a prime >2, then there exists a positive integer $m < p$ such that mp can be written as a sum of four squares, that is, $mp = x_1^2 + x_2^2 + x_3^2 + x_4^2$ for certain integers x_1, x_2, x_3, x_4.

To prove this, we begin by noting that since p is odd and >2, $\frac{1}{2}(p-1)$ is a positive integer. There are $\frac{1}{2}(p+1)$ integers a such that $0 \le a \le \frac{1}{2}(p-1)$. If for each of these a's we divide a^2 by p, we obtain certain remainders r_i which satisfy the condition $0 \le r_i \le p - 1$. These remainders are all distinct from one another. To see this, suppose that two have the same value r. Then there are two integers a_1 and a_2 with $0 \le a_2 < a_1 \le \frac{1}{2}(p-1)$ such that $a_1^2 = q_1 p + r$ and $a_2^2 = q_2 p + r$, so $a_1^2 - a_2^2 = (q_1 - q_2)p$ and p divides $a_1^2 - a_2^2 = (a_1 + a_2)(a_1 - a_2)$. Since a prime which divides a product necessarily divides one of the factors, p must divide $a_1 + a_2$ or $a_1 - a_2$. But this is impossible, since $a_1 + a_2$ and $a_1 - a_2$ are both positive integers $<p$.

We next add 1 to each r_i and subtract the result from p. This gives $\frac{1}{2}(p+1)$ distinct integers s_i such that $0 \le s_i \le p - 1$. These two sets of numbers, the r_i's and the s_i's, must have a common member $r = s$, for otherwise there would be $\frac{1}{2}(p+1) + \frac{1}{2}(p+1) = p+1$ distinct integers among the p numbers $0, 1, \ldots, p - 1$. We know from our construction that there exist integers a and b with $0 \le a$, $b \le \frac{1}{2}(p-1)$ such that $a^2 = q_1 p + r$ and $b^2 = q_2 p + r'$, where $s = p - (r' + 1)$. On adding these three equations we get $a^2 + b^2 + s = (q_1 + q_2 + 1)p + r - 1$; and the fact that $r = s$ enables us to write this as $mp = a^2 + b^2 + 1$, where $m = q_1 + q_2 + 1$. All that remains is to notice that $mp = a^2 + b^2 + 1 = a^2 + b^2 + 1^2 + 0^2$ is a sum of four squares and that m is a positive integer such that

$$m = \frac{a^2 + b^2 + 1}{p} < \frac{1}{p}\left(\frac{p^2}{4} + \frac{p^2}{4} + 1\right) < p.$$

We emphasize that the restriction on m ($1 \le m < p$) is the crux of what has been proved; for if m is allowed to be 0 or p, then mp can be 0 or p^2, and each of these is obviously a sum of four squares.

2. Every prime $p > 2$ can be written as a sum of four squares, that is, $p = x_1^2 + x_2^2 + x_3^2 + x_4^2$ for certain integers x_1, x_2, x_3, x_4.

Let m be the smallest positive integer such that

$$mp = x_1^2 + x_2^2 + x_3^2 + x_4^2 \tag{11}$$

for integers x_1, x_2, x_3, x_4. We know by what we have already proved that $m < p$, and it suffices to show that $m = 1$.

Equation (11) can be written in the form

$$\frac{m}{2}p = \left(\frac{x_1 + x_2}{2}\right)^2 + \left(\frac{x_1 - x_2}{2}\right)^2 + \left(\frac{x_3 + x_4}{2}\right)^2 + \left(\frac{x_3 - x_4}{2}\right)^2. \tag{12}$$

We use this fact to show that m must be odd. For assume that m is even. Then mp is also even, and by (11)—since the square of an even number is even and the square of an odd number is odd—there are three possibilities: (i) the x_i's are all even; (ii) the x_i's are all odd; (iii) two of the x_i's, say x_1 and x_2, are even and the other two are odd. In each of these cases the numbers in parentheses in (12) are integers. But since $m/2$ is also an integer and is $< m$, the expression (12) violates the minimum condition placed on m. We therefore conclude that m must be odd.

We now know that m is odd and we are trying to show that $m = 1$, so we assume that $m \geq 3$ and deduce a contradiction. To accomplish this, we first divide each x_i in (11) by m to obtain a remainder r_i with $0 \leq r_i \leq m - 1$, and then we define y_i to be r_i if $0 \leq r_i \leq \frac{1}{2}(m - 1)$ and to be $r_i - m$ if $\frac{1}{2}(m + 1) \leq r_i \leq m - 1$. Then $x_i = q_i m + y_i$ with $-\frac{1}{2}(m - 1) \leq y_i \leq \frac{1}{2}(m - 1)$; and since $y_i = x_i - q_i m$, (11) yields

$$
\begin{aligned}
y_1^2 + y_2^2 + y_3^2 + y_4^2 &= x_1^2 + x_2^2 + x_3^2 + x_4^2 - 2m(x_1 q_1 + x_2 q_2 + x_3 q_3 + x_4 q_4) \\
&\quad + m^2(q_1^2 + q_2^2 + q_3^2 + q_4^2) \\
&= mp - 2m(x_1 q_1 + x_2 q_2 + x_3 q_3 + x_4 q_4) \\
&\quad + m^2(q_1^2 + q_2^2 + q_3^2 + q_4^2) \\
&= mn,
\end{aligned}
\tag{13}
$$

where n is a non-negative integer. Further, n must be positive. For $n = 0$ would imply that all the y_i's are zero, so all the x_i's would be divisible by m. But then equation (11) in the form

$$m\left[\left(\frac{x_1}{m}\right)^2 + \left(\frac{x_2}{m}\right)^2 + \left(\frac{x_3}{m}\right)^2 + \left(\frac{x_4}{m}\right)^2\right] = p$$

would imply that m divides p, and this is impossible since $1 < m < p$ and p is prime. It is also true that $n < m$, since $mn = y_1^2 + y_2^2 + y_3^2 + y_4^2 < 4\left(\frac{m}{2}\right)^2 = m^2$.

Next, on multiplying (11) and (13) and using the identity (10), we obtain

$$
\begin{aligned}
m^2 np &= (x_1 y_1 + x_2 y_2 + x_3 y_3 + x_4 y_4)^2 + (x_1 y_2 - x_2 y_1 - x_3 y_4 + x_4 y_3)^2 \\
&\quad + (x_1 y_3 + x_2 y_4 - x_3 y_1 - x_4 y_2)^2 + (x_1 y_4 - x_2 y_3 + x_3 y_2 - x_4 y_1)^2. \tag{14}
\end{aligned}
$$

Easy calculations show that each of the squared numbers on the right of (14) is a multiple of m. For instance,

$$
\begin{aligned}
x_1 y_1 + x_2 y_2 + x_3 y_3 + x_4 y_4 &= x_1(x_1 - q_1 m) + x_2(x_2 - q_2 m) \\
&\quad + x_3(x_3 - q_3 m) + x_4(x_4 - q_4 m) \\
&= x_1^2 + x_2^2 + x_3^2 + x_4^2 \\
&\quad - m(x_1 q_1 + x_2 q_2 + x_3 q_3 + x_4 q_4) \\
&= mp - m(x_1 q_1 + x_2 q_2 + x_3 q_3 + x_4 q_4) \\
&= m z_1
\end{aligned}
$$

and

$$
\begin{aligned}
x_1 y_2 - x_2 y_1 - x_3 y_4 + x_4 y_3 &= x_1(x_2 - q_2 m) - x_2(x_1 - q_1 m) \\
&\quad - x_3(x_4 - q_4 m) + x_4(x_3 - q_3 m) \\
&= m(-x_1 q_2 + x_2 q_1 + x_3 q_4 - x_4 q_3) \\
&= m z_2,
\end{aligned}
$$

where z_1 and z_2 are integers; and similarly

$$
x_1 y_3 + x_2 y_4 - x_3 y_1 - x_4 y_2 = m z_3
$$

and

$$
x_1 y_4 - x_2 y_3 + x_3 y_2 - x_4 y_1 = m z_4
$$

for certain integers z_3 and z_4. Substituting these results in (14) and dividing by m^2 yields

$$
np = z_1^2 + z_2^2 + z_3^2 + z_4^2. \tag{15}
$$

Since $1 \le n < m$, (15) contradicts the minimum property of m. The only remaining possibility is that $m = 1$, and Lagrange's theorem is proved.

It is tempting to complete the chain of ideas started in Section B.2 by proving the two squares theorem as well, but the connection with our basic theme of calculus is already fairly remote, and perhaps we have already strained the reader's patience too much in going this far. However, according to the old Armenian proverb, mathematics, love, and home remodeling are the principal human enterprises in which one thing leads to another; and students with the desire to follow this particular road to the end will find an excellent presentation of Fermat's two squares theorem in A. D. Aleksandrov, A. N. Kolmogorov and M. A. Lavrent'ev, *Mathematics, Its Content, Methods, and Meaning* (tr. S. H. Gould), vol. 2, M.I.T. Press, 1964, pp. 225–28.

ANSWERS
TO
PROBLEMS

Section B.2

2. Because $(4n + 2)^2$ is a multiple of 4.

6. If $^n\sqrt{m} = a/b$ where a and b are positive integers with no common factor > 1, then $a^n = mb^n$ and every prime factor of b must be a prime factor of a (why?), so $b = 1$.

7. All n's except $1, 10, 10^2, 10^3, \ldots$.

8. Assume that x is a rational number a/b (with $b > 0$) in lowest terms, and show that $b = 1$ as in Problem 6.

Section B.16

2. (c) $100 = 4 \cdot 25 = 10 \cdot 10$.

3. (a) If N is not prime, then since $(4m - 1)(4n - 1) = 4(4mn - m - n) + 1$, there is no guarantee that N has a prime factor of the form $4n + 1$.

Section B.21

1. $a \sin^{-1} \dfrac{\sqrt{2ay - y^2}}{a} = \sqrt{2ay - y^2} + x.$

4. (a) $\dfrac{dx}{dt} = v_0(1 - \cos\theta), \quad \dfrac{dy}{dt} = v_0 \sin\theta;$

(b) $2v_0$, when P is at the top of the cycloid;

(c) v_0, when $\theta = \pi/2 + 2n\pi$ for some integer n.

6. $\frac{3}{8}\pi a^2$.

7. $6a$.

8. $\frac{12}{5}\pi a^2$.

10. $x = a\left(\dfrac{1}{3}\right)^{3/2}, \quad y = a\left(\dfrac{2}{3}\right)^{3/2}$.

11. $x = 2b\cos\theta + b\cos 2\theta, \quad y = 2b\sin\theta - b\sin 2\theta; \quad \dfrac{16}{3}a$.

12. $x = (a+b)\cos\theta - b\cos\dfrac{a+b}{b}\theta, \quad y = (a+b)\sin\theta - b\sin\dfrac{a+b}{b}\theta$.

14. $x = 3b\cos\theta - b\cos 3\theta, \quad y = 3b\sin\theta - b\sin 3\theta; \quad 12a$.

Section B.25

1. Since

$$M = \left(\frac{4\pi^2}{G}\right)\left(\frac{a^3}{T^2}\right),$$

determine the ratio a^3/T^2 for any particular planet (for instance, the earth) and proceed with the arithmetic.

2. (a) $2\sqrt{2} \cong 2.8$ years; (b) $3\sqrt{3} \cong 5.2$ years; (c) 125 years.

INDEX